The City & Guilds textbook

Bricklaying

LEVEL 2 TECHNICAL CERTIFICATE (7905)
LEVEL 3 ADVANCED TECHNICAL DIPLOMA (7905)
LEVEL 2 & 3 DIPLOMA (6705)
LEVEL 2 APPRENTICESHIP (9077)

Mike Jones
Colin Fearn

HODDER
EDUCATION
AN HACHETTE UK COMPANY

Orders: please contact Hachette UK Distribution, Hely Hutchinson Centre, Milton Road, Didcot, Oxfordshire, OX11 7HH. Telephone: +44 (0)1235 827827. Email education@hachette.co.uk Lines are open from 9 a.m. to 5 p.m., Monday to Friday.
You can also order through our website:www.hoddereducation.co.uk

ISBN: 9781510458147

© The City & Guilds of London Institute and Hodder & Stoughton Limited 2019

First published in 2019 by

Hodder Education,

An Hachette UK Company

Carmelite House

50 Victoria Embankment

London EC4Y 0DZ

www.hoddereducation.co.uk

Impression number   10 9 8 7 6 5 4 3 2

Year   2023

Cover photo © Alekss – stock.adobe.com

City & Guilds and the City & Guilds logo are trade marks of The City and Guilds of London Institute. City & Guilds Logo © City & Guilds 2019.

Typeset in India by Integra Software Services Ltd.

Printed and bound by CPI Group (UK) Ltd, Croydon, CR0 4YY

A catalogue record for this title is available from the British Library.

MIX
Paper | Supporting responsible forestry
FSC™ C104740

Contents

About your qualification v

Acknowledgements vii

About the author viii

Picture credits ix

How to use this book xi

1 Health and safety in the construction industry (Colin Fearn*) 1

Health and safety laws and regulations affecting bricklaying work 2

Dealing with hazards, accidents and emergencies 13

Handling materials and equipment safely 20

Using access equipment and working at height 26

2 Communication and information within the construction industry 33

Communication methods used with colleagues and customers 34

The range of construction documents used and how to interpret them 38

Planning and managing a construction project 53

Calculating resources and quantities of materials 61

3 Principles of construction 71

Company structures and how construction teams work together 72

Types of buildings and how their parts are constructed 79

Environmental and sustainability considerations for buildings 96

4 Setting out masonry structures 108

Preparing the site for setting out procedures 109

Selecting the right tools, equipment and working drawings for setting out the building 113

Methods of setting out and levelling a rectangular structure 118

5 Building cavity walls in masonry 127

Cavity wall design considerations 128

Planning ahead and setting up the work area 135

Positioning and preparing materials for safe and efficient work 146

Setting out and building cavity walls, including forming openings 152

6 Building solid walls, isolated and attached piers 173

Where solid walls can be used 174

Selecting the right materials and DPC for solid walls and piers 176

The masonry bonding arrangements used in solid walls and piers 179

Forming joint finishes and providing weather protection for solid walls and piers 189

7 Maintaining and repairing masonry structures 195

The range of faults that can occur in masonry 196

Repairing masonry faults and replacing components 202

Producing and placing concrete and repairing faults 223

How domestic drainage systems work 228

8 Constructing decorative and reinforced brickwork 236

Methods and techniques used to build decorative brickwork features 237

Setting out and building obtuse- and acute-angle quoins 254

When to reinforce brickwork and how to do it 260

iii

9 Constructing radial and battered brickwork   264
The different types of brick arches and how they are set out   265
Methods and techniques used to build brick arches   273
Setting out and building brickwork that is curved on plan   282
Methods and techniques used to set out and build battered brickwork   286

10 Chimneys, flues and fireplaces   291
The regulations and guidance that control construction of chimneys, flues and fireplaces   292
Components and materials used in chimney, flue and fireplace construction   296
Methods of setting out and building chimneys, flues and fireplaces   305

Glossary   317
Test your knowledge and activities answers   321
Index   327

*Includes some content derived from material originally written by Colin Fearn.

About your qualification

Bricklaying qualifications supported by this book

You are working towards achievement of one of the following qualifications:
- Level 2 Technical Certificate in Bricklaying (7905-20)
- Level 3 Advanced Technical Diploma in Bricklaying (7905-30)
- Level 2 Diploma in Bricklaying (6705-23)
- Level 3 Diploma in Bricklaying (6705-33)
- Bricklayer Trailblazer Apprenticeship (9077).

A Level 2 qualification is considered to be the industry standard that leads to a career as a bricklayer. It prepares you to be able to:
- interpret working drawings
- set out basic structures
- use blockwork in structures (including thin-joint masonry and masonry cladding)
- build solid walls and cavity walls
- build isolated and attached piers.

A Level 3 qualification extends your skills and knowledge to a higher level of craft ability.

It will equip you to work on more challenging bricklaying tasks including:
- repairing and maintaining masonry structures
- constructing radial and battered brickwork
- producing decorative and reinforced brickwork
- building fireplaces, flues and chimneys.

How the different qualifications work

7905

The Level 2 Technical Certificate in Bricklaying is aimed at learners starting a career in the construction industry as a bricklayer. Learners could progress into employment as a bricklayer, onto an apprenticeship, or further their studies by taking other qualifications at Level 3.

The Level 3 Advanced Technical Diploma in Bricklaying is suitable for learners who already have bricklaying experience and wish to make advancement in the industry. Learners could progress to a position supervising others or managing resources.

The Level 2 and Level 3 (7905) qualifications are both assessed using one exam and one practical synoptic assignment.

The exam is in a multiple choice question format with the addition of short written answer questions and extended response questions for Level 3.

The questions are designed to confirm your breadth of knowledge and understanding across the entire content of the qualification, testing your analysis, evaluation and application skills.

For the practical synoptic assignment, a typical Level 2 brief could be to build a right-angled section of cavity wall with an opening and soldier course above to accurately match the brief and the client's specification. This requires you to correctly interpret working drawings and specifications. You will need to draw on your knowledge of setting

v

out a structure and demonstrate competent hand skills when laying and cutting masonry materials, with health and safety requirements satisfied throughout.

A Level 3 synoptic assignment will include more advanced technical elements and may include the requirement to produce accurate drawings or written assignment work.

6705

The Level 2 Diploma in Bricklaying allows candidates to learn, develop and practise the skills required for employment and/or career progression in bricklaying.

The Level 3 Diploma in Bricklaying allows candidates to learn, develop and practise higher level craft skills in bricklaying that involve more technically difficult tasks.

The Level 2 and Level 3 (6705) qualifications are both assessed through a progressive series of practical assignments along with multiple choice tests related to each unit of study.

The practical assignments allow you to demonstrate your developing bricklaying skills. Prior to carrying out each task, you will be provided with specific task instructions by your trainer/instructor. You will be told how the task will be assessed and the time allowed to complete the work.

Your trainer/instructor will observe you as you carry out the tasks and mark your completed work. You can ask your assessor for help in understanding the task instructions, but all of the work must be your own. You must use safe working practices at all times.

The multiple choice tests will cover all of the knowledge outcomes for each unit.

9077

The Bricklayer Trailblazer Apprenticeship is an employer-led programme designed to enable the candidate to achieve competence in a skilled occupation, which is transferable and secures long-term earnings potential, greater security and the capability to progress in the workplace.

The Bricklayer Trailblazer Apprenticeship (9077) is assessed by means of an End Point Assessment (EPA). This assesses how an apprentice can apply their skills, knowledge and behaviours acquired in their apprenticeship. The assessment consists of three parts:
- Knowledge test – consists of multiple choice questions on a computer-based platform
- Skills test – assesses practical skills, knowledge and behaviours acquired throughout the apprenticeship
- Oral questioning – following the skills test, to confirm the learners' understanding and to obtain further evidence of knowledge. It will extend and amplify the ability demonstrated in the skills test.

In order to progress to the end point assessment, you must complete Level 1 English and Maths and attempt Level 2 English and Maths.

As an apprentice, during the on-programme phase of the apprenticeship, you must collect supporting evidence to complete a portfolio of learning. This is not graded but will form the basis of the oral questioning element of the EPA.

Acknowledgements

This book draws on several earlier books that were published by City & Guilds, and we acknowledge and thank the writers of those books:

- Mike Jones
- Colin Fearn
- Clayton Rudman
- Justin Beattie
- Tony Tucker
- Martin Burdfield.

We would like to thank Martin Burdfield for his knowledge and expertise in reviewing this book.

We would also like to thank everyone who has contributed to City & Guilds photoshoots. In particular, thanks to: Andrew Buckle, Paul Reed, Akeem Callum, Frankie Slattery, Wahidur Rahman and all of the staff at Hackney Community College, Mike Jones, Ron Lucock and Pawel Walga of Cardiff and Vale College, Garry Blunt, Ryan Kenneth, Nikesh McHugh, Luke Banneman, Joe Beckinsale, Adeel Qamer and the staff at Central Sussex College, Paul Reed, Akeem Callum, Frankie Slattery, Wahidur Rahman and all of the staff at Hackney Community College.

Contains public sector information licensed under the Open Government Licence v3.0.

From the author

Thanks to Charlie Evans at City & Guilds for suggesting I be considered as author for this new textbook.

Much appreciation to Loren Bowe who has shaped and refined my work in such an encouraging and supportive manner. She cleverly teased out lots of opportunities to expand the value of what I produced.

Also, thanks to Martin Burdfield whose breadth of technical knowledge is so impressive. Thanks for the specific and concise suggestions and amendments to make sure the material is accurate.

Many thanks to my former colleagues Paul Sebburn and Craig Jones for being innovative and imaginative teachers, giving me ideas on how to explain technical construction details clearly.

Stephen Halder and his team at Hodder Education have been so supportive and I have to give special thanks to Imogen Miles for her editorial skill and attention to detail in producing a good-looking book that has real technical value.

Finally, thanks to my lovely wife Sue who has patiently supported and encouraged me throughout. Months of staring at the back of my head as I sat at my desk couldn't have been easy!

Mike Jones, 2019

About the author

Bricklaying and the construction industry have been central to my working life over a long period. I've worked as a skilled tradesman, a supervisor and a site manager on projects ranging from small extensions to multi-storey contracts worth millions of pounds. For a number of years, I employed a small team of skilled workers in my own construction company working on contracts for selected customers.

The skills I developed over the years allowed me to design and successfully build my own family home in rural Wales, which I have always viewed as a highlight of my construction career. Relatively few people are able to have that privilege.

After over 30 years working on construction sites, I was fortunate to move into the education sector, first as an NVQ trainer and assessor for three years and subsequently as a college lecturer. I worked for ten years at Cardiff and Vale College in South Wales in the Brickwork section, where I taught bricklaying and other construction skills and became section leader.

After leaving my post at college, I've maintained my links with training and education producing teaching and learning resources for City & Guilds along with my work as a technical author writing bricklaying textbooks. I am pleased to be currently involved in the writing, reviewing and editing of resources for bricklaying qualification examinations.

Working 'hands-on' in the construction industry followed by my work in the education sector has been very rewarding. My aim during my time teaching others has been to impart to learners the great job satisfaction that can be gained from becoming a skilled practitioner. Put maximum effort into developing your skills and knowledge, and you will be able to take full advantage of career opportunities that come your way.

Mike Jones, 2019

Picture credits

Fig.1.1 © Igor Sokolov (breeze)/Shutterstock.com; Fig.1.3 © Alexander Erdbeer/Shutterstock.com; Fig.1.4 © Avalon/ Construction Photography / Alamy Stock Photo; Fig.1.5 © StockCube/Shutterstock.com; p.8 top left © Darkkong/ Shutterstock.com, top middle © dernarcomedia/Shutterstock.com, top right © phiseksit/Shutterstock.com; Table 1.2 1st © JoePhuriphat/Shutterstock.com, 2nd © Rafa Fernandez/Shutterstock.com, 3rd © mike.irwin/Shutterstock.com, 4th © ArtWell/Shutterstock.com, 5th © 29september/Shutterstock.com; Fig.1.6 courtesy of Axminster Tool Centre Ltd; Fig.1.7 © Virynja/stock.adobe.com; Fig.1.8 © northallertonman / Shutterstock.com; Table 1.3 1st © DenisNata/Shutterstock. com, 2nd, 3rd, 5th, 6th, 7th, 8th, 9th, 11th, 12th, 13th courtesy of Axminster Tool Centre, 4th © Virynja/stock.adobe. com, 10th © Ralph Fiskness/123RF, 14th © Jahanzaib Naiyyer/Shutterstock.com, 15th © Kraska/Shutterstock.com; Fig.1.9 © Dmitry Kalinovsky/Shutterstock.com; Fig.1.11 © Steroplast Healthcare Limited; Fig.1.12 © Health & Safety Executive; Fig.1.14 © Mr.Zach/Shutterstock.com; Fig.1.15 © Alan Stockdale/stock.adobe.com; Fig.1.16 © Jenny Thompson /stock. adobe.com; Fig.1.17 © Hartphotography/stock.adobe.com; Fig.1.18 © Mark Richardson /stock.adobe.com; Fig.1.19 © daseaford/Shutterstock.com; p.20 top left © Kaspri/Shutterstock.com, top middle © Barry Barnes/Shutterstock.com, top right © Tribalium/Shutterstock.com, bottom left © daseaford/Shutterstock.com, bottom right © DeiMosz/ Shutterstock.com; Fig.1.20 © objectsforall/Shutterstock.com; Fig.1.21 courtesy of Axminster Tool Centre Ltd; Fig.1.23 © SergeBertasiusPhotography/Shutterstock.com; Fig.1.24 © sbarabu/Shutterstock.com; Fig.1.26 Vulcascot Cable Protectors Limited; Fig.1.27 © kavalenkau/Shutterstock.com; Fig.1.28 © Rob Kints/Shutterstock.com; Fig.1.29 © Israel Hervas Bengochea/Shutterstock.com; Fig.1.32 © Lyte Ladders & Towers; Fig.1.36 © Gamut/stock.adobe.com; Fig.2.1 © Phovoir/Shutterstock.com; Fig. 2.3 © Michaeljung /stock.adobe.com; Fig.2.4 © Aurielaki/stock.adobe.com; Fig.2.5 © Robert Kneschke/stock.adobe.com; Fig.2.6 © RAGMA IMAGES/Shutterstock.com; Fig.2.7 © .shock/stock.adobe.com; Fig.2.8 © Ionia/stock.adobe.com; Fig.2.9 © Ilizia/Shutterstock.com; Table 2.1 bottom © Al-xVadinska/Shutterstock.com; Fig.2.12 © Corepics VOF/Shutterstock.com; Table 2.6 top right reproduced with permission of the Royal Institution of Chartered Surveyors which owns the copyright; Fig.2.18 © kzww/Shutterstock.com; Fig.2.19 © Sergej Razvodovskij/ Shutterstock.com; Fig.2.20 © Michaeljung/stock.adobe.com; Fig.2.21 © Milanmarkovic78/stock.adobe.com; Fig.2.22 © Biker3/stock.adobe.com; Fig.2.24 © Joe Gough/Shutterstock.com; Fig.2.30 © Jojje11/stock.adobe.com; Fig.2.31 © ivvv1975/Shutterstock.com; Fig.2.32 © Mark Atkins/Shutterstock.com; Fig.2.36 © Andrew Koturanov/Shutterstock.com; Fig.2.39 © Minerva Studio/stock.adobe.com; Fig.2.40 © Goodluz/stock.adobe.com; Fig.2.41 © DragonImages/stock. adobe.com; Fig.2.42 © Gustav/Shutterstock.com; Fig.2.43 © stocksolutions/Shutterstock.com; Fig.2.44 © alexandre zveiger/Shutterstock.com; Fig.3.1 © Duncanandison/stock.adobe.com; Fig.3.2 © Gaid Phitthayakornsilp/123RF; Fig.3.3 © Rawpixel.com/stock.adobe.com; Table 3.1 1st © Rawpixel.com/stock.adobe.com, 2nd © Goodluz/stock.adobe.com, 3rd © Budimir Jevtic/stock.adobe.com, 4th © Sergey Nivens/stock.adobe.com; Fig.3.4 © Patrick J Hanrahan/123RF; Fig.3.5 © Monkey Business/stock.adobe.com; Table 3.2 1st © BillionPhotos.com/stock.adobe.com, 2nd © Pat Hunt/123RF, 3rd © Joyfotoliakid/stock.adobe.com, 4th © Mangostar /stock.adobe.com; Fig.3.8 © MrSegui/Shutterstock.com; Fig.3.9 © Neil Lang/123RF; Fig.3.10 © Paul Wishart/Shutterstock.com; Fig.3.11 © Tony Baggett/stock.adobe.com; Fig.3.12 © Fikmik/123RF; Fig.3.13 © Taina Sohlman/123RF; Fig.3.14 © Enrique del Barrio/stock.adobe.com; Fig.3.15 © Annelie Krause/123RF; Fig.3.22 © TonyV3112/Shutterstock.com; Fig.3.25 © Bubutu/stock.adobe.com; Fig.3.30 © Ungvar/stock. adobe.com; Fig.3.34 © Silvia Crisman/123RF; Fig.3.35 © Stieberszabolcs/123RF; Fig.3.36 © Artursfoto/stock.adobe.com; Fig.3.37 © Artursfoto/stock.adobe.com; Fig.3.38 © Susan Law Cain/Shutterstock.com; Fig.3.47 © SteF/stock.adobe.com; Fig.3.48 © UrbanImages / Alamy Stock Photo; Fig.3.49 © Kasipat/stock.adobe.com; Fig.3.50 © Jane/stock.adobe.com; Fig.3.51 © Dagmara_K/stock.adobe.com; Fig.3.52 © Andrey Popov /stock.adobe.com; Fig.3.53 © Terry Kent/Shutterstock. com; Fig.3.54 © Dmitri Ma/Shutterstock.com; Fig.3.55 © Osmar01/stock.adobe.com; Fig.3.56 © Masauvalle /stock.adobe. com; Fig.3.58 © Smileus/Shutterstock.com; Fig 3.59 © Jesus Keller/Shutterstock.com; Fig.3.61 © Tomislav Pinter/ Shutterstock.com; p.98 top © BanksPhotos/E+/Getty Images, middle © Alena Brozova/Shutterstock.com, bottom © Alena Brozova/Shutterstock.com; p.99 top © SuperFOIL, middle © SueC/Shutterstock.com; p.100 1st © alterfalter/ Shutterstock.com, 2nd © H+H International A/S, 3rd © Ctvvelve/stock.adobe.com, 4th © Anton Starikov/Shutterstock. com; Fig.3.63 © Dario Sabljak/Shutterstock.com; Fig.3.65 © Lineicons freebird/Shutterstock.com; Fig.3.66 © Rparys/ stock.adobe.com; Fig.4.1 © hans engbers/Shutterstock.com; Fig.4.2 © Hoda Bogdan/stock.adobe.com; Fig.4.3 © koldunova/stock.adobe.com; Fig.4.4 © Igor Stramyk/Shutterstock.com; Fig.4.5 © Vadim Ratnikov/Shutterstock.com; Fig.4.6 © kaentian/stock.adobe.com; Fig.4.7 © Gina Sanders/stock.adobe.com; Fig.4.8 © Gundolf Renze/stock.adobe.com;

How to use this book

Throughout this book you will see the following features:

Industry tips and **Key points** are particularly useful pieces of advice that can assist you in your workplace or help you remember something important.

> ### INDUSTRY TIP
> The typical noise level for a hammer drill or a concrete mixer is 90 to 100 dB(A).

> ### KEY POINT
> Remember, if you have heat, fuel and oxygen you will have a fire. Remove any one of these and the fire will go out.

Key terms in bold purple in the text are explained in the margin to aid your understanding. (They are also explained in the Glossary at the back of the book.)

> ### KEY TERM
> **Galvanise:** to cover iron or steel with a protective zinc coating

Health and safety boxes flag important points to keep yourself, colleagues and clients safe in the workplace. They also link to sections in the health and safety chapter for you to recap learning.

> ### HEALTH AND SAFETY
> If you're moving long joists, don't try to transport them alone. It is easier and safer to have two workers to do the job, one at each end of the joist.

Activities help to test your understanding and learn from your colleagues' experiences.

> ### ACTIVITY
> Research 'water level' tool. Explain how this device works and suggest how you could use it for transferring levels for a building.

Values and behaviours boxes provide hints and tips on good workplace practice, particularly when liaising with customers.

> ## VALUES AND BEHAVIOURS
> The smooth running of a construction site of any size is heavily dependent on general operatives who are reliable and trustworthy and who co-operate with colleagues.

Improve your maths items provide opportunities to practise or improve your maths skills.

Improve your English items provide opportunities to practise or improve your English skills.

At the beginning of each chapter there is a table that shows how the main headings in the chapter cover the learning outcomes for each qualification specification.

At the end of each chapter there are some **Test your knowledge** questions. These are designed to identify any areas where you might need further training or revision.

CHAPTER 1

HEALTH AND SAFETY IN THE CONSTRUCTION INDUSTRY

INTRODUCTION

Building sites and construction workshops are varied and rewarding places to work. However, the work environment can present many potential hazards, and, regrettably, each year many construction workers' careers are affected or cut short by injuries. Careful analysis of work activities has led to the development of improved safety regulations, leading to a reduction in accidents and improving working conditions.

Every construction worker has a role to play in creating a safe working environment. This chapter looks at how safety laws and regulations benefit each worker and how you can apply them to contribute to maintaining safety in the workplace. Protecting yourself and those working with you will mean you can look forward to a long and rewarding career in construction.

By the end of this chapter, you will have an understanding of:
- health and safety laws and regulations affecting bricklaying work
- dealing with hazards, accidents and emergencies
- handling materials and equipment safely
- using access equipment and working at height.

The table below shows how the main headings in this chapter cover the learning outcomes for each qualification specification.

Chapter section	Level 2 Technical Certificate in Bricklaying (7905-20)	Level 3 Advanced Technical Diploma in Bricklaying (7905-30)	Level 2 Diploma in Bricklaying (6705-23) Unit 201/601	Level 3 Diploma in Bricklaying (6705-33)	Level 2 Bricklayer Trailblazer Apprenticeship (9077) Module 2
1. Health and safety laws and regulations affecting bricklaying work	N/A	N/A	1.1, 1.2, 1.3, 4.1, 4.2, 8.1, 8.2, 8.3, 8.4, 8.5	N/A	1.1, 2.3, 3.3, 3.4, 3.5
2. Dealing with hazards, accidents and emergencies	N/A	N/A	1.5, 2.1, 2.2, 2.3, 2.4, 2.5, 2.6, 3.1, 3.2, 3.3, 3.4, 3.5, 4.3, 9.1, 9.2, 9.3, 9.4	N/A	1.2, 2.1, 2.2, 3.4
3. Handling materials and equipment safely	N/A	N/A	5.1, 5.2, 5.3, 7.1, 7.2, 7.3, 7.4	N/A	3.1, 3.2, 3.6
4. Using access equipment and working at height	N/A	N/A	6.1, 6.2, 6.3, 6.4	N/A	3.4

1 HEALTH AND SAFETY LAWS AND REGULATIONS AFFECTING BRICKLAYING WORK

The introduction of effective health and safety legislation combined with the efforts of construction workers has made the workplace much safer in recent years. It is the responsibility of everyone involved in the building industry to continue to make it safer – that includes those who write the laws, employers and workers like you.

Many construction industry operatives think that an accident could never happen to them, but the reality is that many workers experience the consequences of an accident each year. Accidents can have a devastating effect on individuals and their families. There can be a significant financial cost due to lost earnings and injury compensation claims and workers can be prosecuted and lose their job if they are found to have broken health and safety laws.

ACTIVITY

Using the HSE website (www.hse.gov.uk), find the most recent health and safety statistics. Of the total number of accidents that resulted in three or more days off work, what proportion (as a percentage) were fatalities during that year?

Health and safety regulations

In the UK there is a great deal of **legislation** that has been put into place to make sure that those working on construction sites, and members of the public who could be affected by construction operations, are kept healthy and safe. If these laws and regulations are not obeyed, then prosecutions may be brought. More concerning, injury and damage to health may have long-lasting consequences.

KEY TERM

Legislation: a law or set of laws made by a government, for example, the Health and Safety at Work etc. Act (HASAWA) 1974

Health and Safety Executive

The Health and Safety Executive (HSE) provides a lot of advice on safety and publishes numerous booklets and information sheets and **Approved Codes of Practice (ACOPs)**. ACOPs have a special legal status and outline preferred or recommended ways of working to ensure both employers and employees comply with regulations and the requirements of the Health and Safety at Work Act. ACOPs are published by official bodies or professional organisations and approved by the HSE. ACOP booklets are available free online from www.hse.gov.uk.

KEY TERM

Approved Code of Practice (ACOP): a set of written directions and methods of work, issued by an official body or professional association and approved by the HSE, that provides practical advice on how to comply with the law

The HSE is a body set up by the UK government and is responsible for workplace health and safety. It provides guidance and advice to employers and workers and has extensive powers to ensure that health and safety law is being followed. The HSE can make spot checks in the workplace, carry out an inspection on the premises and take things away to be examined.

If the HSE finds a problem that breaks health and safety law, it may issue an **improvement notice**, giving the employer a set amount of time to correct the problem. For serious health and safety risks and where there is a risk of immediate major injury, it can issue a **prohibition notice** which will stop all work on site until the health and safety issues are rectified. The HSE can take an employer, employee, self-employed person (subcontractor) or anyone else involved with the building process to court for breaking health and safety legislation.

▲ Figure 1.1 The right equipment and the right mindset are needed to keep safe

Principal health and safety legislation

The principal legislation that relates to health, safety and welfare in the construction industry is as follows:

- Health and Safety at Work etc. Act (HASAWA) 1974
- Control of Substances Hazardous to Health (COSHH) 2002 (amended 2004)
- Reporting of Injuries, Diseases and Dangerous Occurrences Regulations (RIDDOR) 2013
- Construction, Design and Management (CDM) Regulations 2015
- Provision and Use of Work Equipment Regulations (PUWER) 1998
- Manual Handling Operations Regulations 1992 (amended 2002)
- Personal Protective Equipment at Work Regulations 2002
- Work at Height Regulations 2005 (amended 2007)
- Lifting Operations and Lifting Equipment Regulations (LOLER) 1998

- The Electricity at Work Regulations 1989
- Control of Noise at Work Regulations 2005
- Control of Vibration at Work Regulations 2005

A bricklayer is not expected to know every detail of this extensive list of legislation that applies to construction activities. This chapter will look at information that is relevant to the bricklayer's work on site and will provide an awareness of the range of legislation that exists to promote high standards of safety in the workplace.

Health and Safety at Work etc. Act (HASAWA) 1974

The Health and Safety at Work etc. Act (HASAWA) 1974 applies to all workplaces. Everyone who works on a building site or in a workshop is covered by this legislation. This includes employed and self-employed operatives, subcontractors, the employer and those delivering goods to the site. The legislation not only protects those working, it also ensures the safety of anyone else who might be nearby.

KEY POINT

You may hear this legislation referred to as the HSW Act 1974.

INDUSTRY TIP

The Health and Safety at Work etc. Act (HASAWA) 1974 can be accessed at:

www.legislation.gov.uk/ukpga/1974/37/contents

Key employer responsibilities

The employers' general responsibilities are contained in Section 2 of the Act. They are to ensure, 'so far as is reasonably practicable', the health, safety and welfare at work of all their employees, in particular:

- the provision of safe plant and systems of work
- the provision of **personal protective equipment (PPE)** free of charge and ensuring the appropriate PPE is used whenever needed
- the safe use, handling, storage and transport of components, materials and substances
- the provision of any required information, instruction, training or supervision

- a safe place of work, including safe, secure access and exit
- a safe working environment with adequate welfare facilities.

KEY TERM

Personal protective equipment (PPE): this is defined as 'all equipment (including clothing affording protection against the weather) which is intended to be worn or held by a person at work and which protects against one or more risks to a person's health or safety'

KEY POINT

One of the main methods of making sure the workplace and the activities that take place there are as safe as possible is to conduct risk assessments. We'll discuss these in more detail later in this chapter.

Key employee responsibilities

HASAWA defines the key responsibilities of employees and those working as subcontractors as follows:
- Work in a safe manner and always take care.
- Make sure that you do not put yourself or others at risk by your actions or inactions.
- Co-operate fully with your employer regarding health and safety.
- Make full use of any equipment and safeguards (e.g. PPE) provided by your employer.
- Do not interfere or tamper with any safety equipment.
- Do not misuse or interfere with anything that is provided for employees' safety.
- Do not engage in practical jokes and horseplay.

INDUSTRY TIP

Your responsibilities regarding health and safety can be summarised as:
- work safely
- work in partnership with your employer
- report hazards and accidents in accordance with company policy.

Control of Substances Hazardous to Health (COSHH) Regulations 2002 (amended 2004)

The Control of Substances Hazardous to Health (COSHH) Regulations 2002 control the use of potentially dangerous substances such as preservatives, fuel, cement and oil-based paint. These must be moved, stored and used safely without polluting the environment. The Regulations also cover hazardous substances produced while working, such as dust produced when cutting or grinding masonry materials.

INDUSTRY TIP

You can access the COSHH Regulations at: www.legislation.gov.uk/uksi/2002/2677/contents/made

Hazardous substances already present on a project site may be discovered during the construction process. An example is asbestos which is also covered by specific regulations.

HEALTH AND SAFETY

Asbestos was used extensively in construction for many years. It was produced in sheet form and was often used to box around steel girders and columns as well as in fibre form to surround (or 'lag') pipes that carried hot gases or liquids. If you come across what you think might be asbestos when working, do not disturb the material, stop work immediately and inform your supervisor.

When dealing with substances and materials that may be hazardous to health, an employer should do the following to comply with COSHH:
- Carefully read the COSHH safety data sheet which comes with the product. It will outline any hazards associated with the product and the safety measures to be taken.
- Check with the supplier to establish if there are any known risks to health posed by the substance or material.
- Use the trade press to find out if there is any information about this substance or material.
- Use the HSE website, or other websites, to check any known issues with the substance or material.

When assessing the possible risks attached to using a potentially dangerous substance or material, it is important to consider how operatives could be exposed to it. There are a number of ways that a hazardous substance can affect you:

- breathing it in as a gas or mist
- swallowing it
- getting it into your eyes
- through the skin, either by surface contact or through cuts.

You can find out more about COSHH and safety data sheets by reading the leaflet 'Working with substances hazardous to health: A brief guide to COSHH', published by the HSE (www.hse.gov.uk/pubns/indg136.pdf).

Control measures

To manage the risks that can arise from using potentially hazardous substances, established control measures should be applied. A sequence of control measures that can be followed is listed below.

- Eliminate the use of the harmful substance and use a safer one. An example might be to swap high **VOC** oil-based paint for a lower VOC water-based paint.
- Use a safer form of the product. Is the product available ready-mixed? Is there a lower strength option that will still do the job?
- Change the work method to emit less of the hazardous substance. For example, wet masonry grinding or cutting produces less dust than dry masonry grinding or cutting.
- Enclose the work area so that the hazardous substance does not escape. This can mean setting up a tented area or closing doors.
- Use extraction or filtration systems in the work area.

Always use the correct PPE and limit the number of personnel working in the area where the potentially hazardous substance is being used.

▲ Figure 1.2 COSHH symbols

Construction, Design and Management (CDM) Regulations 2015

The Construction, Design and Management (CDM) Regulations 2015 focus attention on the effective planning and management of construction projects, from the conception and design onwards.

The aim of the Regulations is for health and safety considerations to be integrated into the development of a project from the very beginning, rather than simply as an afterthought. The CDM Regulations reduce the risk of harm to those that work on or in a building for the entirety of its use, from construction through to **demolition**.

KEY TERM

Demolition: the process of destruction, when a structure is torn down and destroyed

▲ Figure 1.3 The CDM Regulations play a vital role during demolition

▲ Figure 1.4 CDM Regulations protect those who work on complex structures

The CDM Regulations apply to all projects, with the exception of those arranged by private clients which are not part of a business interest.

ACTIVITY

Under the CDM Regulations, the HSE must be notified about certain projects before they start. Check online and list the requirements for notifying the HSE about a project.

▲ Figure 1.5 HSE inspector

Duty holders

Under the terms of the CDM Regulations, a **duty holder** is a person who is trained and competent, with enough experience and knowledge to know how to avoid specific dangers. The level of competence required will differ for different types of work.

Under the CDM Regulations there may be several duty holders, each with a specific role. Table 1.1 shows how these roles and responsibilities are assigned.

▼ Table 1.1 The roles and responsibilities of duty holders

Duty holder	Roles and responsibilities
Client	This is the person or organisation that wants the work done. The client will check that: ● all the team members are competent ● suitable management is in place ● enough time is allowed for all stages of the project ● welfare facilities are in place before construction starts. HSE-notifiable projects require that the client appoints a CDM co-ordinator and principal contractor and provides access to a health and safety file.
CDM co-ordinator	Appointed by the client, the co-ordinator advises and assists the client with CDM duties. Responsibilities include: ● notifying the HSE before work starts ● co-ordinating health and safety aspects of the design of the building ● ensuring good communication is maintained between the client, designers and contractors.
Designer	At the design stages, the designer: ● identifies hazards and reduces risks ● provides information about the risks that cannot be eliminated ● checks that the client is aware of their CDM duties and that a CDM co-ordinator has been appointed ● supplies information for the health and safety file.
Principal contractor	The principal contractor will plan, manage and monitor the construction and liaise with any other contractors. This involves developing a written plan and site rules before construction begins. The principle contractor ensures that: ● the site is made secure and suitable welfare facilities are provided from the start and maintained throughout construction ● all operatives have site inductions and any further training that might be required to make sure the workforce is competent.
Contractor	Subcontractors and self-employed operatives will plan, manage and monitor their own work and that of employees, ensuring they co-operate with the main contractor in relation to site rules. Contractors will make sure that: ● all operatives have any further training that might be required to make sure they are competent ● all incidents are reported to the principal contractor.
Operatives	As an operative, you must ask yourself: ● Can I carry out the task I have been asked to do safely? ● Have I been trained to do this type of activity? ● Do I have the correct equipment to carry out this activity? You must follow all the site health and safety rules and procedures and fully co-operate with the rest of the team to ensure the health and safety of other operatives and those who may be affected by the work. Any health and safety issues must be reported.

ACTIVITY

Write down what you would do if you spotted any of these hazards:

Site welfare facilities required under the CDM Regulations

The CDM Regulations also give details relating to the welfare of workers on site. Table 1.2 shows the welfare facilities that must be available.

▼ Table 1.2 Welfare facilities that must be provided by employers

Facility	Regulation requirement
Toilets 	Enough suitable toilets should be provided or made available. Toilets should be: • adequately ventilated and lit • maintained in a clean condition • separate for men and women.
Washing facilities 	Enough washing facilities must be available and should include showers if required by the nature of the work. They should be: • in the same place as the toilets and near any changing rooms • supplied with clean hot (or warm) and cold running water, soap and towels • separate for men and women unless the area is for washing hands and face only.
Clean drinking water 	Drinking water must be provided or made available. It must be clearly marked by an appropriate sign. Cups should be provided unless the supply of drinking water is from a water fountain.

▼ Table 1.2 Welfare facilities that must be provided by employers (continued)

Facility	Regulation requirement
Changing rooms and lockers 	Changing rooms must be provided or made available if operatives are required to wear special clothing and if they cannot be expected to change elsewhere. There must be separate rooms for men and women where necessary. The rooms must have seating and include, where necessary, facilities to enable operatives to dry clothing. Lockers should also be provided.
Rest room or rest areas 	There should be enough tables and seating with back-rests for the number of operatives likely to use them at any one time. Arrangements must be made to ensure that meals can be prepared, heated and eaten. It must also be possible to boil water.

ACTIVITY

Look around your workplace or training centre. Make a list of the facilities available and note how they compare to the regulation requirements shown in Table 1.2.

Control of Noise at Work Regulations 2005

Under the Control of Noise at Work Regulations 2005, duties are placed on employers and employees to reduce the risk of hearing damage to the lowest reasonable level achievable. Hearing loss caused by noisy work activities is preventable. Hearing damage is permanent and hearing cannot be restored once lost.

INDUSTRY TIP

You can access The Control of Noise at Work Regulations 2005 at: www.legislation.gov.uk/uksi/2005/1643/contents/made

ACTIVITY

Visit the HSE's website and watch 'The Hearing Video' to find out more about protecting your hearing: www.hse.gov.uk/noise/video/hearingvideo.htm

Employer's duties under the Regulations

Employers must make sure that workers are protected from hearing damage by:
- carrying out a **risk assessment** and identifying who is at risk
- eliminating or controlling exposure to noise in the workplace and reducing the noise to the lowest level possible
- providing suitable hearing protection
- providing health surveillance to those identified as at risk by the risk assessment
- providing information and training about the risks to employees as identified by the risk assessment.

KEY TERM

Risk assessment: a systematic examination of a job or process to identify significant hazards and risks and evaluate what control measures can be taken to reduce the risk to an acceptable level

Employees' duties under the Regulations

Employees must:
- make full and proper use of personal hearing protectors provided to them by their employer
- report any defect in personal hearing protectors or other control measures to their employer as soon as possible.

▲ Figure 1.6 Ear defenders

▲ Figure 1.7 Ear plugs

Specifics about noise

Under the Regulations, specific actions are triggered at specific noise levels. Noise is measured in decibels and for the type of noise likely to be experienced on a construction site, it is shown as 'dB(A)'. The lower and upper action levels are 80 dB(A) and 85 dB(A).

Requirements at 80 dB(A) to 85 dB(A):

- Assess the risk to operatives' health and provide them with information and training.
- Provide suitable ear protection free of charge to those who request it.

Requirements above 85 dB(A):

- Reduce noise exposure as far as achievable by means other than ear protection.
- Set up an ear protection zone using suitable signage and segregation.

- Provide suitable ear protection free of charge to those affected and ensure it is worn.

> **INDUSTRY TIP**
>
> The typical noise level for a hammer drill or a concrete mixer is 90 to 100 dB(A).

Personal Protective Equipment (PPE) at Work Regulations 2002

Employees and subcontractors must work in a safe manner. Not only must they wear the PPE that their employers provide, they must also store it correctly, look after it while using it and report any damage to it. Employers must provide appropriate PPE without charge to employees, including agency workers where they are legally recognised as employed by the contractor.

The hearing and **respiratory** PPE provided for most work situations is not covered by these Regulations because other regulations apply. However, these items need to be compatible with any other PPE provided.

> **KEY TERM**
>
> **Respiratory:** relating to breathing

▲ Figure 1.8 A site safety sign showing the PPE required to work in this area

Table 1.3 shows why it is important to store, maintain and use PPE correctly. Remember that it is also important to check and report damage to PPE.

▼ Table 1.3 Types of PPE used in the workplace

PPE	Correct use
Hard hat/safety helmet	Hard hats must be worn when there is danger of hitting your head or a danger of being hit by falling objects. Most sites insist on hard hats being worn at all times. They must be adjusted to fit your head correctly and must not be worn back to front!
Steel toe-cap boots or shoes	Steel toe-cap boots or shoes are worn on most sites as a matter of course and protect the feet from crushing by heavy objects. Some safety footwear has additional insole protection to help prevent nails going up through the foot.
Ear defenders and plugs	Your ears can be very easily damaged by loud noise. Ear protection will help prevent hearing loss if there is a lot of noise near you. If your ear defenders are damaged or fail to make a good seal around your ears, have them replaced.
High visibility (hi-vis) jacket	This PPE item makes it much easier for other people to see you. This is especially important when plant or vehicles are moving in or near your work area.
Goggles and safety glasses	These protect the eyes from dust and flying debris while you are working. You only get one pair of eyes, so look after them!

▼ Table 1.3 Types of PPE used in the workplace (continued)

PPE	Correct use
Dust masks and respirators	Dust is produced during most construction work and it can be hazardous to your lungs. It can cause serious illnesses that can be life-threatening. A dust mask gives vital protection when correctly fitted. Note: A respirator is used to filter out hazardous gases. Respirators are rated P1, P2 and P3 to show the level of protection they give.
Knee pads	Knee pads provide essential protection to the knee joints when kneeling for extended periods.
Gloves	Gloves protect your hands from hazards such as cuts, abrasions, dermatitis, chemical burns or splinters. Latex and nitrile gloves are good for fine work, although some people are allergic to latex. Gauntlets provide protection from strong chemicals. Other types of gloves provide good grip and protect the fingers. ▲ A chemical burn as a result of not wearing safety gloves
Sunscreen	Sunburn is a risk, especially in the summer months. Overexposure to the sun can cause skin cancer. When out in the sun, cover up and use sunscreen as PPE on exposed areas of your body to prevent burning.

INDUSTRY TIP

A hard hat doesn't last forever. Ultraviolet light from the sun will cause the plastic to deteriorate over time. Solvents, pens and paints can damage the plastic too. Check the date of manufacture and get a replacement hat if it is out of date.

HEALTH AND SAFETY

A range of substances can cause chemical burns to the skin. Visit the HSE website at www.hse.gov.uk/construction/healthrisks/hazardous-substances/cement.htm and list the control measures that should be employed to reduce the risk of cement burns to the skin.

The Control of Vibration at Work Regulations 2005

Vibration white finger (VWF) or hand-arm vibration syndrome (HAVS) is caused by using vibrating tools such as hammer drills or hand-held breakers over a long period of time. The most efficient and effective way of controlling the risk of HAVS is to look for new or alternative work methods which remove or reduce exposure to vibration.

▲ Figure 1.9 Don't use power tools for longer than you need to

Here are some important points that can help reduce the development of HAVS:

- Always use the right tool for each job.
- Check tools before using them to make sure they have been properly maintained and repaired to avoid increased vibration caused by faults or general wear.
- Make sure cutting tools are kept sharp so that they remain efficient.
- Reduce the amount of time you use a tool in one work session. Alternate the work with another task.
- Encourage good blood circulation by:
  - keeping warm and dry (when necessary, wear gloves, a hat and waterproofs, and use heating pads if available)
  - giving up or cutting down on smoking because smoking reduces blood flow
  - massaging and exercising your fingers during work breaks.

INDUSTRY TIP

When using a power tool of any sort, avoid gripping the tool too tightly or forcing the tool into the work piece to try to speed up the job. Let the tool do the work.

Damage from HAVS can cause a worker to have difficulty in performing detailed or intricate work. A worker suffering from HAVS can experience painful finger blanching attacks (when the ends of the fingers go white) when working in cold conditions.

2 DEALING WITH HAZARDS, ACCIDENTS AND EMERGENCIES

The occurrence of accidents and emergencies can be minimised and often prevented by assessing and managing the risks involved in performing a work task in defined conditions. This must be an ongoing process to deal with the changing conditions on site as a project develops.

Managing hazards on site

Using the right tools for any construction task makes it easier to achieve success. Managing hazards successfully will also be made easier by using the right tools. The main tool developed to identify and manage potential hazards is a document known as a risk assessment.

Risk assessments

The Health and Safety at Work etc. Act 1974 requires employers to carry out regular risk assessments to make sure that there are minimal dangers to their employees in a workplace. A risk assessment can be split into three key stages:

1 Identify hazards and potential hazards in the workplace.
2 Assess the risk of harm likely to be caused by these hazards.
3 Establish measures to minimise and control the risk.

Look carefully at the example of a risk assessment and then study the following list of steps that show how the document is created and maintained.

1 Identify the hazards. Consider the environment in which the job will be done. Which tools and materials will be used?

2 Identify who might be at risk. Think about operatives, visitors and members of the public.
3 Evaluate the risk. How severe is the potential injury? How likely is it to happen? A severe injury may be possible but may also be very improbable. On the other hand, a minor injury might be very likely.
4 If there is an unacceptable risk, can the job be changed? Could different tools or materials be used instead?
5 If the risk is acceptable, what measures can be taken to reduce the risk further? This could be training, special equipment or using personal protective equipment (PPE).
6 Keep good records. Explain the findings of the risk assessment to the operatives involved. Update the risk assessment as required – there may be new machinery, materials or staff. Even adverse weather can bring additional risks.

Risk assessment

Activity / Workplace assessed: Return to work after accident
Persons consulted / involved in risk assessment:
Date:
Reviewed on:

Location:
Risk assessment reference number:
Review date:
Review by:

Significant hazard	People at risk Describe who could be harmed (e.g. employees, contractors, visitors etc.)	Existing control measure What is currently in place to control the risk?	Risk rating Use matrix identified in guidance note Likelihood (L) Severity (S) Multiply (L) * (S) to produce risk rating (RR)				Further action required What is required to bring the risk down to an acceptable level? Use hierarchy of control described in guidance note when considering the controls needed	Actioned to: Who will complete the action?	Due date: When will the action be complete by?	Completion date: Initial and date once the action has been completed
			L	S	RR	L/M/H				
Uneven floors	Operatives	Verbal warning and supervision	2	1	2	m	None applicable	Site supervisor	Active now	Ongoing
Steps	Operatives	Verbal warning	2	1	2	m	None applicable	Site supervisor	Active now	Ongoing
Staircases	Operatives	Verbal warning	2	2	4	m	None applicable	Site supervisor	Active now	Ongoing

		Likelihood		
		1 Unlikely	2 Possible	3 Very likely
Severity	1 Slight/minor injuries/minor damage	1	2	3
	2 Medium injuries/significant damage	2	4	6
	3 Major injury/extensive damage	3	6	9

1 Low risk, action should be taken to reduce the risk if reasonably practicable

2, 3, 4 Medium risk, is a significant risk and would require an appropriate level of resource

6 & 9 High risk, may require considerable resource to mitigate. Control should focus on elimination of risk, if not possible control should be obtained by following the hierarchy of control

▲ Figure 1.10 A risk assessment

Method statements

Method statements are used alongside risk assessments. They give a clear, uncomplicated sequence of work to achieve the specified task and can be used to record the specific hazards and potential hazards associated with the task.

An employer will have a method statement written for all trade tasks that will be performed during the construction of a project. Make sure you consult risk assessments and method statements before you start work on your assigned task.

Communicating safety information

Good communication between personnel on site and in construction workshops is essential to keep everyone informed and updated about safety issues.

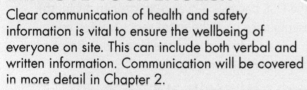

IMPROVE YOUR ENGLISH

Clear communication of health and safety information is vital to ensure the wellbeing of everyone on site. This can include both verbal and written information. Communication will be covered in more detail in Chapter 2.

Inductions

Access to construction sites is controlled to protect members of the public and maintain safety standards. All visitors and new operatives arriving on site will be given an induction session. This will explain:

- the layout of the site
- any site-specific hazards to be aware of
- the location of welfare facilities
- the assembly areas in case of emergency
- site rules, including risks associated with the use of drugs and alcohol
- 'house-keeping' requirements and safe waste disposal
- environmental protection requirements.

During induction, you may be required to provide details of your qualifications so that a record can be kept of skilled personnel working on site.

Toolbox talks

Toolbox talks are short talks given at regular intervals on site at your work location. They give timely safety reminders and outline any new hazards that may have arisen, because construction sites change as they develop. Weather conditions such as extreme heat, wind or rain may also create new hazards.

KEY TERM

Toolbox talks: short talks given at regular intervals onsite at a work location

First aid

Unfortunately, even with constant effort and diligent attention to safety issues, it is possible that first aid will need to be provided at some point in the construction process.

First aid should only be carried out by trained personnel. Remember that even a minor injury can become infected, so it should be cleaned and a dressing applied by a competent person. If any cut or injury shows signs of infection or becomes inflamed and painful, seek proper medical attention.

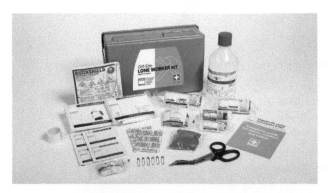

▲ Figure 1.11 A first aid kit

An employer's first-aid needs will be assessed to indicate if a first-aider (someone trained in first aid) is necessary. The minimum requirement is to appoint a person to take charge of first-aid arrangements, whose responsibilities include looking after the first-aid equipment and facilities and calling the emergency services when required. First-aid kits vary according to the size of the workforce. They should not contain tablets or medicines.

HEALTH AND SAFETY

First aid is designed to stabilise a patient for later treatment if required. The casualty may be taken to hospital or an ambulance may be called. In the event of an emergency you should raise the alarm.

ACTIVITY

Find the first aid kit in your workplace or training centre. Compare it with Figure 1.11 and check what it contains. Decide if anything is missing.

Recording accidents

By keeping records of accidents on site, it is possible to see trends that may be emerging, possibly due to bad habits and incorrect work practices. This provides an opportunity to make changes that will improve safety for all workers on site. Keeping records of accidents and reporting them to the HSE is governed by regulations.

IMPROVE YOUR ENGLISH

Written records can provide a wealth of information that can be referred back to, to help improve on-site practices. Records should be ordered and well maintained, using correct written English so the information is easily understood by other employees.

Reporting of Injuries, Diseases and Dangerous Occurrences Regulations (RIDDOR) 2013

The Reporting of Injuries, Diseases and Dangerous Occurrences Regulations (RIDDOR) 2013 state that employers must report to the HSE all accidents that result in an employee needing more than seven days off work. Diseases and dangerous occurrences must also be reported.

INDUSTRY TIP

You can access the Reporting of Injuries, Diseases and Dangerous Occurrences Regulations (RIDDOR) 2013 at: www.legislation.gov.uk/uksi/2013/1471/contents

A dangerous occurrence which has not caused an injury (referred to as a 'near miss') should still be reported to the HSE because if it were to happen again, the consequences could be more serious. Steps must also be taken to minimise the likelihood of such an occurrence happening again.

Here are some examples of injuries, diseases and dangerous occurrences that would need to be reported to the HSE:

- A joiner cuts off a finger while using a circular saw.
- A plumber takes a week off after a splinter in their hand becomes infected.
- A groundwork operative contracts **leptospirosis**.
- An operative develops dermatitis (a serious skin problem) after contact with an irritant substance.
- A scaffold suffers a collapse following severe weather, unauthorised alteration or overloading, but no one is injured.

KEY TERM

Leptospirosis: also known as Weil's disease, this is a serious disease spread by contact with urine from rats and cattle

RIDDOR enables the HSE to investigate serious incidents and collate statistical data. This information is used to help reduce the number of similar accidents happening in future and improve workplace safety.

```
┌─────────────────────────────────────────────────────────────────────────┐
│  ╔═══╗                                                                    │
│  ║HSE║                                             Health and Safety      │
│  ╚═══╝                                             Executive              │
│                                                                           │
│  Health and Safety at Work etc Act 1974                                   │
│  The Reporting of Injuries, Diseases and Dangerous Occurrences Regulations 2013 │
│                                                       Zoom  100   KS  i  ? │
│  F2508 - Report of an injury                                              │
│  About you and your organisation                                          │
│  *Title         *Forename              *Family Name                       │
│  [    ]         [              ]       [              ]                    │
│                                                                           │
│  *Job Title     [              ]    *Your Phone No  [         ]           │
│                                                                           │
│  *Organisation Name  [              ]                                     │
│                                                                           │
│  *Address Line 1  [              ]               (eg building name)       │
│                                                                           │
│  Address Line 2   [              ]               (eg street)              │
│                                                                           │
│  Address Line 3   [              ]               (eg district)            │
│                                                                           │
│  *Town            [              ]                                        │
│                                                                           │
│  County           [              ]                                        │
│                                                                           │
│  *Post Code  [        ]    Fax Number [        ]                          │
│                                                                           │
│  *E-Mail     [              ]                                             │
│                                                                           │
│        □ Remember me    [?]                                               │
│                                                                           │
│  *Did the incident happen at the above address?    □ Yes   □ No           │
│  *Which authority is responsible for monitoring and inspecting health     │
│  and safety where the incident happened?           □ HSE  □ Local Authority  [?] │
│  Please refer to the help for guidance on the responsible authority       │
│                                                                           │
│  [ Next ]  [ Form Preview ]                                               │
│                              Page 1 of 5                                  │
└─────────────────────────────────────────────────────────────────────────┘
```

▲ Figure 1.12 An F2508 injury report form

Although minor accidents and injuries are not reported to the HSE, records must still be kept. Accidents must be recorded in the accident book. This provides a record of what happened and is useful for future reference.

VALUES AND BEHAVIOURS

Everyone on site must work hard to create a safe working environment. Accidents don't just affect the person who has the accident. Work colleagues or members of the public might be affected and so will the employer. Here are some of the potential consequences of an accident on site:

- emotional trauma for workers
- poor company image
- loss of production
- increase in insurance costs
- closure of the site.

Fire safety

Fire safety relies on all personnel on site or in a training centre maintaining awareness of the potential hazards that could lead to an outbreak of fire.

The fire triangle

Fire needs three things to start, so if just one of them is missing there will be no fire. If all are present, then a fire is unavoidable. A fire needs:

- oxygen: a naturally occurring gas in the air that combines with flammable substances under certain circumstances
- heat: a source of fire, such as a hot spark from a grinder or naked flame
- fuel: things that will burn, such as acetone, timber, cardboard or paper.

▲ Figure 1.13 The fire triangle

> **KEY POINT**
>
> Remember, if you have heat, fuel and oxygen you will have a fire. Remove any one of these and the fire will go out.

Preventing the spread of fire

Maintaining a clean and tidy workplace is a key factor in helping to prevent fires from starting and spreading. Think about the following points:

- Wood offcuts should not be left in big piles or standing up against a wall. Instead, useable offcuts should be stored in racks.
- Put waste into the allocated disposal bins or skips.

- Always replace the cap on unused fuel containers when you put them away, otherwise they are a potential source of danger.
- Flammable liquids (not limited to fuel-flammable liquids) such as oil-based paint, thinners and oil must be stored in a locked metal cupboard or shed.
- Never light a cigarette around flammable liquids.
- Wood dust can be explosive, so when doing work that produces wood dust it is important to use some form of extraction and have good ventilation.

Fire extinguishers

To contribute to fire safety, you need to know the location of fire extinguishers and fire blankets in your workplace or training centre and which type can be used on different fires. Study Table 1.4 below to see which type of extinguisher can be used in different situations.

▼ Table 1.4 Types of fire extinguisher and their uses

Classification of fire	Materials within the classification	Type of extinguisher to use
A	Wood, paper, hair, textiles	Water, foam, dry powder
B	Flammable liquids	Foam, dry powder, CO_2
C	Flammable gases	Dry powder, CO_2
D	Flammable metals	Specially formulated dry powder
E	Electrical fires	CO_2, dry powder
F	Cooking oils	Wet chemical, fire blanket

▲ Figure 1.14 CO_2 extinguisher

▲ Figure 1.15 Dry powder extinguisher

▲ Figure 1.16 Water extinguisher

▲ Figure 1.17 Foam extinguisher

▲ Figure 1.18 Fire blanket

ACTIVITY

Check your workplace or training centre and note the types and locations of fire extinguishers.

KEY POINT

Remember, although all fire extinguishers are red, they each have a different coloured label to identify their contents.

HEALTH AND SAFETY

It is very important to use the correct extinguisher when dealing with a fire, as using the wrong one could make the danger much worse. For example, using a water extinguisher on an electrical fire could lead to the user being electrocuted!

ACTIVITY

An additional type of foam fire extinguisher has been introduced called a P50. Do some online research to find out more about this type of fire extinguisher and list the benefits of using it.

Emergency procedures

In an emergency situation, many people have a tendency to panic. It is vital to be prepared so that if an emergency such as a fire does occur, you will know what to do.

It is your responsibility to know the emergency procedures on your work site:

- If you discover a fire or other emergency, raise the alarm.
- Be aware of the alarm signal. It might be a bell, voice or siren.

- Proceed to the designated fire assembly point in an orderly way. Leave all your belongings behind, as they may slow you or others down.
- At the assembly point, there will be someone who will ensure everyone is safely away from the emergency by taking a head count. Co-operate with them.
- Do not re-enter the building or site until you are told by the appointed person that it is safe to do so.

VALUES AND BEHAVIOURS

You should know where the fire assembly point is in advance of a possible evacuation event. If you are not at the fire assembly point when an alarm is sounded and others must look for you, they could be put at risk. Check the route from your work location to the designated fire assembly point. Time how long It takes you to calmly walk there.

▲ Figure 1.19 Fire assembly point

Signs and safety notices

Signs and safety notices in the workplace can contribute to managing potential hazards and reducing accidents and emergencies. There is a range of standardised safety signs used in construction that you should be familiar with. Some give instructions on what you *must* do, while others give instructions on what you must *not* do. Table 1.5 shows some examples.

▼ Table 1.5 Signs and safety notices

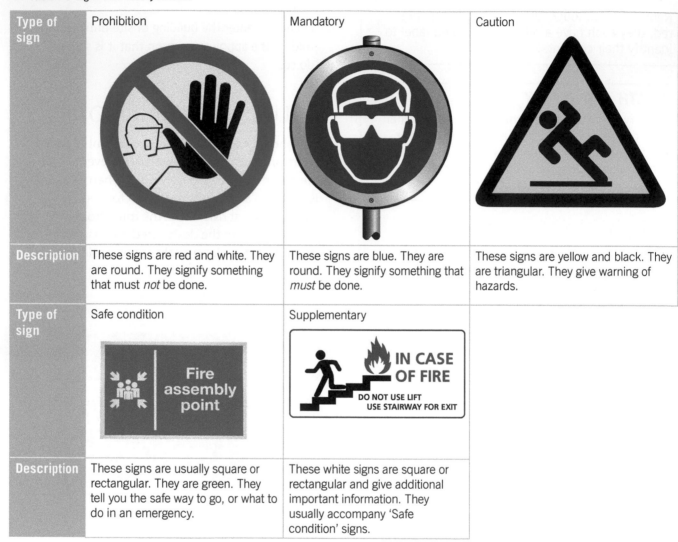

Type of sign	Prohibition	Mandatory	Caution
Description	These signs are red and white. They are round. They signify something that must *not* be done.	These signs are blue. They are round. They signify something that *must* be done.	These signs are yellow and black. They are triangular. They give warning of hazards.
Type of sign	Safe condition	Supplementary	
Description	These signs are usually square or rectangular. They are green. They tell you the safe way to go, or what to do in an emergency.	These white signs are square or rectangular and give additional important information. They usually accompany 'Safe condition' signs.	

3 HANDLING MATERIALS AND EQUIPMENT SAFELY

Health and safety laws and regulations are designed to cover every activity on a construction site. We will now look at some legislation designed to protect you when you're handling materials and using certain types of equipment.

Methods and equipment for handling materials

Using incorrect lifting and handling techniques poses a serious risk to your health. It is very easy to injure your back or other parts of your body, causing pain and discomfort, possibly over an extended period. Using the correct lifting techniques and equipment will reduce the risk of injury and enable you to handle and move materials more comfortably.

Manual Handling Operations Regulations 1992 (amended 2002)

Within reason, employers and training providers must arrange for personnel to avoid manual handling if there is a possibility of injury. If manual handling cannot be avoided, then they must manage and minimise the risk of injury by carrying out a risk assessment.

▲ Figure 1.20 Wheelbarrow

INDUSTRY TIP

Access the Manual Handling Operations Regulations 1992 (amended 2002) at:
www.legislation.gov.uk/uksi/1992/2793/contents/made

HEALTH AND SAFETY

Most workplace injuries occur as a result of manual handling. Remember, pushing or pulling an object also comes under the Manual Handling Operations Regulations.

▲ Figure 1.21 Brick tongs

Lifting and handling

Let's look at some points about correct lifting techniques and the equipment that can be used to safely handle materials. Think about the following:

- Assess the load. Is it too heavy? Do you need assistance or additional training? Is it an awkward shape?
- Can a lifting and transporting aid be used?
- Does the lift involve twisting or reaching?
- Where is the load intended to be moved to? Is there a clear path? Is the intended location clear of obstacles?

For safety and efficiency, most sites will have some means of moving heavy items mechanically, such as a fork lift or crane.

How to lift and place an item correctly

If items cannot be moved by mechanical means and must be moved manually, methods have been developed to move and handle components and materials efficiently and in a way that reduces the risk of injury. One recommended method is called **kinetic** lifting.

Look at Figure 1.22 to see how to use this method.

KEY TERM

Kinetic: relating to, caused by, or producing movement

▲ Figure 1.22 Safe kinetic lifting technique

Always lift with your back straight, elbows in, knees bent and your feet slightly apart. When placing the item, be sure to use the leg muscles, bending your knees. Beware of trapping your fingers when stacking materials. Place the item on levelled, carefully spaced bearers if required.

For very heavy items, get assistance from one or more helpers. It is important that one person in the team is in charge, and that lifting is done in a co-operative way.

ACTIVITY

Consider this list of materials: plywood, cement, aggregates, sawn timber joists, drainage pipes and kerbs. Write a short report stating how you would transport and stack these materials around your place of work. Take account of things such as uneven ground, trip hazards, weather conditions and bright sun reducing visibility.

Provision and Use of Work Equipment Regulations (PUWER) 1998

The Provision and Use of Work Equipment Regulations (PUWER) 1998 place duties on:

- people and companies who own, operate or have control over work equipment
- employers whose employees use work equipment.

INDUSTRY TIP

Access the Provision and Use of Work Equipment Regulations (PUWER) 1998 at: www.legislation.gov.uk/uksi/1998/2306/contents/made

Work equipment is defined as any machinery, appliance, apparatus, tool or installation for use at work (whether exclusively or not). This includes equipment which employees provide for their own use at work. The scope of work equipment is therefore extremely wide.

The use of work equipment is also very widely interpreted and, according to the HSE, means 'any activity involving work equipment including starting, stopping, programming, setting, transporting, repairing, modifying, maintaining, servicing and cleaning.' It includes equipment such as diggers, electric planers, stepladders, hammers and wheelbarrows.

Under the terms of PUWER, work equipment must be:
- suitable for the intended use
- safe to use
- well maintained
- inspected regularly.

Regular inspection is important since a tool which is safe to use when new, may no longer be safe after a considerable period of use. Work equipment must only be used by people who have received adequate instruction and training. Information regarding the use of the equipment must be given to the operator and the equipment must only be used for the designed purpose.

Study the following key points that emphasise what is needed to use equipment safely:
- Protective devices, such as emergency stop buttons, must be used when needed.
- Brakes must be fitted, where appropriate, to slow down moving parts to bring the equipment to a safe condition when turned off or stopped.
- Equipment must have adequate means of isolation.
- Warnings, either by signs or other means such as sounds or lights, must be used as appropriate.
- Access to dangerous parts of the machinery must be controlled.

HEALTH AND SAFETY

Under PUWER, abrasive wheels used for cutting or grinding can only be changed by someone who has received training to do this. Wrongly fitted wheels are a serious danger because they have the potential to explode.

Lifting Operations and Lifting Equipment Regulations (LOLER) 1998

The Lifting Operations and Lifting Equipment Regulations (LOLER) 1998 place responsibility upon employers to ensure that the lifting equipment provided for use at work is:
- strong and stable enough for the particular use and marked to indicate safe working loads
- positioned and installed to minimise any risks
- used safely by making sure the work is planned, organised and performed by competent people
- subject to ongoing thorough examination and, where appropriate, inspection by competent people.

▲ Figure 1.23 Using a scissor lift at height

INDUSTRY TIP

Access the Lifting Operations and Lifting Equipment Regulations (LOLER) 1998 at:
www.legislation.gov.uk/uksi/1998/2307/contents/made

Using electrical equipment

Electricity is a useful energy resource, but it can be very dangerous and must be handled with care. Make sure that you are properly trained before working with electrical equipment. Never dismantle or adjust electrical equipment of any kind – leave it to someone who is trained and competent.

Work involving the use of electrical equipment is regulated by The Electricity at Work Regulations 1989.

INDUSTRY TIP

Access the Electricity at Work Regulations 1989 at:
www.legislation.gov.uk/uksi/1989/635/contents/made

The dangers of electricity

When working with or near equipment powered by electricity always be aware of the main dangers that can arise:

- shock and burns (a 230 V shock can kill)
- electrical faults which could cause a fire
- an explosion where an electrical spark has ignited a flammable gas.

Voltages

The lower the voltage, the safer it is. However, a low voltage will not be suitable to power some machines, so be aware of the voltage ratings of equipment you use or are near. On site, for greater safety, 110 V (volts) is recommended and this is the voltage rating most commonly used in the building industry. This is converted from 230 V by use of a transformer.

230 V (commonly referred to as 240 V) is a domestic voltage that can be used on site for charging batteries. 230 V is often used in workshops. Protection equipment called a residual current device (RCD) should be used when 230 V tools are operated. This will disconnect the supply quickly if a fault or unsafe condition occurs.

410 V (otherwise known as 3 phase) is used for large machinery, such as joinery shop equipment.

INDUSTRY TIP

Voltages are 'nominal', meaning they can vary slightly.

Battery power

Battery power is much safer than mains power. Many power tools are now available in battery-powered versions. They are available in a wide variety of voltages from 3.6 V for a small powered screwdriver all the way up to 36 V for large masonry drills.

▲ Figure 1.24 Battery-powered drill

Wiring

The wires inside a cable are made from copper, which conducts electricity. The copper is surrounded by a plastic coating called insulation that is colour coded. The three wires in a cable are the live (brown), which works with the neutral (blue) to conduct electricity, making the appliance work. The earth (green and yellow stripes) prevents electrocution if the appliance is faulty or damaged.

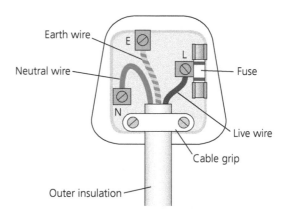

▲ Figure 1.25 A correctly wired plug

Checks on power tools

Power tools should always be checked before use. Make sure that you inform your supervisor if you find a fault. The tool will need to be repaired and until then, the tool needs to be kept out of use. The tool should be removed from the work area, put in a secure location and clearly labelled 'Do not use'.

Table 1.6 shows the list of checks that should be carried out to ensure electric tools are safe to use.

Power tools and equipment should be stored correctly after use. Portable power tools must be returned to carrying boxes and cables need to be wound onto reels or neatly coiled as they can easily become tangled.

▲ Figure 1.27 Cable reel

> **HEALTH AND SAFETY**
>
> Always fully unroll an extension lead before using it. If you leave the cable coiled up during use, it can overheat and cause a fire.

▲ Figure 1.26 Cable protection

▼ Table 1.6 Electrical safety checks

Item	What to check for
Portable Appliance Test (PAT) label	PAT is a regular test programme carried out by a competent person (a qualified electrician) to ensure the tool is in a safe electrical condition. A sticker is placed on the tool after it has been tested. Tools that do not pass the PAT are taken out of use.
Cable	Is it damaged? Is there a repair? Insulation tape may be hiding a damaged cable. Damaged cables must be replaced.
Casing	Is the casing cracked? Plastic casings ensure the tool is double-insulated. This means the live parts inside are safely shielded from the user. A cracked casing will reduce the protection to the user and will require repair.
Electricity supply leads	Are they damaged? Are they creating a trip hazard? You need to place them in such a way that they do not make a trip hazard. Are they protected from damage? If they are lying on the floor with heavy traffic crossing them, they must be covered.

4 USING ACCESS EQUIPMENT AND WORKING AT HEIGHT

Working at height remains one of the biggest causes of fatalities and major injuries on construction sites.

KEY TERM

Working at height: work in any place where, if there were no precautions in place, a person could fall a distance that is liable to cause personal injury

Keeping safe

Employing safe methods of work and using suitable access equipment correctly are vital factors in keeping yourself and others safe when working at height. All equipment should be checked for safe condition before use and inspection tags and notices should be up to date. Never take short cuts, thinking an accident won't happen to you!

Work at Height Regulations 2005 (amended 2007)

The Work at Height Regulations 2005 (amended 2007) place several legal duties upon employers:
- Working at height should be avoided if possible.
- If working at height cannot be avoided, the work must be properly organised and risk assessments carried out.
- Risk assessments should be regularly updated.
- Those working at height must be trained and competent.
- A method statement must be provided.

INDUSTRY TIP

Access the Work at Height Regulations 2005 at: www.legislation.gov.uk/uksi/2005/735/contents/made

▲ Figure 1.28 Workers wearing safety harnesses on an aerial access platform

When planning a work task that involves working at height, consider the following points.
- How long is the job expected to take?
- How would you categorise the work? It could be anything from fitting a single light bulb, through to removing a chimney or installing a roof.

▲ Figure 1.29 A cherry picker can provide a safe working platform

If an access platform is to be used, consider the following:

- How is the access platform going to be reached? By how many people?
- Will people be able to get on and off the structure safely? Could there be overcrowding?
- What are the risks to passers-by? Could debris or dust blow off and injure others in the area nearby?
- What are the prevailing conditions like? Extreme weather, unstable buildings or poor ground conditions need to be considered.

Access equipment and safe methods of use

The means of access should only be chosen after a risk assessment has been carried out and a method statement has been written. There are various types of access equipment suitable for a range of specific jobs.

Ladders

Ladders are usually only used for access onto a working platform. They are not designed to be used as a platform for working from except for light, short-duration work. A ladder should be set at an angle of 75° to vertical. A helpful way of setting up the correct angle is to position the base of the ladder one unit of measurement away from vertical to four units of measurement up the height of the ladder.

The top of the ladder must rest against something secure and stable, not something flexible such as plastic guttering.

There are ladders designed for specific purposes such as accessing a roof surface without causing damage to it. Always use a ladder that's designed for the job.

Stepladders are designed for light, short-term work.

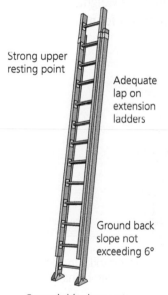

Strong upper resting point

Adequate lap on extension ladders

Ground back slope not exceeding 6°

Ground side slope not exceeding 16°, clean and free of slippery algae and moss

▲ Figure 1.30 Using a ladder correctly

▲ Figure 1.31 Resting ladders on plastic guttering can cause it to bend and break

▲ Figure 1.32 Roof ladder

Stepladders

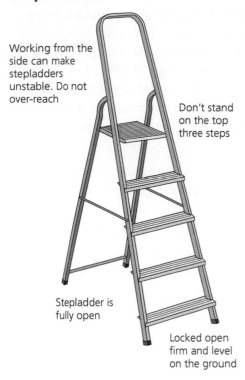

Working from the side can make stepladders unstable. Do not over-reach

Don't stand on the top three steps

Stepladder is fully open

Locked open firm and level on the ground

▲ Figure 1.33 Using a stepladder correctly

▼ Table 1.7 Using ladders and stepladders safely

Table 1.7 shows the key points related to safe use of ladders and stepladders.

HEALTH AND SAFETY

Always complete ladder pre-checks carefully. Check the stiles (the two uprights) and rungs for damage such as splits or cracks. Do not use painted ladders because the paint could be hiding damage!

Trestles

Trestles are part of an access system used for relatively short durations. This access equipment is not designed to carry heavy loadings of materials, so a bricklayer using trestles would need to limit the weight placed on the platform when stacking materials for a work task.

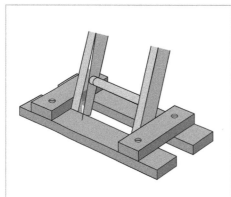

A ladder secured at the base

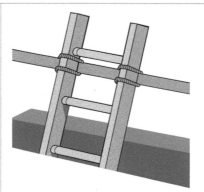

A ladder secured at the top

Access ladders should extend 1 m above the landing point to provide a strong handhold

Certain step ladders are unsafe to work from the top three rungs

Don't over-reach. Remain on the same rung during a task

Grip the ladder when climbing and remember to keep three points of contact

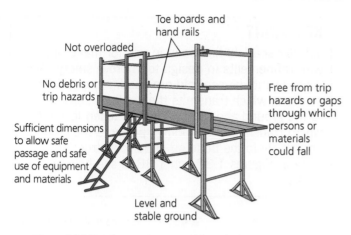

▲ Figure 1.34 Trestles used as a working platform

Labels in Figure 1.34:
- Toe boards and hand rails
- Not overloaded
- No debris or trip hazards
- Free from trip hazards or gaps through which persons or materials could fall
- Sufficient dimensions to allow safe passage and safe use of equipment and materials
- Level and stable ground

Tower scaffold

Tower scaffold systems are manufactured from galvanised steel or lightweight aluminium alloy. They must be erected by someone competent in accordance with the manufacturer's instructions.

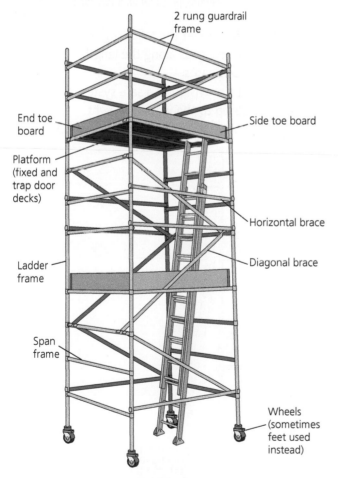

Labels in Figure 1.35:
- 2 rung guardrail frame
- End toe board
- Side toe board
- Platform (fixed and trap door decks)
- Ladder frame
- Horizontal brace
- Diagonal brace
- Span frame
- Wheels (sometimes feet used instead)

▲ Figure 1.35 Parts of a tower scaffold

To use a tower scaffold safely:
- Always read and follow the manufacturer's instruction manual.
- Only use the equipment for its designed purpose.
- The wheels or feet of the tower must be in contact with a firm surface.
- Outriggers should be used to increase stability.
- The maximum height given in the manufacturer's instructions must not be exceeded.
- The platform must not be overloaded.
- The platform should be unloaded (and reduced in height if required) before it is moved.
- Never move a tower scaffold, even a small distance, if it is occupied.

INDUSTRY TIP

Remember, a mobile access tower should always have guard rails and toe boards fitted when in use to prevent you and your materials from accidentally falling from the working platform.

Tubular scaffold

Tubular scaffold must only be erected by specialist scaffolding companies and often requires structural calculations. The scaffold is assembled using steel tubes and clips which bolt together tightly in a specified pattern.

Some types of tubular scaffold are designed to be assembled using specially shaped cups on the upright tubes which allow horizontal tubing to be slotted into them and wedged together tightly.

▲ Figure 1.36 Safe scaffolding being set up

Never alter tubular scaffold yourself. There can be a temptation to remove handrails to allow materials to be placed more conveniently on the working platform by forklift or crane. However, to do so would be dangerous to yourself and others working on the scaffold. Don't do it! When stacking materials on tubular scaffold, make sure the load is spread evenly to distribute the weight carried by the structure.

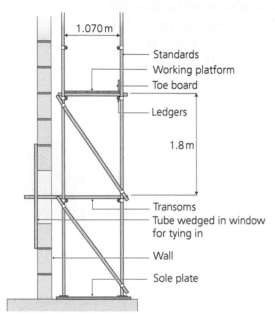

▲ Figure 1.37 Independent tubular scaffold

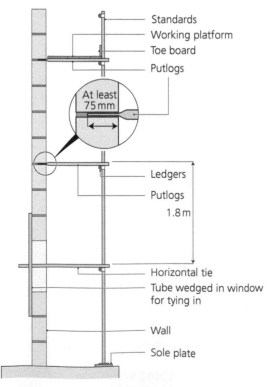

▲ Figure 1.38 Putlog tubular scaffold

KEY POINT

Tubular scaffold has been in use for many years with refinements in design to improve safety. Two types of tubular scaffold are used: independent scaffold, which transmits the loadings carried by it to the ground and putlog scaffold (used less often) which transmits the loadings partly to the ground and partly to the building under construction. Look at Figures 1.37 and 1.38 to see how this works.

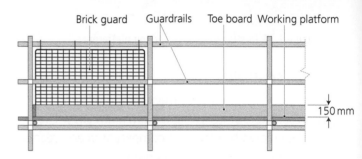

▲ Figure 1.39 A safe working platform on a tubular scaffold

All scaffolding must meet the following requirements:
- There must be no gaps along the length of the handrail or toe boards.
- There must be provision of a safe system for lifting any materials up to the working height.
- There must be a safe system of debris removal.

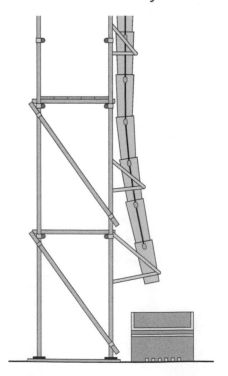

▲ Figure 1.40 A debris chute for scaffolding

HEALTH AND SAFETY

If a debris chute is not installed, don't be tempted to simply throw waste materials off the working platform of a scaffold. Place the materials safely on a pallet or in a bin for controlled removal by a fork lift or crane.

Summary

This chapter has covered the relevant health and safety laws and regulations that govern the work of a bricklayer on site. Of course, the principles identified also apply to activities in a training centre or college.

The aim of this chapter is to keep you and those working with you safe and to increase your awareness of the potential for accidents and how to prevent them. Review the information contained in this chapter frequently. Discuss what you learn with others and work with them to develop and apply your understanding of health and safety in the workplace.

Test your knowledge

1 Which document must be completed to manage safe working practices?

   a Accident report

   b Improvement notice

   c Prohibition notice

   d Risk assessment

2 Which shape and colour of sign inform you of something you *must* do?

   a Blue circle

   b Green circle

   c White square

   d Yellow triangle

3 Two parts of the 'fire triangle' are heat and fuel. What is the third?

   a Carbon dioxide

   b Hydrogen

   c Hydrogen sulphide

   d Oxygen

4 Which type of fire extinguisher would be used on an electrical fire?

   a Blanket

   b Foam

   c Powder

   d Water

5 Which specific legislation protects an operative when using solvents and adhesives?

   a Construction Design and Management Regulations

   b Control of Substances Hazardous to Health Regulations

   c Control of Vibration at Work Regulations

   d Manual Handling Operations Regulations

6 What is the correct angle at which to set a ladder against a wall?

   a 70°

   b 75°

   c 80°

   d 85°

7 What does 'VOC' stand for as a measure of how much pollution a product will release?

   a Varying organic compound

   b Virtual organic compound

   c Visual organic compound

   d Volatile organic compound

8 What is the upper exposure value in decibels above which an employer must provide hearing protection?

   a 75 dB(A)

   b 80 dB(A)

   c 85 dB(A)

   d 90 dB(A)

9 After a construction accident, how many days must an operative be off work before the accident must be reported to the HSE?

   a More than 5 days

   b More than 7 days

   c More than 14 days

   d More than 21 days

10 What voltage rating is referred to as '3 phase'?

   a 230 V

   b 240 V

   c 410 V

   d 420 V

11 Why should a report be made if a dangerous occurrence happens on site that results in a near-miss, but no one is hurt?

12 Describe the difference between an independent tubular scaffold and a putlog tubular scaffold.

13 State the key points required to lift and position an item safely using the kinetic lifting method.

14 Describe the characteristics of 'safe condition' signs.

15 List the information that a site induction session will provide.

CHAPTER 2

COMMUNICATION AND INFORMATION WITHIN THE CONSTRUCTION INDUSTRY

INTRODUCTION

Working in the construction industry will often mean working as part of a varied team, ranging from small groups of bricklayers to large projects with many skilled personnel. Teamwork is rewarding, but it can present challenges, particularly associated with your ability to communicate effectively with your colleagues and managers when working together.

Communication takes many forms, from verbal to written, and even drawn information presented in a range of documents. In this chapter, we will look at methods of communication to establish and maintain good working relations between personnel in the workplace. We will also consider how you gather and interpret information from a range of documentary sources to achieve success when working on a construction project.

By the end of this chapter, you will have an understanding of:
- communication methods used with colleagues and customers
- the range of construction documents used and how to interpret them
- planning and managing a construction project
- calculating resources and quantities of materials.

The table below shows how the main headings in this chapter cover the learning outcomes for each qualification specification.

Chapter section	Level 2 Technical Certificate in Bricklaying (7905-20) Unit 201	Level 3 Advanced Technical Diploma in Bricklaying (7905-30) Unit 301	Level 2 Diploma in Bricklaying (6705-23) Unit 202/602	Level 3 Diploma in Bricklaying (6705-33) Unit 301/701	Level 2 Bricklayer Trailblazer Apprenticeship (9077) Module 3
1. Communication methods used with colleagues and customers	1.3, 2.1	5.1	16.4, 16.5, 16.6, 16.7	14.1, 14.2, 14.3, 14.4, 14.5	1.1, 1.2, 5.1, 5.2
2. The range of construction documents used and how to interpret them	2.1, 2.2, 2.3	1.2, 3.2, 6.1, 6.2,	10.1, 10.2	10.1, 10.2, 10.3, 10.4, 10.5	3.1, 3.2, 3.3, 3.4, 3.5
3. Planning and managing a construction project	1.2, 2.1, 3.1, 3.3	3.1, 3.3, 4.1, 6.3	16.3	13.1, 13.2, 13.3, 13.4	3.7, 4.1, 5.1, 5.2
4. Calculating resources and quantities of materials	1.3, 2.1	3.2, 3.3	12.6, 13.3, 13.7, 14.2, 14.3	12.1, 12.2, 12.3, 12.4, 12.5, 12.6, 12.7, 12.8	3.6, 6.1, 6.2

Note: for 7905-20, Unit 201:
Partial content for Topic 1.3 and total content for Learning outcomes 4 and 5 are covered in Chapter 3.
Total content for Topics 2.4 and 3.2 is covered in Chapter 1.
Note: for 7905-30, Unit 301:
Content for Topics 1.1 and 4.2 is covered in Chapter 1.
Note: for 6705-23, Unit 202/602:
Content for Assessment criterion 10.3 is covered in Chapter 4.

1 COMMUNICATION METHODS USED WITH COLLEAGUES AND CUSTOMERS

Using effective communication skills is crucial to developing and maintaining good working relationships and to achieving success in the workplace. Communication involves sharing thoughts, information and ideas between people.

A breakdown of communication can be the cause of misunderstandings and mistakes that can lead to wasted materials, time and effort. Poor communication can also lead to accidents.

Ways to communicate

Every day we naturally use different communication methods without thinking about it. We might use **verbal communication**, for example, when talking to others face to face or on a telephone. Body language and facial expressions can communicate our mood or feelings in a non-verbal way. We can use many forms of written communication.

KEY TERM

Verbal communication: talking to others to transmit information

Let's look at the different types of communication that we can use when working in construction and how they can be applied.

Written communication methods

Written communication forms a permanent record of information passed to others. If it is read and interpreted carefully, it can reduce the chance of misunderstandings and enable relevant information to be referred to at a later date. Written information commonly includes letters and emails.

Letter

Sometimes it is necessary to communicate a complaint or a concern in writing. To check that the written information has been received, a letter can be sent that requires the recipients to sign to say they have received it. Letters are rarely handwritten now and are almost exclusively produced using a word-processor.

▲ Figure 2.1 Operative typing a letter

A letter should be set out with the following information:
- Your address – also known as the 'return address', this ensures a reply can be made if required.
- The date – needed as a point of reference to show the timescale over which the communication takes place.
- The recipient's name and address – this is important to confirm the intended audience for the information.
- The greeting – must be appropriate to the recipient. It could be 'Dear Sir/Madam', 'Dear Mr Jones' or 'Dear Bob'. The greeting can set the tone of the letter.
- The subject – this is often helpful to the recipient to identify details such as an order number or the job referred to.
- The text of your letter, which should be:
  - concise – letters that are clear and brief can be understood quickly
  - authoritative – letters that are well written and professionally presented have more credibility and are taken more seriously
  - factual – accurate and informative letters enable the reader to see immediately the relevant details, dates and requirements
  - constructive – letters with positive statements encourage action and quicker decisions
  - friendly – letters with a considerate, co-operative and complimentary tone are prioritised because the reader responds positively to the writer and wants to help.
- The closing phrase, your name and signature – insert after the body of text. Your letter should end

with an appropriate closing phrase such as 'Yours sincerely'.

- Leave several blank lines after the closing phrase (so that you can sign the letter after printing it), then type your name. You could put your job title and company name on the line beneath this.

VALUES AND BEHAVIOURS

The 'tone' of written communication is often more formal than spoken communication. Keep in mind that although communicating person-to-person is more informal, you should still maintain a respectful tone. See 'Communication with customers' later in this chapter on spoken communication.

KEY POINT

If your letter is addressed to someone whose name you know (for example, 'Dear Mrs Smith'), end it with 'Yours sincerely'.

If your letter is addressed to someone whose name you don't know (for example, 'Dear Sir/Madam'), end it with 'Yours faithfully'.

IMPROVE YOUR ENGLISH

Draft a letter to a supplier informing them that two pallets of bricks out of 14 were badly damaged on the bottom two layers and that replacements are required.

INDUSTRY TIP

In the past, memos were often used to communicate information between office staff on site. A memo is a brief note informing or reminding the receiver of something that must be done. Memo is short for 'memorandum', which is from the Latin for 'to be remembered'. Memos have now mostly been replaced by emails.

Memo

Hi Clare, just to let you know that the client will be joining us at the review meeting on site next Monday.

Kind regards, Martin.

▲ Figure 2.2 A memo

Email

While email is an electronic means of communicating, to be effective it still requires competent writing skills and an ability to present information in a clear and logical way. Emails have the advantage of almost instant delivery and allow documents and drawings to be sent as attachments.

Here are some points that should be kept in mind when communicating by email:

- Help the person or persons receiving the email to understand the urgency or importance of your message by including a clear 'subject' line.
- Keep your emails short and make sure your message is clear by avoiding phrases that could easily be misinterpreted.
- Use spell check before sending your message to avoid potentially embarrassing errors.
- Carefully read your email before you hit the 'send' button. Would you be comfortable with someone other than the intended person reading it?
- Remember that emails can be easily forwarded to others, so be careful not to write anything that could offend another reader.

Verbal communication methods

Talking face to face or on the phone is probably the most common form of verbal communication. Verbal methods of communicating include the use of site radios.

Keep in mind that mistakes can easily be made while communicating verbally. The person giving the information might not make the matter clear enough or the person receiving the information might misunderstand something. Often on site, there will be a lot of background noise which can lead to information being mis-heard.

▲ Figure 2.3 Verbal communication can occur in many ways

Since there is rarely a record of such conversations, it's sensible to remember these important points:

- Think before you speak, so that you get across exactly what you want to say.
- Be clear and concise in what you say.
- Ask for confirmation that what you have said has been understood.

INDUSTRY TIP

Confusion and misunderstanding can be minimised by recording important conversations for future reference, but make sure that the other person in the conversation is aware that recording is taking place and that they accept this. For greater certainty that there is clear understanding of the verbal communication, you could confirm important details by sending a brief confirmation email.

Body language as communication

Body language refers to the different forms of non-verbal communication shown below and can be a powerful indicator of a person's mood or feelings. Body language can also reveal an unspoken message. Examples include:

- rolling your eyes – can express frustration or mean 'here we go again!'
- yawning – indicating boredom
- hands in pockets – indicating lack of interest
- crossed arms – indicating discontent with what is being said
- smiling – indicating happiness
- frowning – indicating doubt or disagreement.

▲ Figure 2.4 Types of body language

VALUES AND BEHAVIOURS

Everyone has the right to be treated with respect. Aggressive speech or body language breeds aggression and leads to poor working relations, reduced levels of productivity and a poor image to customers. Good relationships in the workplace will grow if you treat others as you would like to be treated yourself.

Communication with customers

Your body language forms a significant part of interacting and communicating with others on any construction project. The way you communicate with customers will play a huge part in whether they are satisfied with the service you provide, and whether they will recommend you to others and return to you in the future.

Your style and manner of interacting and communicating with customers should be professional, respectful and promote the feeling that their property is safe in your care. Customers should feel like their needs are being listened to and that you are working towards satisfying their requirements.

Demonstrating a slovenly, disrespectful attitude in the way you conduct yourself will not contribute to building and establishing a good reputation. Here are some important points to keep in mind when dealing with customers on a personal level.

- Behave in a respectful manner, looking after the customers' property and protecting their environment (for example, by sheeting over property that could easily be damaged and keeping the site clean and tidy).
- Speak to customers in a tone that shows politeness and don't cause confusion by using highly technical terms about the job.
- Promote a professional image by being reliable, on time and clean and neat in your appearance.
- Be professional in the way you plan and organise your work and organise your tools and equipment. Store materials neatly and position them considerately.

- Maintain good morale by dealing with other trades and personnel in a considerate way. Deal with problems in a mature way, knowing when to refer matters to your line manager.
- Be honest and trustworthy.

Providing good customer service will establish a positive reputation for you and the company you work for, possibly leading to more work in the future.

▲ Figure 2.5 Maintain good communication with the customer

Communication at meetings

Communication on site often takes place as a formally arranged meeting. If there are problems to resolve, a review of progress is needed or there are questions about planning ahead, a meeting provides an opportunity for the subject matter to be clearly communicated to all those involved.

A meeting will mainly involve verbal communication, but some written documentation may be provided to support the matters being discussed.

Agenda

An **agenda** is a written document that sets out the points for discussion in sequence. This provides a focus for discussions and allows the meeting to progress in an orderly manner. Each point on the agenda should be resolved before moving on to the next.

The agenda should be provided well before the arranged time and date of the meeting to allow all those who will attend to prepare their contribution to the discussions, if appropriate. The agenda may be circulated by email for efficiency.

> **KEY TERM**
>
> **Agenda:** a written document that sets out the points for discussion in sequence

> **ACTIVITY**
>
> Progress on site is being slowed down by a shortage of skilled bricklayers. Your supervisor has asked you to write an agenda for a site meeting to discuss what can be done to resolve the problem. Draw up an agenda that includes at least four points for discussion. You could, for example, include a discussion on how the existing workforce is being used, where extra workers can be found, etc. Discuss your agenda with someone else to see if they can add to it.

Minutes

The minutes of a meeting form a record of decisions made and actions planned by those in attendance. The minutes from past meetings are used to make sure that actions decided on previously have been carried out.

▲ Figure 2.6 Set of minutes

The minutes of meetings usually include:
- the time, date and place of the meeting
- a list of the people attending
- a list of absent members of the group (their names are entered as 'apologies')

- approval of the previous meeting's minutes, and any matters arising from those minutes
- for each item in the agenda, a record of the principal points discussed, the decisions taken and who is responsible for actioning any discussed items
- any other business (often abbreviated to 'AOB')
- the time, date and place of the next meeting
- the name of the person taking the minutes.

Communicating through toolbox talks

A toolbox talk is a short presentation to communicate information to workforce members at their work location, usually on a single aspect of health and safety. The toolbox talk may be prepared by a company's safety officer, a manufacturer's representative to discuss a product being used or a site supervisor.

Toolbox talks are a convenient and effective way of keeping personnel informed of important information. Since they are conducted in the workplace, the minimum amount of production time is lost and the information can be applied directly to the work in hand.

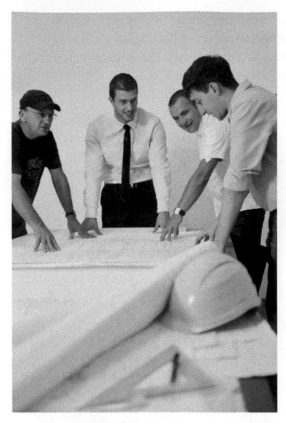

▲ Figure 2.7 Site meeting

2 THE RANGE OF CONSTRUCTION DOCUMENTS USED AND HOW TO INTERPRET THEM

Even a relatively small-scale project will require many different documents and drawings to provide the information needed for successful completion. We'll now look at the range of documents you will need to know about in order to play your part in the construction process.

Drawings used for construction

Drawings are documents that are required at every stage of building work and are an efficient way of providing a great deal of clear information, without the need for lots of potentially confusing text. After the building has been designed and the details are agreed with the client, drawings are required in order to apply for planning consent and Building Regulations approval.

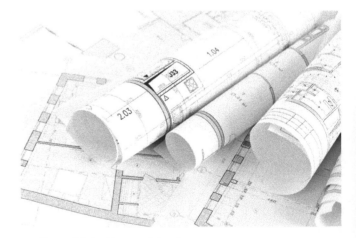

▲ Figure 2.8 Architectural drawings

These drawings will show the size, position and general arrangement of the proposed construction and allow the local authority planning authority and Building

Control officers to decide whether approval should be given and whether the proposed building design meets the current Building Regulations.

ACTIVITY

Research what documents are required to submit a planning application by visiting:
www.planningportal.gov.uk

INDUSTRY TIP

Planning authorities are involved in approving the design and regulation of projects at various stages of development. The first step is to obtain outline planning permission, which, as the name suggests, shows the broad design of the proposed structure. This is followed by detailed planning permission covering the full design. Planning authorities will also be involved in regulating work on listed buildings.

Architectural drawings are drawn according to a set of 'conventions', which include particular views, sheet sizes and units of measurement.

When producing drawings, an architectural technician or a draughtsperson will draw a building to **scale**. This means that large structures can be represented on a document that is much smaller and more manageable than the real thing. (We'll look at how architectural drawings are produced later in this chapter.)

Scale is shown by using a **ratio**, such as 1 to 10. This would usually be written in the form '1:10' and means that if we wanted to draw a real wall that is 1 m (or 1000 mm) long, it would be shown as a drawing 100 mm long. This is because 100 mm is 1/10 of 1000 mm.

KEY TERMS

Ratio: the amount or proportion of one thing compared to another
Scale: when accurate sizes of an object are reduced or enlarged by a certain amount

INDUSTRY TIP

When reading drawings, always work to the written dimensions. Although it is possible to take dimensions off a scale drawing using a scale rule, they should not be relied on as accurate.

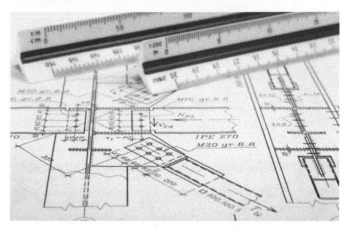

▲ Figure 2.9 Scale rule on a drawing

IMPROVE YOUR MATHS

Show your understanding of how scale works by filling in the gaps in the table.

Scale size on drawing (mm)	Scale	Actual size (mm)
10	1:10	100
25	1:20	a
b	1:50	300
50	1:200	c

Types of drawings

You will now look at the different types of drawings, the scales in which they are presented and what they communicate to a skilled operative.

General location drawings

The local authority will use the following drawings to consider planning approval applications.

▼ Table 2.1 Types of drawings

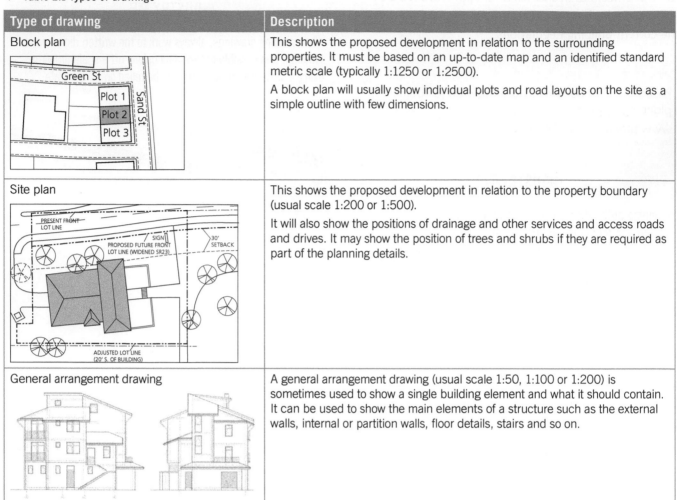

Type of drawing	Description
Block plan	This shows the proposed development in relation to the surrounding properties. It must be based on an up-to-date map and an identified standard metric scale (typically 1:1250 or 1:2500). A block plan will usually show individual plots and road layouts on the site as a simple outline with few dimensions.
Site plan	This shows the proposed development in relation to the property boundary (usual scale 1:200 or 1:500). It will also show the positions of drainage and other services and access roads and drives. It may show the position of trees and shrubs if they are required as part of the planning details.
General arrangement drawing	A general arrangement drawing (usual scale 1:50, 1:100 or 1:200) is sometimes used to show a single building element and what it should contain. It can be used to show the main elements of a structure such as the external walls, internal or partition walls, floor details, stairs and so on.

Construction drawings

These drawings focus on specific parts or elements of a structure that will be used to establish compliance with Building Regulations.

▼ Table 2.2 Specific drawings

Type of drawing	Description
Floor plans	Used to show the layout of the internal walls, doors and stairs and, in a dwelling, the arrangement of special use rooms such as bathrooms and the kitchen. These will be drawn at a scale of 1:50 or 1:100.

▼ Table 2.2 Specific drawings (continued)

Type of drawing	Description
Elevations 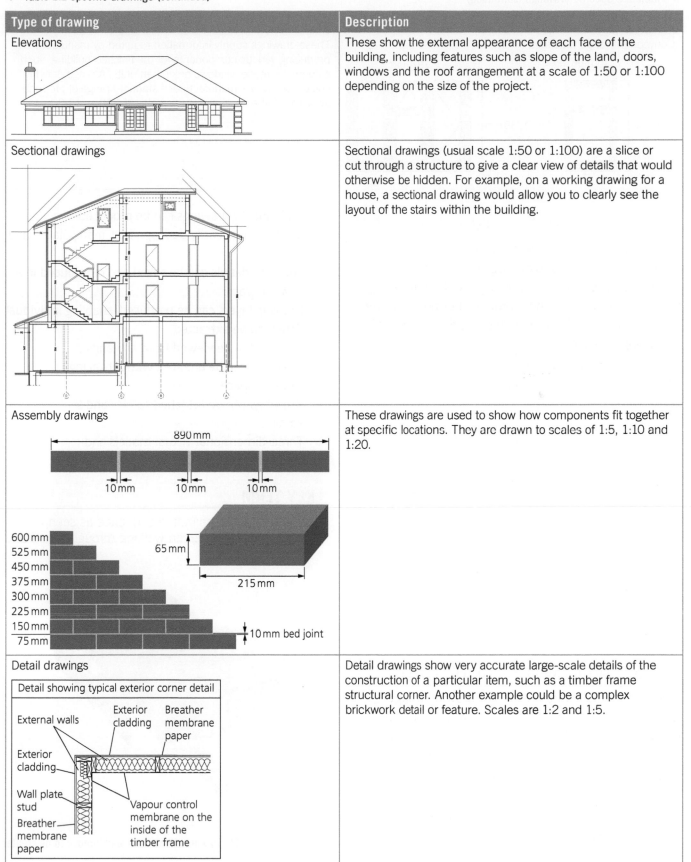	These show the external appearance of each face of the building, including features such as slope of the land, doors, windows and the roof arrangement at a scale of 1:50 or 1:100 depending on the size of the project.
Sectional drawings	Sectional drawings (usual scale 1:50 or 1:100) are a slice or cut through a structure to give a clear view of details that would otherwise be hidden. For example, on a working drawing for a house, a sectional drawing would allow you to clearly see the layout of the stairs within the building.
Assembly drawings	These drawings are used to show how components fit together at specific locations. They are drawn to scales of 1:5, 1:10 and 1:20.
Detail drawings	Detail drawings show very accurate large-scale details of the construction of a particular item, such as a timber frame structural corner. Another example could be a complex brickwork detail or feature. Scales are 1:2 and 1:5.

▼ Table 2.2 Specific drawings (continued)

Type of drawing	Description
Component/range drawings 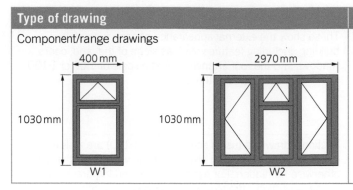	These drawings supply information required by manufacturers producing various components for the finished building, such as purpose-made windows or kitchen units. A range drawing could also show a manufacturer's standard range of products available off the shelf. Drawn to scales of 1:10 and 1:20.

How drawings are produced

In the past, drawings were produced by a draughtsperson by hand on large parallel-motion drawing boards. Some drawings are still produced this way, but most are now produced using computer aided design (CAD), so we'll focus on this method of producing drawings.

▲ Figure 2.10 Parallel-motion drawing board

Producing drawings using CAD

CAD is a sophisticated method of drawing using specialist software packages which require training to use effectively. A competent user can produce drawings in two or three dimensions and CAD programs are used in the design of many of the items we use every day.

Some of the advantages of using a CAD system to produce drawings are described below.

- High quality drawings can be produced.
- Drawings can easily be magnified, manipulated and amended.
- Standard details can be saved and reproduced in a new contract.
- Drawing layers can be produced, enabling individual details to be extracted.
- Objects can be viewed from any angle.
- Drawings can be archived without taking up valuable space.
- Drawings can be attached as files and sent via email.
- Three-dimensional virtual models and **walkthroughs** can be created.

KEY TERM

Walkthrough: an animated sequence as seen at the eye level of a person walking through a virtual structure

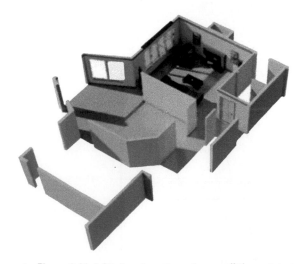

▲ Figure 2.11 A 3D drawing allows for a walkthrough to be created

While there are significant advantages in producing drawings using a CAD system, there are some disadvantages that should be considered:

- The high initial outlay for the equipment required including a high specification computer and visual display unit (VDU).
- The cost and time needed to train staff to use the software.
- The cost of purchasing and updating software.

▲ Figure 2.12 Typical CAD workstation with large screens

INDUSTRY TIP

There are software packages for producing CAD drawings that can be downloaded free of charge, some of which can be used to produce 3D work to a professional standard.

ACTIVITY

Go online and search for free 3D drawing programs (hint: Try Sketchup). Download a program and try it out. See if you can produce a 3D model of a simple structure. Be warned – these programs can be addictive once you become familiar with how they work!

Standard paper sizes

To produce drawings that are consistent and can be formatted to suit particular applications, they are laid out on standard-sized sheets of paper. Look at Table 2.3 and Figure 2.13 to see how standard paper sizes relate to each other.

▼ Table 2.3 Sheet sizes

Standard drawing paper sizes	
Name	Size (mm)
A0	1189 × 841
A1	841 × 594
A2	594 × 420
A3	420 × 297
A4	297 × 210
Less frequently used standard paper sizes	
A5	210 × 148
A6	148 × 105
A7	105 × 74
A8	74 × 52

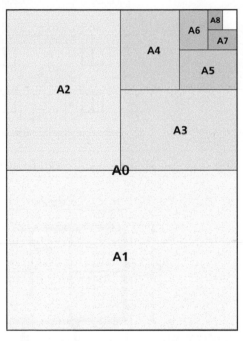

▲ Figure 2.13 Standard drawing sheet sizes

Format and layout of drawings

Drawings for construction activities are produced using what are known as 'conventions'. This means that the format and layout of the drawing follow agreed standards that allow the information they contain to be consistently understood. When drawings are produced, they can be laid out in a number of ways.

Two more common methods of presenting drawings are:

- **orthographic projection**
- isometric projection.

Orthographic projection

Orthographic projection is a two-dimensional method of laying out a drawing, where the front elevation of a structure has the plan view directly below it. The side or end elevations are shown directly each side of the front elevation.

This method of drawing allows views of all elevations to be looked at in relation to each other to gain a good understanding of the overall layout of the structure. It is easy to see the measurements of a building, the position of doors and windows, the shape of the roof structure and much more.

The most commonly used type of orthographic projection is called 'first angle projection'. Look at Figure 2.15 to see how this type of projection works. This shows how the views are projected onto flat planes surrounding the object. The surround is then folded flat showing how the views are arranged in first angle projection on a drawing.

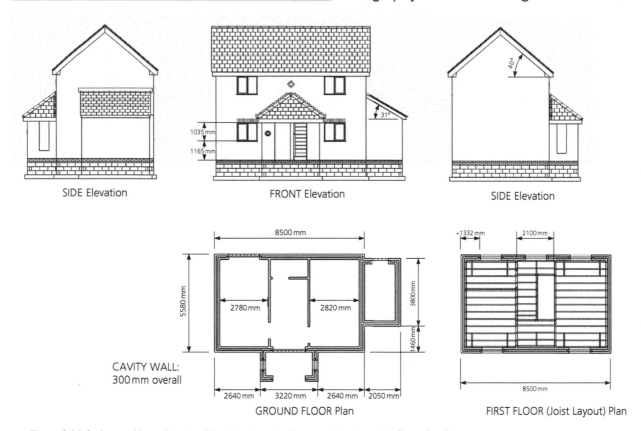

SIDE Elevation · FRONT Elevation · SIDE Elevation

1035 mm
1165 mm
40°
31°

8500 mm
5580 mm
2780 mm · 2820 mm
3800 mm
1460 mm
CAVITY WALL: 300 mm overall
2640 mm · 3220 mm · 2640 mm · 2050 mm
GROUND FLOOR Plan

~1332 mm · 2100 mm
8500 mm
FIRST FLOOR (Joist Layout) Plan

▲ Figure 2.14 Orthographic projection. The plan view in this example shows the floor plan layout

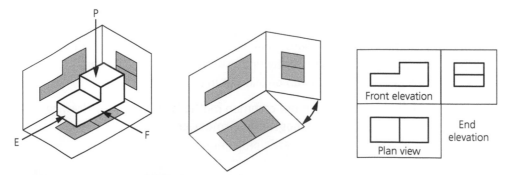

P
E
F
Front elevation
End elevation
Plan view

▲ Figure 2.15 First angle orthographic projection

Isometric projection

Isometric projection can be described as a pictorial method of presenting information on a drawing. This type of drawing layout can add greater clarity, giving the viewer a more life-like representation of the building. Isometric projection has the structure drawn at specified angles, with one corner represented as closer to the person viewing the drawing.

Although it doesn't provide a true three-dimensional view (since it doesn't include **perspective**), the more 'realistic' view it presents maintains accurate scaled dimensions on the drawing. In isometric projection, the vertical lines in the structure will be drawn at 90° to the horizontal (or bottom edge of the page) and the horizontal lines of the structure will be drawn at 30° to the horizontal on the page.

▲ Figure 2.16 An isometric projection drawing

> ### KEY POINT
>
> A less common layout method for drawings is **oblique projection**. In this drawing method, the front elevation is drawn to its actual size and shape, with the third dimension (or depth) shown drawn back to either the left or right as required. The lines are drawn horizontally, vertically or at a 45° axis to the left or right.

Symbols used on construction drawings

Standardised symbols are used on drawings as a means of passing on information in a simpler way. If all the parts of a building were labelled in writing, the drawing would soon become very crowded with text and potentially confusing. Materials can be represented in a sectional drawing by symbolic drawings and diagrams called hatchings.

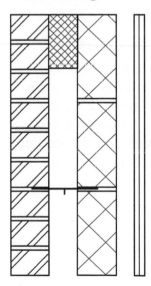

▲ Figure 2.17 A cross-section of a cavity wall showing hatchings

It is important to use symbols and hatchings that conform to a standard so that everyone using the drawing can interpret the information consistently and accurately. Look at Table 2.4 to see the meaning of symbols shown on a construction drawing.

▼ Table 2.4 Basic drawing symbols and hatchings

Sink	Sinktop	Wash basin	Bath	Shower tray
WC	Window	Door	Radiator	Lamp
Switch	Socket	North symbol	Sawn timber (unwrot)	Concrete
Insulation	Brickwork	Blockwork	Stonework	Earth (subsoil)
Cement screed	Damp proof course (DPC)/membrane	Hardcore	Hinging position of windows — Side / Top / Bottom	Stairs up and down — Stairs up 1234567 / Stairs down 1234567
Timber – softwood. Machined all round (wrot)	Timber – hardwood. Machined all round			

Administration documents

While drawings are an effective visual method of communicating a great deal of important information, a range of text documents are necessary to support the efficient administration of a construction project. These documents are used to ensure clear and accurate recording and transmitting of information between operatives and supervisory staff.

Study Tables 2.5, 2.6 and 2.7 to familiarise yourself with the range of text documents that are used on site. The documents have been split into categories to make it easier to see where they apply to your work activities. Table 2.5 shows documents that give specific information about the work task.

▼ Table 2.5 Documents specific to your task

Type of document	Description
Specification	The specification is a contract document, which means it is legally binding. It is used in conjunction with the drawing. Giving written instructions in a separate document prevents the drawing from becoming cluttered with text. The example shown here includes a lot of information about the bricks to be used for a project, including: ● brick bond ● mortar type ● joint finish ● insulation.

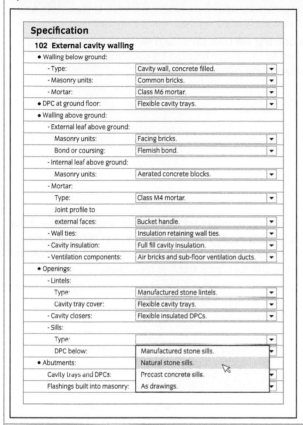

Type of document	Description
Method statement	This is a detailed description of how a task is to be completed safely, including all the resources required. (For more on method statements and how they integrate with risk assessments, see page 15.)

▼ Table 2.5 Documents specific to your task (continued)

Type of document	Description
Schedule	A schedule is mainly used on larger sites where there are several house designs, with each type having different components and fittings. It is usually used in conjunction with drawings. A typical drawing will have doors and windows labelled D1, D2, W1, W2 and so on. These labelled components are then listed in the schedule along with other information.
Variation order	A variation order is a document that is used when a change to the original contract is made that affects the design of the structure. An example might be when a client instructs that a door position is moved or a different brick finish is requested. Alterations like these may incur a cost not covered by the contract and must only be undertaken with written instructions from the architect or client's agent.

Master Internal Door Schedule

Ref:	Door size:	S.O. width	S.O. height	Lintel type	FD30	Self closing	Floor level
D1	838 × 1981	900	2040	BOX	Yes	Yes	GROUND FLOOR
D2	838 × 1981	900	2040	BOX	Yes	Yes	GROUND FLOOR
D3	762 × 1981	824	2040	BOX	No	No	GROUND FLOOR
D4	838 × 1981	900	2040	N/A	Yes	No	GROUND FLOOR
D5	838 × 1981	900	2040	BOX	Yes	Yes	GROUND FLOOR
D6	762 × 1981	824	2040	BOX	Yes	Yes	FIRST FLOOR
D7	762 × 1981	824	2040	BOX	Yes	Yes	FIRST FLOOR
D8	762 × 1981	824	2040	N/A	Yes	No	FIRST FLOOR
D9	762 × 1981	824	2040	BOX	Yes	Yes	FIRST FLOOR
D10	762 × 1981	824	2040	N/A	No	No	FIRST FLOOR
D11	686 × 1981	748	2040	N/A	Yes	No	SECOND FLOOR
D12	762 × 1981	824	2040	BOX	Yes	Yes	SECOND FLOOR
D13	762 × 1981	824	2040	100 HD BOX	Yes	Yes	SECOND FLOOR
D14	686 × 1981	748	2040	N/A	No	No	SECOND FLOOR

CPF Building Co
Variation order

Project Name: Penburthy House, Falmouth, Cornwall

Reference Number: 80475   **Date:** 13/11/19

From: : _____   **To:** _____

Reason for change:	Tick
Customer requirements	✔
Engineer requirements	☐
Revised design	☐

Instruction:
Entrance door to be made from Utile hardwood with brushed chrome finished ironmongery (changed from previous detail, softwood with brass ironmongery).

Signature _____

As a bricklayer, you may not often see the complete range of documents used to order and manage materials, but it's important to know about them so that you can contribute to efficiency on site when appropriate.

KEY TERM

Requisition: a written request or order for supplies

INDUSTRY TIP

On larger sites, you may be required to complete a **requisition** to submit to a central store to obtain some or all of the materials you need for a job. It's very likely that you will be required to take delivery of materials from time to time, so you should know how to deal with delivery notes effectively.

▼ Table 2.6 Documents specific to your materials

Type of document	Description
Bill of quantities	A bill of quantities is produced by the quantity surveyor. The bill of quantities is prepared using a standardised rule book called *New Rules of Measurement 2*, which came into force in 2013.

BILL OF QUANTITIES

Number	Item Description	Unit	Quantity	Rate	Amount £	p
	CLASS A: GENERAL ITEMS					
	Specified Requirements					
	Testing of Materials					
A250	Testing of recycled and secondary aggregates	sum				
	Information to be provided by the Contractor					
A290	Production of Materials Management Plan	sum				
	Method Related Charges					
	Recycling plant/Equipment					
A339.01	Mobilise; Fixed	sum				
A339.02	Operate; Time-Related	sum				
A339.03	De-mobilise; Fixed	sum				
	CLASS D: DEMOLITION AND SITE CLEARANCE					
	Other Structures					
D522.01	Other structures; Concrete	sum				
D522.02	Grading/processing of demolition material to produce recycled and secondary aggregates	m³	70			
D522.03	Disposal of demolition material offsite	m³	30			
	CLASS E: EARTHWORKS					
	Excavation Ancillaries					
E542	Double handling of recycled and secondary aggregates produced from demolition material	m³	70			
	Filling					
E615	Importing primary aggregates for filling to structures	m³	15			
E619.1	Importing recycled and secondary aggregates for filling to structures	m³	15			

The bill of quantities includes:
- information about the client and architect
- description of materials
- measured quantities related to each task
- details of required components.

Requisition order	A requisition order is filled out to order materials or components from a supplier or central store on site. This document will usually need authorisation from a supervisor or line manager before it is sent to the supplier or central store.
	The requisition order will be filed to allow cross checks to be made between items ordered and items delivered.

CPF Building Co
Requisition order

Supplier Information: Construction Supplies Ltd   **Date:** 9/12/19

Contract Address/Delivery Address: Penburthy House, Falmouth, Cornwall

Tel number: 0207294333

Order Number: 26213263CPF

Item number	Description	Quantity	Unit/Unit Price	Total
X22433	75mm 4mm gauge countersunk brass screws slotted	100	30p	£30.00
YK7334	Brass cups to suit	100	5p	£5.00
V23879	Sadikkens water based clear varnish	1 litre	£20.00	£20.00
Total:				£55.00

Authorised by: Denzil Penburthy

▼ Table 2.6 Documents specific to your materials (continued)

Type of document	Description
Invoice **Davids & Co** **Invoice** Invoice number: 75856 Date: 2nd January 2019 PO number: 4700095685 **Company name and address:** Davids & Co, 228 West Retail Park, Ivybridge, Plymouth **Customer name and address:** CPF Building Co, Penburthy House, Falmouth, Cornwall VAT registration number: 663694542 For: 	Item number	Quantity	Description	Unit Price
---	---	---	---
BS3647	2	1 ton bag of building sand	£30
CM4324	12	25 kg bags of cement	£224	 Subtotal £2748.00 / VAT 20% / Total £3297.60 Please make cheques payable to Davids & Co Payment due in 30 days	Sent by a supplier. It lists the services or materials supplied along with the price the contractor is requested to pay. There will be a time limit within which the invoice must be paid. Sometimes there will be a discount for early payment or penalties for late payment.
Delivery note **Construction Supplies Ltd** **Delivery note** **Customer name and address:** CPF Building Co, Penburthy House, Falmouth, Cornwall Delivery Date: 16/12/19 Delivery time: 9am Order number: 26213263CPF 	Item number	Quantity	Description	Unit Price	Total
---	---	---	---	---
X22433	100	75 mm 4 mm gauge countersunk brass screws slotted	30p	£30.00
YK7334	100	Brass cups to suit	5p	£5.00
V23879	1 litre	Sadikkens water based clear varnish	£20.00	£20.00 | Subtotal £55.00 / VAT 20% / Total £66.00 Discrepancies: .... Customer Signature: .... Print name: .... Date:	A delivery note accompanies a delivery of materials or components to site. An invoice for the delivered items will be sent by the supplier at a later date. The goods being delivered must be checked carefully for quantity and good condition before the delivery note is signed. If it's confirmed that the materials or components are damaged or not the items ordered, the delivery can be refused and the items returned to the supplier.

There are some document types that you will use that apply specifically to the work you do. For example, a bricklayer could be paid an hourly rate or daily rate (called day work). It's especially important that documents like time sheets are completed accurately when they are used on a construction project. If they're not accurate, your wages might be calculated wrongly!

▼ Table 2.7 Documents specific to your work activity

Type of document	Description
Job sheet	A job sheet simply gives details of a job to be carried out. It sometimes gives information about: ● location of the work on site ● materials needed for the job and the allocated hours to complete the job in line with the work schedule ● specific details for a job, to avoid confusion and misunderstandings.

CPF Building Co
Job sheet

Customer name: Henry Collins	Date: 9/12/19
Address: 57 Green St Kirkham London	

Work to be carried out
Finishing joint work to outer walls

Instructions
Use weather struck and half round

Type of document	Description
Timesheet	A timesheet is used to record the hours of work completed daily. It can be used to: ● calculate an operative's wages based on hourly rates ● calculate the total labour costs for a job to check if the project is staying within budget ● add to information needed to estimate costs for future projects a company gives a price for.

Timesheet

Employer: CPF Building Co.	Employee Name: Louise Miranda	Week starting: 1/6/19

Date: 21/6/19

Day	Job/Job Number	Start Time	Finish Time	Total Hours	Overtime
Monday	Penburthy, Falmouth 0897	9am	6pm	8	
Tuesday	Penburthy, Falmouth 0897	9am	6pm	8	
Wednesday	Penburthy, Falmouth 0897	8.30am	5.30pm	8	
Thursday	Trelawney, Truro 0901	11am	8pm	8	2
Friday	Trelawney, Truro 0901	11am	7pm	7	1
Saturday	Trelawney, Truro 0901	9am	1pm	4	
Totals				43	3

Employee's signature:_____

Supervisor's signature: _____

▼ Table 2.7 Documents specific to your work activity (continued)

Type of document	Description
Permit to work **PERMIT TO WORK** 1. Area 2. Date 3. Work to be Done 4. Valid From / 5. Valid To 6. Company 7. Person in Charge 8. No. of Persons 9. Safety Precautions 10. Safety Planning Certificate (cancelled if alarm sounds) I have inspected the above job which has been safely prepared according to requirements of a safety planning certificate Signed .... 11. Approval of Permit to Work I am satisfied that this permit is properly authorised, that safe access is provided, and that all persons affected by this job have been informed Signed .... 12. Electrical Equipment All power has been isolated/locked/tagged/tried* Circuits are live for troubleshooting only Signed .... 13. Acceptance of Permit to Work I/we* have read and understood the above precautions and will observe them. All equipment complies with relevant standards. I understand the site emergency plan. Signed .... 14. Completion of Permit to Work I/we* certify that this job is complete/incomplete*, all guards have been replaced and secured and all equipment has been removed. The job site has been left clean and tidy. Signed .... 15. Renewal of Permit to Work (same day only) Approved until .... Signed .... Approved until .... Signed .... If the alarm sounds 1. Stop Work 2. Make equipment safe 3. Leave the building by the nearest exit 4. Make your way to the main car park If you discover a fire 1. Break fire point 2. Leave the building 3. Ring 999 and give name, position, description etc 4. Report to incident controller in Main car park **Do not re-enter any building until you are told it is safe**	The permit to work is a documented procedure that gives authorisation for certain personnel to carry out specific work within a specified time frame. It sets out the procedures to complete the work safely, based on prior risk assessment.

Most of the documents listed will be available to site managers to monitor and check that things are proceeding as they should, and many will be filed in a manager's office. A site diary will be kept and entries made each day recording anything of note that happens on site, such as deliveries, absences of personnel or things that might affect progress such as delays due to bad weather.

Site managers will also be careful to check that manufacturer's instructions are kept and filed. They will ensure that instructions are followed to maintain safety on site and to protect the product or equipment being used.

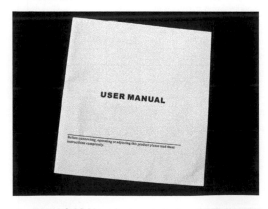

▲ Figure 2.19 Manufacturer's instructions

▲ Figure 2.18 Site diary

③ PLANNING AND MANAGING A CONSTRUCTION PROJECT

To ensure that work can be completed on schedule and within budget, careful planning of the construction process is required. Good management is vital for a successful build. The quality and effectiveness of management can determine whether a company makes a profit so that the company can remain in business.

▲ Figure 2.20 Careful planning is needed to manage a construction project effectively

A major part of managing a project so that a profit can be made is the process of calculating or estimating the cost of materials, equipment and labour requirements before work starts. For a larger project, it may be that several contractors will compete to get the job.

Each contractor will make calculations of resource requirements and their profit margins for a project, which they will submit to a client for consideration. This process is referred to as a '**tender**'. The client then decides which contractor they wish to use, considering the reputation of the contractor as well as the tendered costings.

The contract is then awarded so that detailed planning for the work can begin. The contract may include incentives to complete the work to schedule in the form of a 'penalty clause' which will be applied if the work is not completed on time.

To ensure that the quality of work is maintained, the client may hold back a proportion of payments to the contractor in the form of 'retention'. This is an amount that is held to cover any work that may have to be re-done after completion of the contract, if standards have not been met.

> **KEY TERM**
>
> **Tender:** the process where a contractor will make calculations of resource requirements and their profit margins for a project, which they will submit to a client for consideration

> **KEY POINT**
>
> Tenders are referred to as 'open' or 'closed'. An open tender (sometimes called a competitive or public tender) will allow all qualified bidders to enter. The sealed bids are opened in public for scrutiny and are chosen on the basis of price and quality.
>
> A closed tender is where only invited contractors can submit a bid for consideration by the client.

How quantities of resources are calculated will be considered later in this chapter. First let's look at the stages of planning that lead to successful management of a project.

Stages of planning

The stages at which management planning decisions have to be made are as follows:

- Pre-tender planning (before the tender is prepared). This includes deciding whether the contractor has the resources and personnel to take on the project and whether there is the capability to achieve it within the timescale.
- Pre-contract planning (before the contract is written). This occurs after the contract has been won but before the build starts. The contractor may have up to six weeks to plan the commencement of work on site. During this time, planning for the following occurs:
  - placing of orders for subcontractors
  - planning of the site layout in terms of temporary site accommodation, storage of resources, traffic routes and positioning plant
  - laying on of temporary services
  - preparation of the work programme
  - production of method statements
  - sourcing of suppliers and labour.
- Contract planning (after the contract is finalised) is required to arrange the work activities in a logical

sequence, determine labour, resource and plant requirements and ensure work progresses as planned to meet the handover date.

Getting things right at the early stages of a project is vital to achieving a successful outcome.

▲ Figure 2.21 Planning takes place in stages

VALUES AND BEHAVIOURS

The planning of a large project may be something that a bricklayer will not be directly involved with. However, your approach to supporting and co-operating with managers on site will have an impact on how successful they can be in following the project planning that has taken place during the writing of tenders and contracts. If you can demonstrate your skills and knowledge by contributing valuable thoughts and ideas to your managers, you may well progress quickly, being given greater responsibilities or a more challenging career role.

Planning the site layout

On a large site, planning the site layout will form part of the pre-contract planning, while on smaller sites this will be planned by the site manager. As a bricklayer, you may be involved in the planning or setting up of the site layout for smaller projects.

The purpose of site planning is to ensure that the layout contributes to productivity and efficiency. The planning will establish the best location and position of temporary site offices, welfare facilities and storage areas. Routes across the site for plant and personnel are set out with convenience and safety in mind.

Positioning of secured fenced storage areas (known as 'compounds') and routes should be planned to minimise the movement of materials to save double handling.

Important points to consider during the planning stage are:

- the effect of site work on neighbouring properties from dust and other pollution
- noise considerations
- parking for site personnel
- positioning of waste management and recycling facilities
- protection of the natural environment.

Good planning will support the protection of the public near the site, protection of employees on the site and protection of materials, tools and equipment used during the construction process.

▲ Figure 2.22 Planning the site layout

KEY POINT
If planning is not done with care, unnecessary expense can be caused if portable offices or materials must be relocated during the construction process.

VALUES AND BEHAVIOURS

Planning should take into account the effect the construction process can have on the public and nearby residents. Noise and dust should be kept to a minimum out of consideration for the comfort and convenience of others. To prevent long-term noise nuisance, a Noise Abatement Order can be issued. If short-term noise and dust cannot be avoided, inform those nearby in advance so that they can be prepared to deal with the potential problems caused.

A site plan should be produced showing the planned position of offices, site routes and storage compounds. Fencing or hoarding may be erected around the site perimeter for safety and security. The hoarding must have warning signs on display at the entrance providing information to visitors and workers alike.

ITEMS included on SITE LAYOUT PLAN

Site security fencing

Entrance gates

A Welfare facilities

B Site offices

C Stores – lock up

D Storage racking for finishing materials

E Brick storage area on hardstanding

F Formwork/reinforcement fabrication areas

G General hardstanding area – formed up on commencement of contract

H Area for subcontractor's accommodation and storage

I Bagged aggregates and cement storage

● Mortar mixing area

⊠ Position of tower crane

Car parking spaces

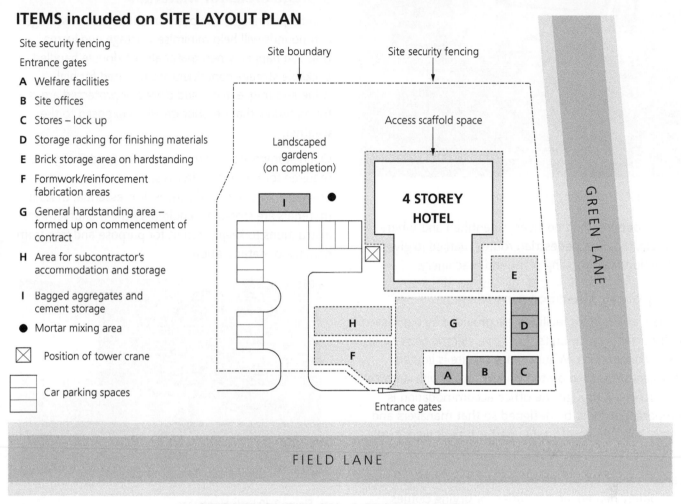

SITE LAYOUT PLAN: 4 STOREY HOTEL

▲ Figure 2.23 Typical site layout plan

▲ Figure 2.24 Safety notices

Access onto the site

Consideration should also be given to how the site is accessed from surrounding roads. In some cases, temporary roads must be constructed to allow access to the site by plant and delivery lorries. Permission must be obtained from the local authority for access over a public footpath, since roads and paths for public use are known as 'adopted' routes and are maintained by official bodies.

▲ Figure 2.25 Temporary road access

Traffic routes should be clearly identified and, where possible, separate pedestrian routes marked to give protection from moving plant and machinery.

Temporary site accommodation

Offices and welfare facilities are provided by purpose-built portable cabin units that have electric power and water connections. Where space on site is limited, they can be stacked on top of each other with suitable stair access. Although the office accommodation is temporary, it must be designed so that managers and supervisors can comfortably perform their work in all weathers for the full duration of the project.

Canteen facilities should be suitably furnished and adequately heated, and facilities for drying clothes must be provided. Larger sites may have a separate facility for storing and drying clothes. The toilets should be cleaned and disinfected daily.

Storage areas for materials

Careful planning of the position and layout of storage compounds will help minimise wastage and losses that can happen when materials are double handled. Lockable storage containers are required for high-value and fragile items and closed or protected areas for materials that deteriorate when exposed to the weather.

Open storage areas with easy access for delivery are required for heavy bulk items such as timber, bricks, drainage pipes and roof trusses. It is essential that all materials are stored as required by the manufacturers' instructions to keep them fit for purpose and free from damage or deterioration.

▲ Figure 2.30 Lockable storage

▲ Figure 2.26 Site manager's office

▲ Figure 2.27 Typical site accommodation set up as rest facilities

▲ Figure 2.28 Typical clothes storage and drying facilities

▲ Figure 2.29 Typical site accommodation set up as a meeting room facility

Stationary plant

Some sites will work more efficiently if a crane is installed for moving materials more easily. The crane is often centrally positioned on the site to allow for the whole site to be accessed within its radius of operation.

▲ Figure 2.31 Construction crane

Planning the sequence of work

Planning the sequence or programme of work is often undertaken using a form of bar chart. Each company is likely to have its own variation, but traditionally this bar chart is referred to as a Gantt chart (developed by Henry Gantt in the early twentieth century).

Gantt chart

Planning carefully is a key to efficiency and productivity on site. A Gantt chart is a clear and easy-to-understand documentary tool that helps site personnel to follow a set sequence of activities. It shows clearly if work on site is progressing to schedule. The chart will show:

- the start date
- the sequence in which the building operations will be carried out
- an estimated time for each operation
- the labour required
- the plant required
- when materials require delivering to site
- the contract end date
- any public holidays.

▲ Figure 2.32 Materials being delivered on schedule

▼ Table 2.8 Sequence of work activities on site

Operation number	Description	Trade	Comment
1	Site preparation and setting out	Carpenter, Labourer	
2	Excavation/concrete to foundations and drains	Labourer	
3	Brickwork to DPC	Bricklayer, Labourer	Start to DPC
4	Backfill and ram	Labourer	
5	Hardcore and ground floor slab	Labourer	

▼ Table 2.8 Sequence of work activities on site (continued)

Operation number	Description	Trade	Comment
6	Brickwork to first lift	Bricklayer, Labourer	DPC to watertight
7	Scaffolding	Subcontractor, Labourer	
8	Brickwork to first floor	Bricklayer, Labourer	
9	First floor joisting	Carpenter, Labourer	
10	Brickwork to eaves	Bricklayer, Labourer	
11	Roof structure	Carpenter, Labourer	
12	Roof tiles	Subcontractor, Labourer	
13	Windows fitted	Carpenter, Labourer	
14	Carpentry first fix	Carpenter, Labourer	Internal work and finishing
15	Plumbing first fix/second fix	Subcontractor, Labourer	
16	Electrical first fix/second fix	Subcontractor, Labourer	
17	Services	Subcontractor, Labourer	
18	Plastering	Subcontractor, Labourer	
19	Second fix carpentry	Carpenter, Labourer	
20	Decoration	Subcontractor, Labourer	
21	External finish/snagging	Subcontractor, Labourer	

The sequenced activities can now be entered on a Gantt chart to set out the timescale that will be allotted for each one (Figure 2.33). Experience gained from past projects together with information provided by method statements can be used to work out how much time should be allotted to individual tasks.

This method of presenting information gives a clear view of overlapping activities and helps to identify when labour and plant requirements will need to be met. Notice that the section for each task in the numbered sequence is split into two rows. This is to show actual progress against the planned activities. The top row will show the planned activity timings and the bottom row will be filled in a different colour to show actual activity timings.

This allows easy monitoring and identification of any 'slippage' in the programme (perhaps caused by bad weather) and steps can be taken to make adjustments to bring work back on schedule. Study Figure 2.33 and note how the chart shows when various trade activities are happening at the same time.

ACTIVITY

Use the chart in Figure 2.33 to establish how many weeks in total the following trades are required on site.

1 Bricklayers
2 Carpenters

Critical path analysis (CPA)

Critical path analysis (CPA) is unlikely to be used on smaller projects. It is a very specialised system of planning management and requires a high level of analysis skills to create. CPA tends to be used on long-running complex projects such as high-rise commercial buildings incorporating highly technical systems.

KEY TERM

Critical path analysis (CPA): a network analysis technique of planning complex working procedures

CPA uses similar principles to a Gantt chart in that it shows what has to be done at specific times in a sequence. However, CPA shows this information as a series of circles called 'event nodes'.

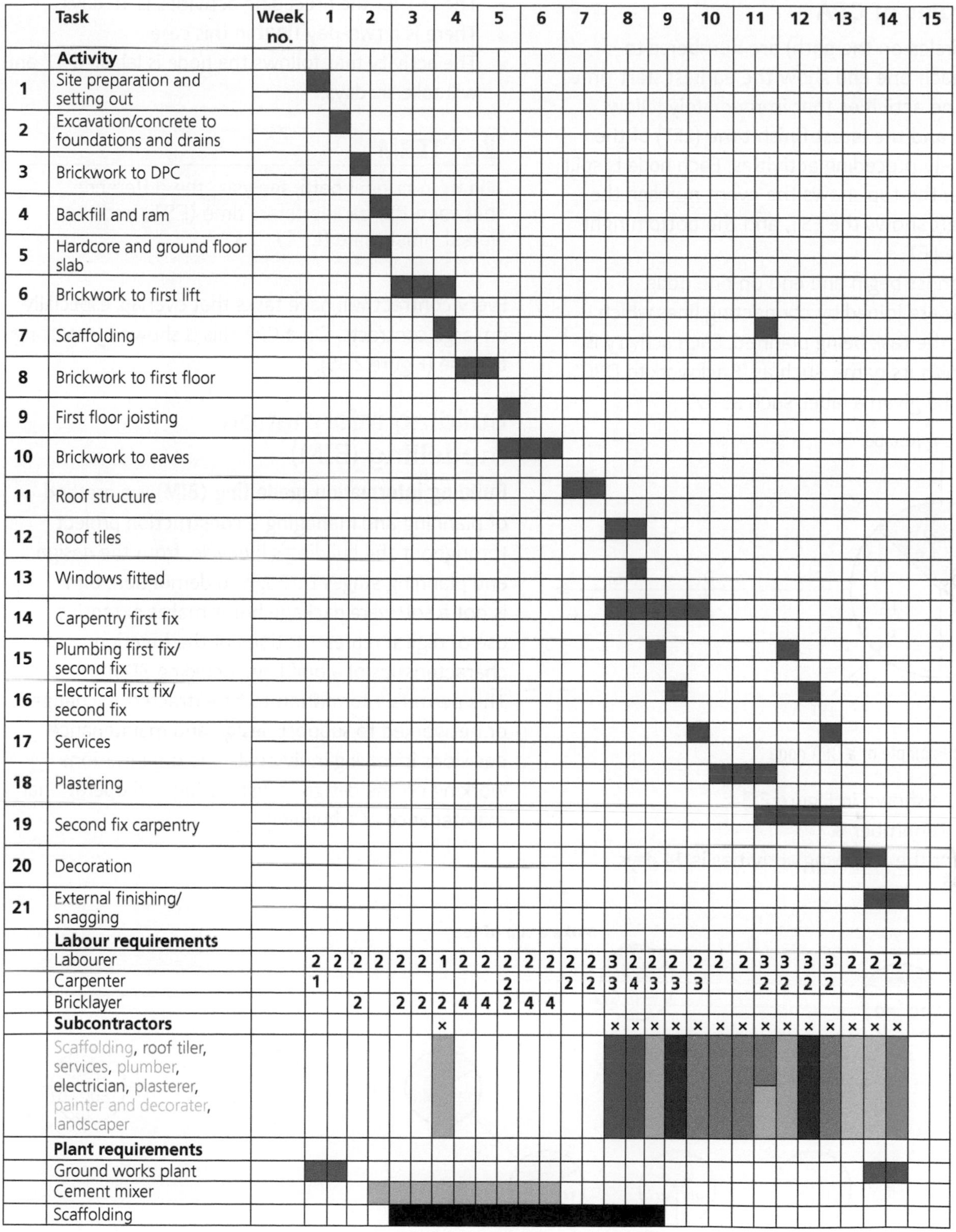

	Task	Week no.	1	2	3	4	5	6	7	8	9	10	11	12	13	14	15
	Activity																
1	Site preparation and setting out		■														
2	Excavation/concrete to foundations and drains			■													
3	Brickwork to DPC			■													
4	Backfill and ram				■												
5	Hardcore and ground floor slab				■												
6	Brickwork to first lift				■												
7	Scaffolding												■				
8	Brickwork to first floor					■											
9	First floor joisting						■										
10	Brickwork to eaves						■										
11	Roof structure								■								
12	Roof tiles									■							
13	Windows fitted									■							
14	Carpentry first fix									■	■						
15	Plumbing first fix/second fix										■			■			
16	Electrical first fix/second fix										■			■			
17	Services										■			■			
18	Plastering											■	■				
19	Second fix carpentry												■	■			
20	Decoration														■		
21	External finishing/snagging															■	
	Labour requirements																
	Labourer		2 2	2 2	2 2	1 2	2 2	2 2	2 2	3 2	2 2	2 2	2 3	3 3	3 2	2 2	
	Carpenter		1				2		2 2	3 4	3 3	3		2 2	2 2			
	Bricklayer			2	2 2	2 4	4 2	4 4									
	Subcontractors				×					×	×	×	× ×	× ×	× ×	× ×	×
	Scaffolding, roof tiler, services, plumber, electrician, plasterer, painter and decorater, landscaper																
	Plant requirements																
	Ground works plant		■													■	
	Cement mixer				■ ■	■ ■	■ ■	■ ■									
	Scaffolding				■ ■	■ ■	■ ■	■ ■	■ ■								

▲ Figure 2.33 Planned activities, labour and plant requirements shown on a Gantt chart

The key rules of CPA

- Nodes (circles on the path) are numbered to identify each one and show the earliest start time (EST) of the activities that immediately follow the node, and the latest finish time (LFT) of the immediately preceding activities. Each node is split into three: the top shows the event number, the bottom left shows the EST, and the bottom right shows the LFT.
- The CPA must begin and end on one node.
- The nodes are joined by connecting lines which represent the task being planned. Each activity is labelled with its name, such as 'Brickwork to DPC', or it may be given a label, such as 'D'.

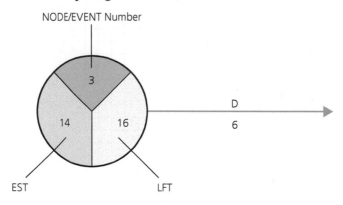

▲ Figure 2.34 Example of a CPA node

In the example shown in Figure 2.34:
- The node is number 3.
- The EST for the following activities is 14 days.

- The LFT for the preceding activities is 16 days.
- There is a two-day **float** in this case.
- The activity that follows the node is labelled 'D' and will take six days.

KEY TERM

Float: in critical path analysis, the difference between the earliest start time (EST) and the latest finish time (LFT)

Every contract will have tasks that overlap, especially on large contracts. On a CPA this is shown by splitting the line (Figure 2.35).

Building information modelling (BIM)

Building information modelling (BIM) is a method of planning and managing a construction project throughout the building's lifecycle, from the design and planning stages through to demolition. BIM is not a software package but it makes extensive use of digital representations of the functions and characteristics of a building, including 3D modelling. This permits information to be extracted, exchanged or networked to support design and maintenance decisions involving individuals and organisations working on the design, construction and operational maintenance of a building.

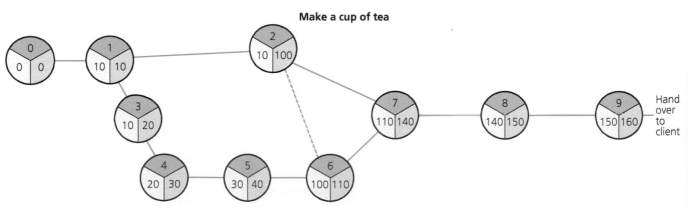

▲ Figure 2.35 CPA diagram

▲ Figure 2.36 BIM is a system that improves the planning and management of complex buildings

Although BIM is not something that a bricklayer is likely to have an active part in, it's important to know about it since the Government is promoting greater use of this method of planning and managing construction projects.

It's an intelligent system that integrates the professional skills of architects, engineers and construction personnel to work more efficiently together. The construction industry is closely regulated by a number of official bodies to ensure that a building is safe and durable as well as environmentally friendly, and BIM allows the efficient combined use of official information sources such as British Standards, Building Regulations and Planning Regulations.

> **KEY TERM**
>
> **Building information modelling (BIM):** a method of planning and managing a construction project throughout the building's lifecycle from the design and planning stages through to demolition

> **ACTIVITY**
>
> Research the history of BIM and write a short report about when the system was invented and what sort of projects it has been used on.

> **KEY POINT**
>
> You could summarise the uses and advantages of using BIM to include:
> - 3D modelling
> - change management
> - building simulation
> - data management
> - building operation.

4 CALCULATING RESOURCES AND QUANTITIES OF MATERIALS

Earlier in this chapter, we looked at bills of quantities and specifications as documents that provide important information for the job. These documents play a significant role in the process of calculating resources and quantities of materials for a contract.

Estimating and quoting

The documents for calculating costs will be used together with relevant drawings to extract information that will be used to produce **estimates** or quotes for the cost of a project.

> **KEY POINT**
>
> It is important to understand the difference between an estimate and a quote.
>
> An estimate is essentially a 'best guess' of how much specific work will cost, or how long it will take to complete. This means that the job could cost more or less than the estimated cost.
>
> A quote is a promise to do work at an agreed price and sets out exactly what work will be done for that agreed price. Acceptance of this quote by the client or their representative creates a binding agreement between the two parties. Additional work requested by the client will be costed separately.

KEY TERM

Estimate: essentially a 'best guess' of how much specific work will cost, or how long it will take to complete

▲ Figure 2.37 Quotes

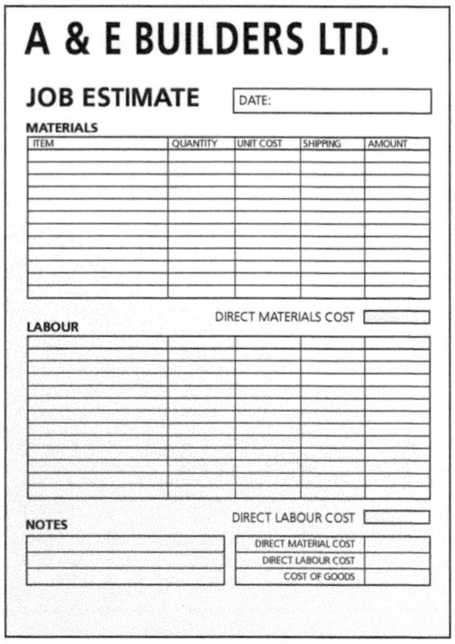

▲ Figure 2.38 An estimate

The costing process

There are a number of professional members of the building team who will contribute to the costing process.

KEY POINT

The building team is made up of craft workers such as bricklayers, operatives such as the workers assisting the bricklayers and professionals such as the quantity surveyor who has a major part in the planning and costing of a project.

Quantity surveyor

The quantity surveyor will have produced the bill of quantities, which will be used with the specification and drawings to extract information about resources and materials, including:

- quantity
- colour
- dimensions
- location
- installation details
- manufacturer.

The bill of quantities will contain standard definitions of specific information:

- Preliminaries – these may be time-related costs that have to be included, such as management and supervision costs or **setting-up costs**, rather than the costs related to the actual building work.
- Preambles – these are statements in the bill of quantities referring to the tender. They give descriptions of the requirements for construction such as the quantity, type and characteristics of the materials required.
- Provisional sum – this refers to work for which there is no accurate estimate or quote. Descriptions must be given of the type of work, how the work is to be done and any time limitations likely to be encountered.
- Prime cost (PC) – these are sums that cover work undertaken by nominated subcontractors, nominated suppliers and **statutory undertakings**.

▲ Figure 2.39 Quantity surveyor

Estimator

The estimator (sometimes referred to as the cost planner) will use the bill of quantities and other details to work out how much it will cost to deliver a completed building to a client. This is a complex job and requires experience in weighing up all the information about the building work, including materials, labour and plant costs and the time needed to complete the work.

▲ Figure 2.40 The estimator has experience in weighing up all the information

An estimator is concerned with establishing the best price that will win the contract in a competitive bidding situation, while ensuring that the contract can be carried out profitably. There are many factors that must be taken into account:

- Labour rates – includes costs such as annual holiday costs, employers' National Insurance contributions, pension contributions, sickness pay, bonuses, travelling time.
- Overheads – includes such things as electricity, gas, phone, insurance, plant and equipment, depreciation of equipment and buildings.
- Contingencies – a small percentage added to cover unforeseen costs that may arise during the contract.
- Profit – the amount of money gained over and above direct costs. The rate applied will depend on several factors, including current workload, the competition and the complexity of the project.
- VAT on materials.

Other costing methods

Traditionally, many companies use a building price book, which is a complete guide for estimating, checking and forecasting building work. The figures in these books are established rates which are updated on a yearly basis. To calculate the cost of proposed work you simply find the description of work requiring costing and apply the figures to the current contract.

There are also many software packages that have been developed to produce costings for complex projects. The software has a database of prices and costs for materials and labour which must be kept up to date. Using this information together with entered details specific to the project, the program can calculate costings for the contract.

Choosing suppliers

To ensure stability of supply and consistency of prices throughout a project, suppliers may be identified as 'preferred' during the planning and costing stages of a contract. The quantity surveyor and estimator may have been approached by a supplier with details of what they can offer, or they may have been recommended (or nominated) by another contractor.

If there are several suppliers from which you can source the same materials locally, the preferred supplier may be one that consistently gives the best price or the one that is most reliable and always delivers on time.

How to calculate quantities

As a bricklayer, you will rarely be involved in the complex costing processes used for a large construction project. However, you may be involved in calculating quantities of resources and materials for smaller projects, so it's important to understand methods used to determine the resource requirements for completion of a project.

▲ Figure 2.41 You may be involved in calculating quantities of resources

Units of measurement

The construction industry uses metric units as standard, but there will often be occasions when you will hear older measures called imperial units used in trade discussions. Material sizes are often still referred to in imperial units, even though they are now sold in metric units. An example of this is 8 ft × 4 ft sheets of ply, where the correct size is 2440 mm × 1220 mm.

Look at Table 2.9 to compare metric units of measurement with imperial units.

▼ Table 2.9 Units of measurement comparison

Units for measuring	Metric units	Imperial units
Length	millimetre (mm) metre (m) kilometre (km)	inch (in) or " e.g. 6 in or 6" foot (ft) or ' e.g. 8 ft or 8'
Liquid	millilitre (ml) litre (l)	pint (pt)
Weight	gram (g) kilogram (kg) tonne (t)	pound (lb)

▼ Table 2.10 How metric units are expressed

Units for measuring	Quantities	Example
Length	There are 1000 mm in 1 m There are 1000 m in 1 km	1 mm × 1000 = 1 m 1 m × 1000 = 1 km 6250 mm can be shown as 6.250 m 6250 m can be shown as 6.250 km
Liquid	There are 1000 ml in 1 l (litre)	1 ml × 1000 = 1 l
Weight	There are 1000 g in 1 kg There are 1000 kg in 1 t (tonne)	1 g × 1000 = 1 kg 1 kg × 1000 = 1 t

ACTIVITY

Look online to find answers to the following:
1 What other imperial units are still commonly used?
2 How many millimetres are there in an inch?
3 How many litres are there in a gallon?

Calculation methods

There are four basic mathematical operations used in construction calculations, which you will frequently use.

Addition

The addition of two or more numbers is shown with a plus sign (+).

Example:

A stack of bricks is 3 bricks long and 2 bricks high. It contains 6 bricks.

3 + 3 = 6

More examples of addition:

5 + 2 = 7

19 + 12 = 31

234 + 105 = 339

Subtraction

The reduction of one number by another number is shown with a minus sign (–).

Examples:

5 – 2 = 3

19 – 12 = 7

234 – 105 = 129

Multiplication

The scaling of one number by another number is shown with a multiplication sign (×).

Example:

A stack of bricks is 3 bricks long and 2 bricks high. It contains 6 bricks.

3 × 2 = 6

More examples:

5 × 2 = 10

19 × 12 = 228

234 × 105 = 24,570

In the last example, the comma (,) is used to show the number is in the thousands. In words we would say, twenty-four thousand, five hundred and seventy.

▲ Figure 2.42 Accurate calculations are needed to order the right quantities of materials

Division

Sharing one number by another number in equal parts (how many times does it go into the number) is shown with a division sign (÷).

Examples:

5 ÷ 2 = 2.5

19 ÷ 12 = 1.583

234 ÷ 104 = 2.25

A decimal point (.) allows us to show parts of whole numbers (decimal fractions). A decimal fraction representing a ½ is shown as 0.5. A decimal fraction representing a ¼ is shown as 0.25.

IMPROVE YOUR MATHS

Try these sums:
1   29 + 51
2   79 – 23
3   54 × 76
4   23 ÷ 4
5   Show three quarters as a decimal fraction.

Applying calculation methods

Calculation methods can be applied in several ways to establish the amounts of materials required for a job. Some categories we can use are described as:

- linear measurement
- area
- volume
- percentages.

We also use ratios when calculating amounts of some resources for construction tasks. (We'll discuss ratios and how they're used for calculating material quantities in more detail on pages 143 and 151.)

Linear measurement

Linear means how long a number of items would measure from end to end if laid in a straight line. Examples of things that are calculated in linear measurements are:

- skirting board
- lengths of timber
- foundations
- drainage systems.

▲ Figure 2.43 The quantity of skirting required is calculated using linear measurement

IMPROVE YOUR MATHS

Linear measurement

A site carpenter has been asked how many metres of skirting are required for the room below.

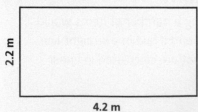

2.2 m

4.2 m

They can add all the sides together:
2.2 + 4.2 + 2.2 + 4.2 = 12.8 m
Or, they can multiply each side by 2, and add them together:
(2.2 × 2) + (4.2 × 2) = 12.8 m
Either way, 12.8 m is the correct answer.

KEY POINT

Adding all the sides of the rectangle together gives you a total measurement called the 'perimeter', which is defined as 'the continuous line forming the boundary of a closed geometrical figure'.

IMPROVE YOUR MATHS

A rectangular building is 5.75 m long and 4.35 m wide.
Calculate the total length of DPC required for the outer wall.

Area

Area is the way we describe the measurement of a two-dimensional surface such as the face of a wall. To find the area of a surface of a wall, you need to multiply its length by its height (L × H). If we calculate the area of the face of a wall in square metres (m²), we can use a simple formula to calculate how many bricks or blocks we need.

KEY TERM

Area: the measurement of a two-dimensional surface such as the face of a wall

KEY POINT

It is accepted that there are 60 bricks or 10 blocks in a square metre. This is not an absolutely accurate figure, but in practical terms is close enough to calculate the quantities that are required to build a wall.

The area of floors is calculated in a similar way to the area of a wall, but this time we multiply the length by the width (L × W).

IMPROVE YOUR MATHS

Find the area using the following measurements:
1 2.1 m × 2.4 m
2 0.9 m × 2.7 m
3 250 mm × 3.4 m

This is simple in the case of a rectangle but a room with an irregular shaped floor would be more difficult. In this instance, the area can be calculated by splitting the floor into sections and adding the results for each section together.

▲ Figure 2.44 Irregularly-shaped rooms can be split into sections to calculate the floor area

Look at the examples of a rectangular room and an irregularly-shaped room and follow the workings out used to calculate the floor areas.

IMPROVE YOUR MATHS

Area: Example 1

A bricklayer has been asked to work out the area of the floor below.

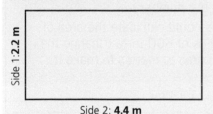

Side 1 × side 2 = floor area
2.2 × 4.4 = 9.68 m²
The total floor area is 9.68 m².

IMPROVE YOUR MATHS

Area: Example 2

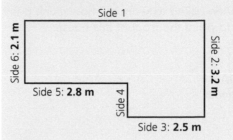

Step 1

Divide the area into two parts, and then calculate the area of each part. The easiest way to do this is to divide it into two smaller sections:

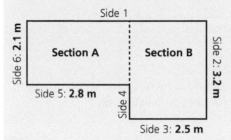

Step 2

Work out the area of section A and section B:
Section A = 2.1 × 2.8 = 5.88 m²
Section B = 2.5 × 3.2 = 8 m²

Step 3

Add the areas of section A and section B together:
section A + section B = total floor area
5.88 + 8 = 13.88 m²
The total floor area is 13.88 m².

Sometimes it is necessary to calculate the area of a non-rectangular shape such as the triangle that forms a gable in the end elevation of a house or other structure. Calculating the area in m² will allow us to work out the number of bricks or blocks required.

Other trades will also use this method. An example could be a painter calculating the amount of paint required.

Look at the example and take note of the method used to calculate the area of the triangular section.

IMPROVE YOUR MATHS

Area: Example 3

A painter has been asked to work out how much paint will be needed to paint the front of this house.

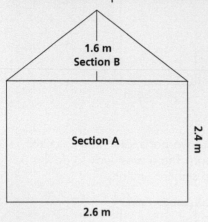

1.6 m
Section B

Section A

2.4 m

2.6 m

Step 1

Divide the area up into a rectangular section (section A) and a triangular section (section B).

Step 2

Find the area of section A:

2.4 × 2.6 = 6.24 m²

The area of section A is 6.24 m².

Step 3

Find the area of section B.

The area of a triangle can be found by multiplying the base by the height, then dividing by 2.

(base × height) ÷ 2 = area

2.6 × 1.6 = 4.16

4.16 ÷ 2 = 2.08 m²

The area of section B is 2.08 m².

Step 4

Area of section A + area of section B = total wall area

6.24 + 2.08 = 8.32 m²

The total wall area is 8.32 m².

IMPROVE YOUR MATHS

Using the diagram, calculate the number of bricks required for the whole elevation. (Remember – there are 60 bricks to 1 m².)

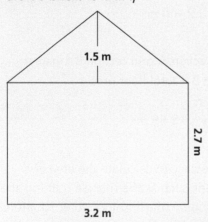

1.5 m

2.7 m

3.2 m

More rarely, it may be necessary to calculate the area of a circle. If you are required to do this, remind yourself of the formula and method of calculation before you start.

KEY POINT

To calculate the area of a circle we use a formula including the mathematical symbol 'π' (or 'pi'). This symbol is used to represent the number 3.142.

The formula is π × radius². (The small figure 2 means that the number is multiplied by itself. This is called squaring the number.)

As an example, we could calculate the area of a circle with a radius of 600 mm. (Change the radius from millimetres to metres to make it easier.)

3.142 (pi) × 0.6² = 1.13 m²

Volume

The volume of an object is the total space it takes up. Examples could be a foundation for a wall or the capacity of a concrete mixer. To find the volume of an object you must multiply length by width by height to give you cubic metres (m³).

So, volume = length × width × height (L × W × H).

Look at the example to see how this works. Each individual cube shown has faces representing 1 m². Count each cube and then follow the calculations shown in the example to check the total volume.

IMPROVE YOUR MATHS

Volume

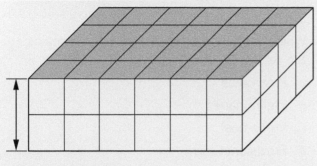

Area = 6 × 4 = 24
Volume = area × 2
Volume = 24 × 2 = 48 cubic units

IMPROVE YOUR MATHS

A brick wall needs a concrete foundation with the following dimensions:
Length 3.5 m, width 600 mm and depth 300 mm
Calculate the volume of concrete required in m³.

Percentage

Careful calculation of quantities for materials is an essential factor in keeping a project financially within budget. However, producing an amount of waste during the construction process is inevitable (and has a financial effect).

Examples could be a small amount of surplus mortar left over at the end of the working day or an amount of bricks that have been damaged and must be disposed of.

When planning and costing a project, allowance must be made for this inevitable waste and this is usually expressed as a **percentage** (or '%') of the amount required.

KEY TERM

Percentage: parts of a hundred

To find a percentage of a number, divide it by 100 and then multiply it by the percentage required. Look at the example to see how this works.

IMPROVE YOUR MATHS

Percentage

Increase 19 m by 12%
19 ÷ 100 = 0.19
0.19 × 12 = 2.28
19 + 2.28 = 21.28 m
Total required = 21.28 m

IMPROVE YOUR MATHS

Calculate the percentages for the following:
1 Increase 49 m by 10%
2 Increase 27 m by 20%
3 Decrease 22 m by 5%

Summary

Constructing a building is a complex process involving many different organisations. Good communication is essential for effective and efficient organisation. In this chapter, we have considered different methods of communicating between personnel and the different textual and drawn documents that are used to communicate information clearly.

We've also looked at how a project is planned and costed, and the documents and methods used when calculating quantities of materials and resources needed to successfully complete a project within specified timescales. The information in this chapter will remain useful to you throughout your working career as you extend your understanding of ways to apply it.

Test your knowledge

1 Which document is used to authorise a client's request for a change to the work?

   a Memorandum

   b Method statement

   c Specification

   d Variation order

2 How can safety information on a new product be communicated to workers at their work location on site?

   a Alarm

   b Email

   c Safety signs

   d Toolbox talk

3 Which of these is an advantage of producing drawings using CAD?

   a Drawings are produced automatically

   b Drawings can be amended quickly

   c No architect is needed

   d No software is required

4 Which of the following scales is used to produce a block plan?

   a 1:50

   b 1:250

   c 1:500

   d 1:2500

5 Which of the following is the largest standard paper size?

   a A0

   b A1

   c A3

   d A4

6 Which of the following is a pictorial drawing method which shows horizontal lines drawn at 30°?

   a Isometric

   b Oblique

   c Orthographic

   d Perspective

7 Which document is a promise to carry out a contract for a specific sum?

   a Bill of quantities

   b Invoice

   c Permit to work

   d Quote

8 Which section of a bill of quantities contains the setting up costs?

   a Preambles

   b Preliminaries

   c Prime cost

   d Provisional sum

9 Which professional team member produces the bill of quantities?

   a Architectural technician

   b Contractor

   c Estimator

   d Quantity surveyor

10 For which resource would linear measurement be used when calculating quantities?

   a Blocks

   b Bricks

   c Cement mortar

   d Skirting boards

11 State one advantage of using written communication on site.

12 List three expressions of body language that can reveal an unspoken message. Explain what your selected expressions mean.

13 Explain the purpose of minutes, which are used as a record of discussion at site meetings.

14 Explain orthographic projection as a method of laying out drawings.

15 State the reason for using standardised symbols to represent features on a working drawing.

CHAPTER 3

PRINCIPLES OF CONSTRUCTION

INTRODUCTION

Building safe and durable structures that work efficiently is a complex process that requires the application of established principles. In this chapter, we will look at how an organised team of construction professionals, technicians and operatives apply ideas and rules within established principles to control the construction process.

We will consider traditional and modern methods of building and see how principles of construction are being refined to support sustainability and protection of the environment. By developing your understanding of construction principles and the methods used to control the creation of different parts of a building, you can contribute more effectively to an industry that has a significant impact on everyday life.

IMPROVE YOUR ENGLISH

A principle can be defined as 'a basic idea or rule that controls how something happens or works'.

By the end of this chapter, you will have an understanding of:
- company structures and how construction teams work together
- types of buildings and how their parts are constructed
- environmental and sustainability considerations for buildings.

The table below shows how the main headings in this chapter cover the learning outcomes for each qualification specification.

Chapter section	Level 2 Technical Certificate in Bricklaying (7905-20) Unit 201	Level 3 Advanced Technical Diploma in Bricklaying (7905-30) Unit 301	Level 2 Diploma in Bricklaying (6705-23) Unit 202/602	Level 3 Diploma in Bricklaying (6705-33) Unit 301/701	Level 2 Bricklayer Trailblazer Apprenticeship (9077)
1. Company structures and how construction teams work together	1.1, 1.3	N/A	16.1, 16.2	N/A	Module 3: 2.1, 2.2, 2.3, 2.4, 2.5
2. Types of buildings and how their parts are constructed	4.1, 4.2, 5.1, 5.2, 5.3, 5.4, 5.5	N/A	12.1, 12.2, 13.6, 14.1, 15.1, 15.2	N/A	Module 1: 1.1, 1.2
3. Environmental and sustainability considerations for buildings	N/A	2.1, 2.2, 2.3	11.1, 11.2, 11.3, 11.4, 11.5	11.1, 11.2, 11.3, 11.4, 11.5	Module 1: 2.1, 2.2

Note: for 6705-23, Unit 202/602:

Content for Assessment criteria 12.3 to 12.5 is covered in Chapter 4.

Content for Assessment criteria 13.1, 13.2, 13.4 and 13.5 is covered in Chapter 5.

1 COMPANY STRUCTURES AND HOW CONSTRUCTION TEAMS WORK TOGETHER

Construction companies are structured in different ways that allow for flexibility in meeting the requirements of individual customers. They range in size and scale from single (or 'sole') traders who take on small projects and employ subcontractors as and when required, to large organisations that take on multi-million-pound contracts and employ hundreds or even thousands of workers.

There are many companies that work between these two extremes and they are referred to as small and medium-sized enterprises (SMEs).

Project types

Construction companies may choose to specialise in one sector of the industry (**public sector** or **private sector**) and arrange their management structure and construction teams to suit the needs of that sector:

- Public sector – projects that are funded by central government and local government
- Private sector – projects that are independently financed, as well as public projects with private finance.

The range of construction work

Construction work is undertaken in a range of situations and formats. We will now look at the various types of construction work and how each is described.

ACTIVITY

As you travel around, list all the forms of construction work you observe in a day and compare your list with the types of work described in this section.

New-build

The name makes the work activity easy to identify. New buildings, from domestic work constructing a new dwelling or an extension, to **commercial** and **industrial** work constructing factories, offices and retail facilities, are seen in many locations.

New housing construction is currently a stated government priority in many parts of the UK and the intention is that construction of this type of new-build work will increase.

KEY TERMS

Commercial: when referring to buildings, those used for business activities to generate a profit, e.g. an office block

Industrial: when referring to buildings, those used for processing materials and manufacturing goods, e.g. an insulation manufacturer

Private sector: projects that are independently financed, as well as public projects with private finance

Public sector: projects that are funded by local and central government

▲ Figure 3.1 New-build housing construction site

Renovation

Sometimes termed 'remodelling', **renovation** of a building is the process of renewing outdated or damaged structures to make them useful again. This can be a challenging process since a structure may be difficult to access within an existing built-up area and the building may have been unused or derelict for some time.

Renovation work can be undertaken on structures that are for commercial or **residential** use and some construction companies specialise in this type of work.

Maintenance

This type of construction work is undertaken on buildings that are in current use and may have to be done while the occupants continue to use the property. Maintenance work is important to prevent deterioration of a building and, if performed at the right time, can prevent the need for expensive repairs to a structure or replacement of failed components. (For more on maintenance and repair methods used for masonry structures, see Chapter 7.)

Restoration

This is an important type of work that returns a historic building to useable condition and restores the appearance it had at a specific point in its history. Some structures of special architectural interest or historic significance may be protected as 'listed' buildings.

Listed buildings can only be restored, altered or extended if consent is given under government planning guidance. They are graded to indicate the level of special interest or significance that they have.

The nature of this type of construction work may require many specialist skilled workers and can require materials that are not easy to obtain. This can lead to a project being expensive to complete.

All of these different types of construction activities require the skills and abilities of an extensive team of personnel to bring a contract to completion. We will now look at how the team is structured and organised.

▲ Figure 3.2 Restoration work on the Houses of Parliament

Job roles within the construction team

Personnel in a construction team can be categorised according to the type of work they perform. We can split the job roles as follows:

- Professionals
- Technicians
- Trade operatives
- Specialists
- General operatives

Remember, you work on site as a team and each member has an important role in applying the principles of construction that will be discussed later in this chapter.

▲ Figure 3.3 The construction team

Professionals

These construction personnel are trained and qualified to perform specific tasks. The training may involve many years of study to gain a recognised qualification.

Look at Table 3.1 to see the roles and responsibilities of each qualified professional.

▼ Table 3.1 Professional personnel

Job title	Roles and responsibilities
Architect	Creates the concept and design of a building in accordance with the client's requirements. Advises the client on practical matters related to the project.
Quantity surveyor	Calculates materials, time and labour costs. Prepares tenders and contracts. (For more about costings and tenders, see pages 53 and 61.)
Surveyor	Makes exact measurements and determines property boundaries. Calculates heights, depths, relative positions, property lines and other characteristics of the terrain.
Civil engineer	Plans, designs and oversees construction and maintenance of building structures and infrastructure, such as roads, railways, bridges and water and sewerage systems.

ACTIVITY

Choose one of the professional personnel in Table 3.1 and research online their roles and responsibilities. Write a short report expanding on the description of responsibilities given in the table.

Technicians

Technicians provide support for professionals. They are trained to understand the technical aspects of whichever job they take on. For example, an architectural technician assists an architect in producing drawings for a project, which frees the architect to concentrate on concept and design matters and dealing with the client.

The broad term 'building technician' covers jobs such as:
- estimating different costs for use in contract tenders
- negotiating with suppliers on the cost of materials, equipment and labour
- drafting construction plans using CAD software (For more on CAD applications, see page 42.)
- monitoring build progress against completion dates.

(For more on CAD applications, see page 42.)

INDUSTRY TIP

A technician could move into construction project management after gaining experience and additional training. Project and site managers must have a range of technical understanding in order to effectively supervise skilled trade operatives on site.

Trade operatives

A skilled tradesperson, a chargehand (or foreman) and a supervisor can be categorised as trade operatives. A range of skills must be competently performed for a construction project to be successful. In the past, an apprentice learned a trade or craft over a period of up to five years. Nowadays the training period for most trade skills has been reduced to two or three years, depending on the level of competence the apprentice or trainee achieves.

VALUES AND BEHAVIOURS

Trade operatives form the 'backbone' of the construction industry. As a bricklayer, you are continuing a long history of skilled workers who have made a significant contribution to the built environment we see around us today. Always strive to improve your skills and abilities and have pride in the way your work contributes to the modern built environment.

Specialists

Specialists can be used to install items such as ventilation systems, renewable energy sources or methods of installing cavity **insulation** such as injection (a method discussed in more detail in Chapter 5).

Some traditional skilled trades have been split into specialist sections of activity, for example, plastering. In the past, a plasterer laid floor **screeds** as well as producing wall finishes. Today, on many projects, if a floor screed is specified, it will be installed by a specialist operative using a product designed to be pumped to the work location.

KEY TERMS

Insulation: materials used to reduce heat transfer in a building. Can also be used to control sound transmission

Screed: a levelled layer of material (often sand and cement) applied to a floor or other surface

Other specialists might be involved in restoration projects such as repairing or replacing a thatched roof. Restoration and conservation projects may require a range of skills that are less common today and that have become specialist trades.

▲ Figure 3.4 A section of thatched roof that has been renewed by a specialist

General operatives

General operatives perform a range of semi-skilled and non-skilled work. They might perform essential tasks such as driving plant or machinery, working with the bricklayers mixing mortar or looking after site storage facilities.

VALUES AND BEHAVIOURS

The smooth running of a construction site of any size is heavily dependent on general operatives who are reliable and trustworthy and who co-operate with colleagues.

▲ Figure 3.5 A general operative loading a mixer

Other key personnel

Other personnel who are important to the success of a construction project may not be part of the company structure but are closely linked to the construction process during the planning stage and when managing quality standards throughout the project. Table 3.2 outlines the roles and responsibilities of these key personnel.

▼ Table 3.2 Other key personnel

Job title	Roles and responsibilities
Building control officer 	Ensures that new buildings, alterations and extensions meet Building Regulations from design to completion. They may also monitor safety, sustainability and accessibility.
Planning officer 	Planning officers mostly work for a local authority planning department. Their duties may include: ● processing planning applications ● conducting site visits to determine whether developments are proceeding in accordance with permissions ● preparing reports for planning committees and making recommendations.
Contracts manager 	Oversees projects from the start through to completion, managing and monitoring contracts to ensure that the project is completed on time and within budget. A contracts manager needs to be skilled in: ● understanding legal matters ● negotiation ● developing relationships.

▼ Table 3.2 Other key personnel (continued)

Job title	Roles and responsibilities
Clerk of works 	Provides an independent assessment of the works undertaken, checking the quality of the build on behalf of the client. Depending on the size of the project, they are either on site continuously or make regular site visits.

Organising the construction team

Although all the members of the construction team play important roles in working towards the successful completion of a project, some roles are more demanding and carry greater responsibility.

This is reflected in what is termed the 'management **hierarchy**'. Figure 3.6 shows a management hierarchy chart, with the highest level of management in the red boxes at the top and the skilled operatives in green boxes at the bottom.

> **KEY TERM**
>
> **Hierarchy:** a system in which roles are arranged according to their importance or level of responsibility

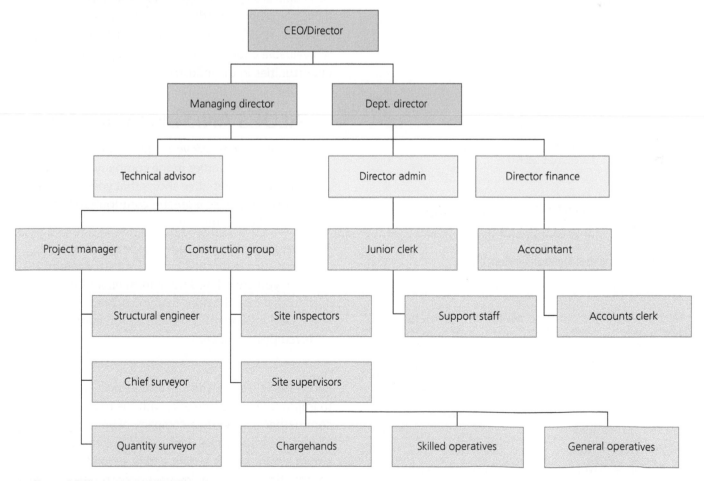

▲ Figure 3.6 Management hierarchy

You may hear the term 'line management' used on site. This refers to the arrangement of responsibilities and relationships between the members of the team. The greater the responsibility, the higher up the 'line' the role will be placed. Each person on the line of management is responsible for managing those below his or her stated position.

Although they may not be viewed as a member of the construction team, we can place the customer or client at the top of a management chart showing levels of responsibility. They have commissioned and funded the work project and everyone on the line of management below them is accountable to them.

Look at the simplified example of a line management chart in Figure 3.7 and notice the various responsibilities positioned on it. It will never be a simple straight line but will always have branches to link roles with similar levels of responsibility or oversight to each other.

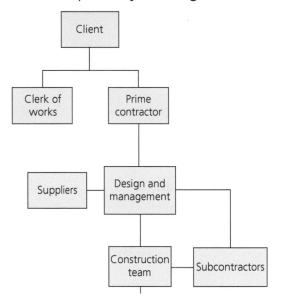

▲ Figure 3.7 A simple line management chart

ACTIVITY

In the simple line management chart, notice the line extending below the 'Construction team' box. Copy the chart and extend the line, adding boxes to show the relative positions of general operatives, technicians, specialists and trade operatives to each other. Discuss the positions you've chosen with others and then select in which box the following roles should be entered.

- Stores manager
- Dumper driver
- Roofer
- Foreman bricklayer

VALUES AND BEHAVIOURS

Skilled workers are in great demand in construction and the personnel needs of the industry are difficult to meet. Although you may become skilled in your chosen trade and sought after by busy contractors, always co-operate with your supervisors and managers and recognise their authority. Many of them have long and varied experience in the industry and have studied to improve their career progression, so they should be given due respect in their responsible positions.

Career progression

The construction industry offers opportunities for career advancement to workers who wish to develop their skills and abilities. Many companies invest in their workforce by encouraging further education and study to gain valuable qualifications.

You could study for a Higher National Certificate (HNC) or progress further to achieve a Higher National Diploma (HND) in construction. A construction degree is a valuable qualification which can open many career opportunities in the industry.

VALUES AND BEHAVIOURS

Don't be afraid to ask your supervisors about training opportunities in the company you work for. Taking the initiative shows that you're keen to progress and make a greater contribution to the success of the company.

There are many opportunities for a bricklayer to build a successful construction industry career. When you are offered training, keep a record of formal qualifications that you achieve. This is called a record of continuous professional development (CPD).

Becoming self-employed can also lead to opportunities to improve your income and, with the necessary commitment, to set up a company of your own. Many skilled operatives have progressed to become employers in their own right and have made a valuable contribution to the employment opportunities of others.

② TYPES OF BUILDINGS AND HOW THEIR PARTS ARE CONSTRUCTED

There is a range of building types, with individual and sometimes distinctive design appearances. Architectural design and types of construction are often categorised as belonging to a distinctive time period or **era,** which is named according to the king or queen living at that time.

> **KEY TERM**
>
> **Era:** a period of time distinguished by particular characteristics

Buildings from different eras

Some examples of eras in which building features and designs are easily identified are described below.

- Elizabethan – houses were often black and white, featuring structural frames built using large timbers with plaster infill between them known as wattle. They featured tall chimneys, overhanging first floors and porches with pillars.

- Georgian – often built square in brick or stone, two or three storeys high, with a large panelled entrance door. This was centrally placed in **detached** houses or to one side in **terraced** houses. In detached houses, the roof was often 'hipped', which means it sloped from all sides of the building. (Roof shapes are discussed later in this chapter.)

▲ Figure 3.9 Georgian house

> **KEY TERMS**
>
> **Detached:** when referring to a house, a stand-alone free-standing residential building
>
> **Terraced:** when referring to houses, a row of similar dwellings joined together by the side (or party) walls

▲ Figure 3.8 Elizabethan oak framing

▲ Figure 3.10 Detached house

- Victorian – houses predominantly built in terraces using brick. Flemish bond was frequently used (for more on this brick bond see page 181). This architectural style featured bay windows to the front of the building which were built with their own roof covering. With a fireplace in every room, there were many chimney pots for each dwelling.
- Edwardian – many houses in this era were built as detached or **semi-detached** residential properties in brick. They often featured porches with balconies above them decorated with patterned timber or iron railings.

▲ Figure 3.11 Victorian terraced houses

KEY TERM

Semi-detached: sometimes referred to as a 'semi', a house that is joined to another house on one side by a shared (party) wall

In addition to designing buildings as detached, semi-detached or terraced structures, construction of flats or apartments has become more popular as a means of maximising the number of dwellings that can be provided on increasingly expensive building land.

For each type of building, new materials and different methods of work have been developed over time to speed up the construction process and reduce environmental impact.

▲ Figure 3.12 Apartment block

ACTIVITY

Draw up a table listing the different building eras described in this section. As you travel around your area, add a tick every time you identify a building from an era in your list. Which building era examples do you see the most?

Modern construction methods

New construction methods have been developed that modify or enhance traditional methods with new ideas and versatile materials. Some methods have been widely adopted, whereas others are used more rarely for specific applications.

Structural methods

Timber frame construction methods are discussed later in this chapter and thin-joint masonry construction is discussed on page 167. Other methods of producing a structure are described below.

Modular

This describes the method of using factory-produced pre-engineered units (or modules) that are built to combine major elements of a structure. Modules can be set up for use as specific rooms, such as a bedroom module, a kitchen module and so-on. They are delivered to site ready for assembly to produce a finished structure relatively quickly. A modular system can be produced with **services** already installed, allowing fast setup and occupation.

▲ Figure 3.13 Building modules ready for delivery

KEY TERM

Services: systems installed in buildings to make them comfortable, functional, efficient and safe

KEY POINT

A relatively new modular method of construction uses what are referred to as 'pods'. This type of construction method uses robust, low maintenance materials that are lightweight and easy to use in mass-produced products. The concept reduces the reliance on multiple skilled workers on site, since the technical work (such as electrical and plumbing) is completed in the factory, sometimes using automated assembly processes.

Monolithic

This term applies to structures that are created by carving or excavating them from a single piece of material. Some ancient structures were produced from rock in this way. A modern application of this idea is to pour a material such as concrete into a mould which is constructed to produce walls, floors and staircases at one time. Concrete domes have been produced monolithically by spraying concrete onto the inside of a mould to produce a structure without joints or seams.

Elements of a structure

Each type of building has parts (or elements) that it shares in common with other types. Let's look at the construction principles used for each element and its purpose within a structure.

Foundations

A foundation supports the building and transfers the loadings produced by and within the structure to the natural foundation (the ground on which it sits). The loadings produced are described as **dead loads** and **imposed loads**. A foundation is part of the design known as the **substructure**.

KEY TERMS

Dead load: the weight of all the materials used to construct the building

Imposed load: additional loads that may be placed on the structure, such as the weight of people and furniture or the effects of wind and snow; also referred to as live load

Substructure: all brickwork, blockwork and structural materials below finished floor level (FFL)

ACTIVITY

Make a list of heavy items, such as a piano, that could be contained in a building as an imposed load.

Foundations must be designed to support the building, regardless of any potential movement in the ground on which the building will sit. Ground conditions can vary widely, so a survey of the land is necessary.

Soil testing

Soil samples will be taken by specialist surveyors before the design stage to help decide on the type of foundation to use. Testing usually takes the form of bore holes dug or drilled around the site and samples taken are sent for testing in a laboratory. The results will identify:

- the soil condition (clay or sandy)
- the depth of the soil above bedrock
- the depth of the water table
- if any contamination is present.

▲ Figure 3.14 Soil test drilling rig

The soil condition is important because:

- clay soil doesn't drain well and will expand (or heave) when waterlogged or shrink during prolonged dry conditions
- sandy soils drain well but can be unstable as a natural foundation.

Based on the test and survey results, a foundation that is suitable for the ground type and load of the building will be designed.

KEY POINT

When clay soils become saturated and expand or heave, the forces generated are powerful enough to move a building upwards, causing potentially serious damage to the structure.

Materials for foundations

Most foundations are formed in concrete. Concrete is made from fine **aggregate** (sand) and coarse aggregate (crushed stone) mixed with cement and water. Water produces chemical reactions in the cement called **hydration**, which causes it to harden and lock the aggregates together.

KEY TERMS

Aggregate: the coarse mineral material, such as sharp sand and graded, crushed stone (gravel), used in making mortar and concrete

Hydration: chemical reactions in which products of hydration bond individual sand and gravel particles together to form a solid mass

Concrete is very strong under compression (when weight or pressure is placed on it) but is weak in tension (when subjected to stretching or pulling forces). Tension can be created in a foundation when it must bridge softer sections of ground or when ground conditions are unstable. To prevent failure of a concrete foundation, it can be strengthened or reinforced with steel bars or mesh cast into the concrete before it hardens.

▲ Figure 3.15 Steel reinforcing mesh

Strength and durability of concrete

Since it will be buried underground, a concrete foundation is likely to be subjected to constant water saturation. Chemicals called sulphates that dissolve in water can affect the strength of cement, so sulphate-resisting cement can be used to maintain the strength of a concrete foundation in wet conditions.

The amount of cement added to a mix will determine its working strength. Look at Table 3.3 to see the different concrete mix ratios and the applications they can be used for.

▼ Table 3.3 Concrete mix ratios and uses

Concrete type	Ratio	Uses
C7.5 (low strength)	1:3:6 or 1:3:7 (cement/sand/ coarse aggregate)	For general non-structural use such as fixing fence posts
C10 to C15 (medium strength)	1:4:5 or 1:4:6 (cement/sand/ medium aggregate)	General use in typical house foundations
C20 (strong)	1:2:4 (cement/ sand/medium aggregate)	Used for foundations on softer ground

▼ Table 3.3 Concrete mix ratios and uses (continued)

Concrete type	Ratio	Uses
C25 (stronger)	1:2:3 (cement/sand/medium aggregate)	Used for foundations for larger houses and for creating suspended floors
C30 (very strong)	1:1½:3 (cement/sand/aggregate with stone dust)	General purpose, strong concrete for applications such as retaining walls
C35 (industrial strength)	1:1½:2½ (cement/sand/fine aggregate)	Structural concrete

ACTIVITY

Neighbours have asked you to build a garden wall along the boundary of their property. The wall is to be built at the bottom of a slight slope alongside a public footpath. Select the strength of concrete that would be suitable for the foundation of the wall.

There may be reasons for speeding up or slowing down the hardening process after concrete has been positioned. Chemical additives known as an **accelerant** and a **retarder** can be added to the mix. An accelerant may be used to prevent frost damage occurring if newly laid concrete won't have hardened when the temperature drops. A retardant may be used to allow a large area of concrete to have a surface finish produced before it hardens.

KEY TERMS

Accelerant: a substance added to speed up a process

Retarder: a substance added to slow down the rate of a chemical change

INDUSTRY TIP

If you are required to use additives when mixing concrete, follow the manufacturer's instructions carefully. Using the wrong quantities or proportions of a product may affect the final strength of the concrete.

Types of foundations

After considering the type of building and the ground conditions it will be built on, the most suitable type of foundation can be designed. An engineer will often be employed at the design stage to produce structural calculations that will confirm the ability of the proposed foundation to perform as intended.

Each type of foundation has different characteristics and performance capabilities. Let's look at the different types that can be used.

Strip foundation

Strip foundations are widely used for housing and small commercial developments. They are formed by digging a trench to the required width and depth, determined by the soil conditions and the weight of the structure. A strip of concrete is poured into the trench to a minimum thickness of 150 mm to suit the design. More commonly the thickness will be increased to 225 mm.

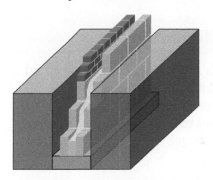

▲ Figure 3.16 Traditional strip foundation

KEY POINT

As a bricklayer, you will often work on buildings with strip foundations. You will be required to work in the foundation trench to build the masonry known as the **footings** up to finished floor level (FFL) or **damp proof course (DPC)** level. (For more on DPC, see Chapter 5.)

KEY TERMS

Damp proof course (DPC): a barrier designed to prevent moisture travelling through a structure

Footings: the masonry constructed from the top of the foundations up to finished floor level (FFL)

The masonry built on the foundation is constructed in bricks or blocks which must be resistant to moisture and capable of supporting the loadings produced by the complete structure. (For details on characteristics of materials like bricks and blocks, see page 137.)

It may be decided that the full depth of trench will be filled with concrete (called 'trench-fill') if there are trees nearby. The roots of trees can undermine a foundation and they also extract water from the ground, causing the soil to shrink. This can lead to failure of the foundation and the structure of the building will be severely affected.

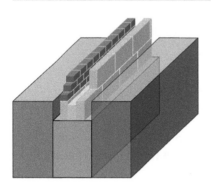

▲ Figure 3.17 Trench fill foundation

Where ground conditions are unstable, a wider strip may be specified which can be reinforced with steel to add strength.

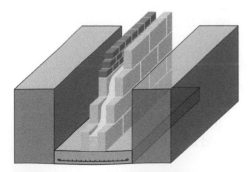

▲ Figure 3.18 Wide strip foundation

If a strip foundation is used on sloping ground, the concrete can be formed in steps to follow the angle of the slope.

Pad foundation

A pad foundation is used to support a load focused in one point, such as a brick pier or a column in a steel-framed building. When it is used to support steel work it will often have bolts cast into the top ready for fixing the steel column. The dimensions and depth of the pad foundation can be adjusted to suit different ground conditions.

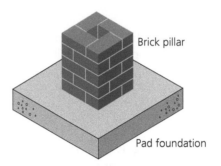

Brick pillar

Pad foundation

▲ Figure 3.19 Pad foundation

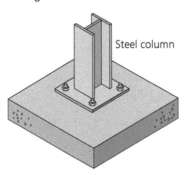

Steel column

▲ Figure 3.20 Pad foundation with bolts

Pile foundation

Piles are essentially long cylinders of a strong material such as steel or concrete.

▲ Figure 3.21 A cylindrical pile for a foundation

Piles are used to transfer the load of a building through soft or unsuitable soil layers into the harder layers of ground below, even down to rock if required. This type of pile foundation is known as 'end bearing' and is effective when a building has very heavy, concentrated loads, such as in a high-rise structure or a bridge.

▲ Figure 3.22 High-rise building

A second type of pile foundation is known as 'friction pile'. Support for a building is provided by the full height of the pile creating **friction** with the soil it stands in. The deeper into the ground the pile is driven, the greater the friction and **load bearing** capacity.

Both these types of pile foundation require specialist equipment and trained personnel to install them successfully.

KEY TERMS

Friction: the resistance that one surface or object encounters when moving over another
Load bearing: relating to the carrying of a load

Raft foundation

A raft foundation is often laid over an area of softer soil that would be unsuitable for a strip foundation or where a pile foundation would be too expensive. It consists of a reinforced slab of concrete covering the entire base of the building, spreading the weight over a wide area. The edge of the slab is usually thickened as a support for load bearing walls around the face line of a building.

If any minor movement takes place due to poor ground conditions, the building is protected since the whole foundation can move slightly as a unit.

When foundations are constructed, provision should be made for the entry of services to and from the building. There is more on services later in this chapter.

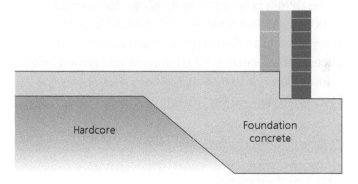

Hardcore

Foundation concrete

▲ Figure 3.23 Raft foundation

INDUSTRY TIP

When you build the masonry for footings, check the entry position of services since you may have to create an opening for them at a specific location in the building.

Floors

Floors are designed to be load bearing so that they can carry the weight of occupants and items of equipment and furniture contained in a building. The loadings they carry can be transferred through the walls on which they bear or are attached to, down to the foundation and ultimately to the ground on which the building sits.

There are various methods used to construct floors. The design choice depends on the type of building and the potential load that will be imposed. As well as being load bearing, floors can be designed to perform other functions, for example:

- prevent heat loss
- reduce or prevent transfer of sound
- prevent moisture penetration.

Ground floors

Floors constructed within the substructure can be formed in a solid or suspended configuration. They can be constructed using a range of materials.

Solid concrete floors

Solid floors are laid on **hardcore** and have a **damp proof membrane (DPM)** under them to prevent damp rising through the floor. The DPM is usually laid on sand to prevent the hardcore from puncturing it. The thickness of the DPM is referred to as its gauge, for example '1000 gauge'. This floor design is termed 'ground bearing' because the weight of the slab and loads placed upon it bears on the hardcore and ground below it. Insulation is also installed either under or over the floor, depending on the design, to reduce heat transfer.

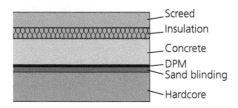

Screed
Insulation
Concrete
DPM
Sand blinding
Hardcore

▲ Figure 3.24 Sectional view of a concrete floor

KEY TERMS

Damp proof membrane (DPM): a sheet of strong waterproof material

Hardcore: solid materials of low absorbency used to create a base for load-bearing concrete floors, for example, crushed stone

Suspended concrete floors

Some concrete slab floors at ground level are suspended on walls at their edge and on any central walls of the masonry footings. They are reinforced with steel mesh to allow them to bridge from wall to wall. This type of floor is used where the material under the floor is too soft to support it. The weight of the suspended concrete slab and the loadings on it are transmitted through the walls to the foundation.

▲ Figure 3.25 The weight of the reinforced concrete floor will be carried by the footing masonry

'Block and beam' is a method of producing a suspended floor that can speed up the process of construction. This uses factory manufactured concrete beams that span across the walls of the masonry footings. The shaped beams are carefully spaced to allow dense concrete blocks to be positioned between them. Look at Figure 3.26 to see how this works.

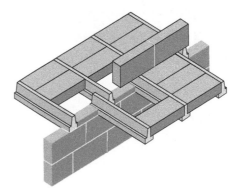

▲ Figure 3.26 Suspended concrete floor (block and beam)

Suspended timber floors

A traditional method of constructing ground floors is to use timber beams called **joists** spanning across the outer walls of a structure with 'sleeper' walls supporting them at intermediate positions. Timber floor boarding or timber sheet material is fixed across the joists to form the floor surface.

KEY TERM

Joists: parallel timber beams spanning the walls of a structure to support a floor or ceiling

For this type of floor arrangement, it is important that adequate ventilation is provided to keep the underfloor space dry. This reduces the chance of rot in the timber components. (For more on rot in timber, see Chapter 7.)

> **KEY POINT**
>
> The structure from the ground floor level upwards is termed the **superstructure**. A suspended timber ground floor could be viewed as part of the superstructure rather than the substructure if it is installed above the level of the DPC.

> **KEY TERM**
>
> **Superstructure:** the upwards part of a building that begins where the substructure ends

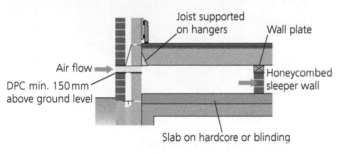

▲ Figure 3.27 Suspended timber floor

Upper floors

Upper floors are suspended, in that they are attached to or bear on the walls of a structure. There are several ways of linking the timber components to the walls to produce a solid and stable result. Galvanised steel connectors and hangers in a range of shapes and sizes allow quick and easy installation of timber beams when constructing upper floors.

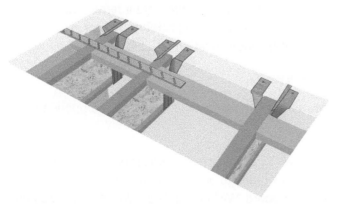

▲ Figure 3.28 i-beam floor joists bearing on a load-bearing masonry leaf using steel hangers

> **ACTIVITY**
>
> Search online for 'joist hangers'. Choose a website and count how many different shapes and sizes are on offer on just one webpage. There's quite a range available!

Solid timber joists can be prone to movement and bending due to changes in moisture levels in a structure. As a result, new products, referred to as 'timber engineered' components, have emerged, which are more stable in use and lighter to use by the installer.

> **INDUSTRY TIP**
>
> Two timber engineered components that are increasingly used have the trade names Posi-Joist and i-beam. (For important points about handling and storing these components see page 149.)

Walls

There are two basic types of masonry wall design that a bricklayer will build:

- cavity walls
- solid walls.

You'll learn more about these types of masonry walls built in brick and block (or a combination of both) in Chapters 5 and 6. In this chapter, we'll look at some alternative and more modern methods of constructing walls.

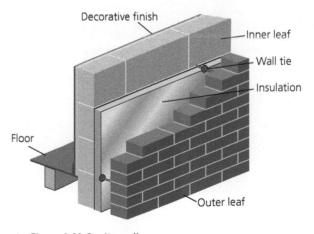

▲ Figure 3.29 Cavity wall

Timber frame

As mentioned earlier in this chapter, Elizabethan houses were often constructed using timber framing, so this is not a new method of constructing walls. However, the dimensions of timber used in those older buildings would not make the method cost effective in modern times.

Modern timber frame homes are precision engineered, strong and durable. They are commonly constructed from factory-made panels which are transported to site ready for assembly. The modern arrangement is to have a timber core that carries the structural loadings and an outer leaf (or skin) of masonry or **cladding** such as treated timber, to waterproof the structure.

KEY TERM

Cladding: the application of one material over another to provide a skin or layer

▲ Figure 3.30 A timber frame structure

A timber frame structure can be well insulated to reduce heat transfer and, if resources from properly managed forests are used, is a sustainable building method using renewable materials. Sustainability is covered in more detail later in this chapter.

Timber frame is a method that speeds up the building process because the timber core can be quickly erected and weatherproofed. This means that internal work can start at the same time as the masonry outer leaf is being constructed.

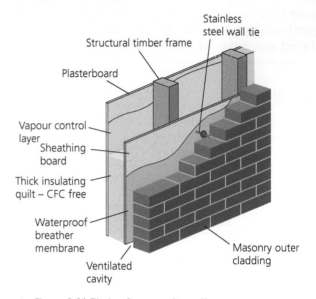

Stainless steel wall tie

Structural timber frame

Plasterboard

Vapour control layer

Sheathing board

Thick insulating quilt – CFC free

Waterproof breather membrane

Ventilated cavity

Masonry outer cladding

▲ Figure 3.31 Timber frame cavity wall

KEY POINT

The principles of constructing a building with a load-bearing core in timber which is tied to an outer leaf of masonry can also be applied to steel frame buildings.

Internal walls

Walls within a structure can be load bearing if they support floors above them, or non-load bearing if used simply as partitions to divide large internal spaces into smaller ones. They can be **prefabricated** or assembled on site at the work location.

KEY TERM

Prefabricated: factory-made units or components transported to site for easy assembly

Walls can be constructed in timber or metal as a framework ready to be covered in a sheet material such as plasterboard. Wall elements produced in this way are commonly referred to as 'stud' walls.

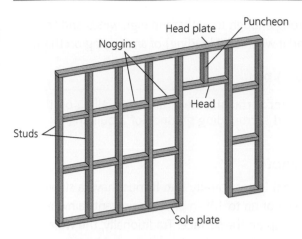
▲ Figure 3.32 Timber stud wall

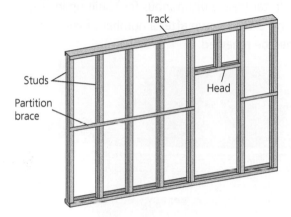
▲ Figure 3.33 Metal stud wall

Wall finishes

A face brick wall, when completed, provides a finish that is durable and long lasting. Other types of wall construction may require additional materials to give a suitable finish.

External finishes

There is an extensive range of finishes suitable for external walls. Some frequently used types are described below.

- Render systems – can be applied as a sand/cement mix coating applied by hand.
- Coatings – silicone-based products that can be applied by hand or machine.
- External wall insulation – installed as insulation sheets glued and mechanically fixed to the outside of a wall with a smooth or textured thin coat render applied by hand or machine. The render can be coloured if required.

Paint systems can be applied to all of these wall finishes when specified.

▲ Figure 3.34 External insulation ready for the finishing coat

Internal finishes

Internal walls (and ceilings) can be plastered, which means they are covered with a thin layer of gypsum plaster over plasterboard. This gives a very smooth surface which is then usually finished with emulsion paint or papered coverings.

▲ Figure 3.35 Plaster gives a smooth finish

ACTIVITY

Do some online research about gypsum. Write a short report on what gypsum is and discuss with others what products can be made from it for construction purposes.

An alternative method of providing a wall finish using plaster-based products is known as dry-lining (or dry-wall). This method involves fixing plasterboards with tapered edges to the wall structure and attaching a paper tape to the joints between the boards which is then finished with a specialised filler. The plasterboards are screwed to a stud wall or glued to a masonry wall using special adhesive.

▲ Figure 3.36 Fixing plasterboards to a wall

The paint finish provided by emulsion paint will be more durable on new plaster if the surface is first sealed with a diluted coat of paint called a mist coat followed by two undiluted coats.

In areas that will be subjected to higher levels of moisture, such as bathrooms and kitchens, the plastered surface can be finished in ceramic tiles which are waterproof. Alternatively, ceramic tiles can be fixed directly to an un-plastered surface with the appropriate adhesive.

▲ Figure 3.37 Applying ceramic tiles as a wall finish

Roofs

Roofs are designed to protect the structure below them by providing a weatherproof surface that directs rain to **storm (or surface) water** drainage systems. They must be strong enough to withstand high winds and the potential weight (or loading) of snow lying on them.

Flat roofs

A flat roof is not literally flat. It must have a slope or incline of up to 10° in order to prevent rainwater building up on the surface. Traditionally, the waterproof coating of a flat roof has been provided by felt material covered in tar (called bituminous felt) built up in several layers. This breaks down after a number of years and must then be renewed.

▲ Figure 3.38 House with a flat roof

An improved, more durable covering is provided by layers of glass fibre sheets impregnated with a special resin.

Both these systems are bonded to timber sheet decking fixed to beams (or joists) spanning across the walls of the structure. The slope (or fall) can be created by fixing tapered lengths of timber, called firrings, to the top of the joists.

▲ Figure 3.39 Flat roof

Pitched roofs

A pitched roof has a weatherproof surface or surfaces that slope at more than 10°. They are constructed using timber components called **rafters** and have a range of different designed shapes.

The simplest design of pitched roof is the 'lean-to', where a single roof surface literally leans against an adjoining wall. If a single pitched roof surface covers the entire building, this is referred to as a 'mono-pitch' roof.

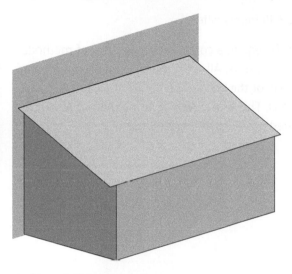

▲ Figure 3.40 Lean-to roof

'**Gabled**' roofs with two sloping surfaces are a very common design, with triangular walls closing each end of the roof up to the **ridge**.

KEY TERMS

Gable: the triangular wall at the end of a ridged roof

Rafter: a beam forming part of the internal framework of a roof

Ridge: the highest horizontal line on a pitched roof where sloping surfaces meet

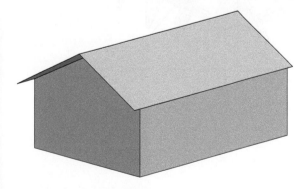

▲ Figure 3.41 Gabled roof

INDUSTRY TIP

The whole end wall of a structure, including the triangular section at roof level, is often referred to as a gable end.

To protect the gable from heavy rain, the roof structure can be extended over the masonry and lengths of timber known as bargeboards can be fitted along the roof edge.

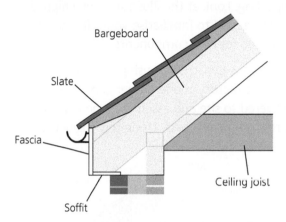

▲ Figure 3.42 Weatherproofing a gable

A 'hipped' roof has no gables and the weatherproof surfaces slope down from the ridge to the top of the walls on all sides. These are more complex and therefore more expensive to construct.

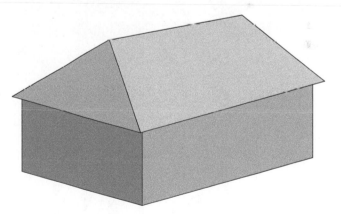

▲ Figure 3.43 Hipped roof

ACTIVITY

Refer to page 45 and review the section on isometric projection drawings. Produce your own ruler-assisted isometric projection drawing of each type of roof in this section. (Note: Don't just copy the illustrations in this section – they are not true isometric drawings.)

Roof components

Traditionally, roofs have been constructed on site from individual lengths of timber. These are referred to as 'cut' roofs since the timber is cut to the required lengths and angles to suit the design. This is a process that requires carpentry skills and a good understanding of geometry.

For a large roof, relatively heavy timbers might be used because, until it is completed, the structure needs to be self-supporting. Look at the illustration of a hipped cut roof in Figure 3.44 to familiarise yourself with the terminology used for roof components.

Most roofs are now constructed using 'trussed rafters'. This system uses factory-made timber components that are delivered to site and assembled more quickly than a cut roof, reducing costs. Using this engineered system means that lighter timbers can be used, making additional cost savings.

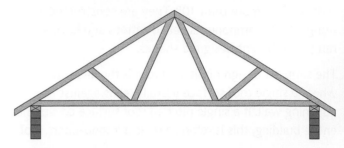

▲ Figure 3.45 Trussed rafter

Some roofs require a combination of cut roof methods and trussed rafters. All roofs must be securely anchored to the walls of the building to conform to Building Regulations. This is achieved by using mild steel straps at specified positions in the structure.

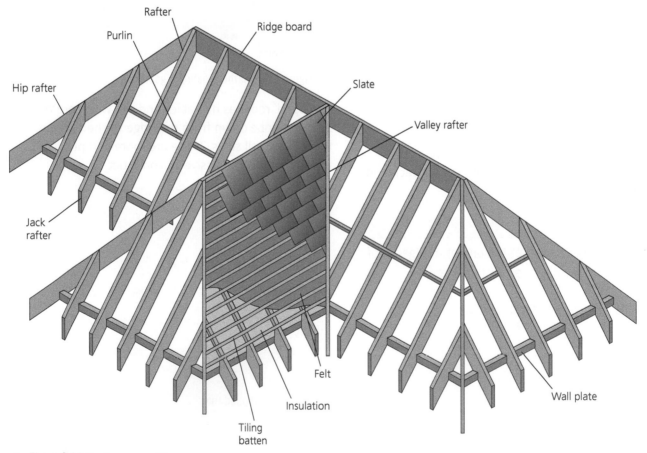

▲ Figure 3.44 Roof components

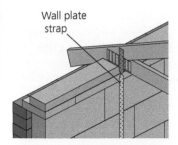

▲ Figure 3.46 Restraining strap

Roof covering materials

The materials selected for use as roof coverings must be highly water resistant and durable.

Natural slate

Traditionally, slate has long been a favoured material for use as a roof covering. This is a natural product in the form of a mineral that can be readily split into thin sheets by skilled workers. Slate continues to be used extensively as a roof covering despite its relative expense.

▲ Figure 3.47 Natural slate roof tiles

Fibre cement slate

A cheaper alternative to natural slate is fibre cement slate. The material is manufactured from a mixture of cement, sand and cellulose. Fibre cement slate has a more uniform appearance than natural slate but can be manufactured with features that imitate the visual characteristics of the natural product.

▲ Figure 3.48 Fibre cement slate

Clay and concrete tiles

Clay tiles are manufactured from a natural material and are produced in a range of sizes and shapes to suit different applications. They produce a finished roof covering that is attractive and very durable.

Concrete tiles are manufactured in different sizes, shapes, textures and colours. They are relatively quick to install and are often designed to interlock with each other to form a wind resistant covering.

Both clay tiles and concrete tiles are heavy components requiring a roof structure that is capable of carrying the loadings adequately.

▲ Figure 3.49 Roof tiles

Lead sheet is commonly used to waterproof specific areas of a roof, for example where a chimney projects through the roof covering. Used in this location the lead weatherproofing is called 'flashings'. (For more on the use of lead to weatherproof specific parts of buildings, see Chapters 7 and 10.)

▲ Figure 3.50 Flashings providing waterproofing

The construction sequence

Managers and operatives must apply the principles of construction in a recognised sequence so that the work progresses in an orderly and efficient manner. There are two expressions that are used to describe the programmed construction timing of certain elements of a structure:

- **First fix**
- **Second fix**.

Let's discuss what these terms mean in relation to constructing and finishing the project.

First fix

The following components and elements are part of the first fix stage of construction:

- Partitions – the internal walls that separate rooms.
- External door and window frames – if not installed when the walls are being built, they are fitted during first fix.
- Internal door lining – the initial component of the internal door frame.
- Stairs – fitted before the wall finishes start to be applied. The stairs will need temporary protection during subsequent construction operations.
- Service runs – cables for electricity, pipework for heating and water and, if appropriate, pipework for gas. They will not usually be 'live' until later stages of completion are reached.

▲ Figure 3.51 Plumbers fitting underfloor central heating pipes

Second fix

Second fix categories are:

- Finishes – plaster or dry-lined finishes, skirting boards and architrave.
- Sanitary ware – shower, sinks and toilets.
- Internal doors – these will need to be protected from subsequent operations such as applying paint finishes.
- Kitchen units – also require temporary protection from damage by subsequent operations such as wall tiling.

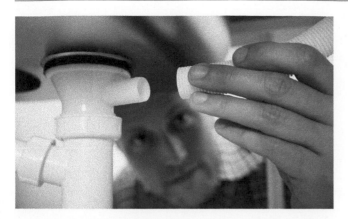

▲ Figure 3.52 Fitting a sink

Services

As previously discussed, the installation of services is an important factor to consider when constructing buildings that are designed for human occupation. We'll now look at how services are integrated into the construction process.

Water

Water is brought into a building through pipework, which must enter the building at a suitable underground depth to give protection from frost. Supply pipes entering the building are usually made of plastic, with internal plumbing pipework being made from plastic or copper. The plumber is skilled in using a variety of pipework fittings to create watertight routes to supply bathrooms, kitchens and central heating systems.

Electricity

Electricity is an important service to provide for lighting, heating and cooking. It is brought into a

building by means of cables which can be installed underground or overhead.

Gas

Gas is brought into a building through pipework which must be installed in accordance with strict regulations. Only trained and certified operatives can work on gas installations in order to maintain safety. Gas is a fuel that has been used for heating and cooking for many years.

Drainage systems

Drains are installed to carry waste products and rainwater away from a building.

New structures may be designed to include drainage systems that store rainwater in tanks, often buried underground to allow heavy high-capacity tanks to be used. This is called water harvesting and the water stored can be used to flush toilets in what is known as a 'grey water' system.

> **KEY POINT**
>
> Domestic drainage systems are mostly connected to large pipes that take the toilet (foul) waste to treatment plants. These must be carefully designed to protect the environment from pollution. If it is not harvested, rainwater will either be channelled to an excavated hole in the ground called a soakaway or be taken through storm drains to rivers or the sea. (For more on drainage systems, see page 228.)

▲ Figure 3.53 Copper pipe for plumbing

▲ Figure 3.54 Plastic waste water pipe

▲ Figure 3.55 Electrical installation

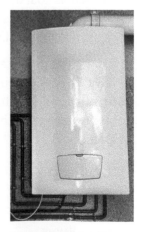

▲ Figure 3.56 Pipework to a gas boiler

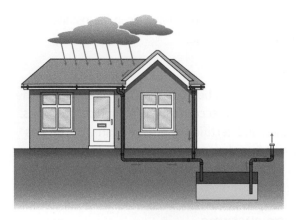

▲ Figure 3.57 Rainwater stored in an underground tank

Other services

Other provisions that must be planned for in most modern buildings are telephone and data services.

③ ENVIRONMENTAL AND SUSTAINABILITY CONSIDERATIONS FOR BUILDINGS

Increasingly, there is concern about how construction activities and work practices have an impact on the environment. A great deal of effort is being put into making construction methods and completed buildings more environmentally friendly. We'll now look at sustainability principles used in building design and construction.

Sustainability

Constructing and operating buildings uses significant amounts of energy. To produce this energy, power stations often burn fossil fuels such as coal and oil. However, reserves of these fuels are declining, so continuing to use them as an energy source is unsustainable. In addition, the process of burning fossil fuels such as coal or oil combines carbon with oxygen to introduce carbon dioxide (CO_2) into the atmosphere, which is leading to **climate change**.

It is therefore important to make greater use of renewable energy sources such as solar and wind, both in the manufacture of construction materials and components and in the way buildings are designed to operate.

KEY TERM

Climate change: a large-scale, long-term change in the Earth's weather patterns and average temperatures

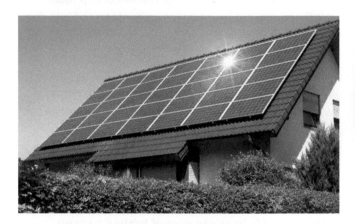

▲ Figure 3.58 Solar panels

▲ Figure 3.59 Wind turbines

Official guidance

The Government has supported a number of initiatives to provide guidance on sustainable construction. The Building Research Establishment (BRE) has produced the Home Quality Mark, a national standard for new homes, and has developed a structured method of assessing a building's impact on the environment, referred to as BREEAM (Building Research Establishment Environmental Assessment Methodology).

INDUSTRY TIP

The Home Quality Mark has replaced the Code for Sustainable Homes. To find out more, go to: www. homequalitymark.com/what-is-the-hqm

Building Regulations also play a part in controlling construction activities to support sustainability, and statutory guidance is provided in 'Conservation of fuel and power: Approved Document L'. This standard applies to construction projects that are new, extended, renovated, refurbished or involve a change of use. Among other things, the guidance states the minimum levels of insulation required in buildings to control heat transfer.

INDUSTRY TIP

You can find out more about 'Conservation of fuel and power: Approved Document L' at: www.gov.uk/government/publications/conservation-of-fuel-and-power-approved-document-l

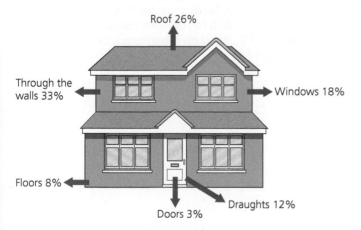

Roof 26%

Through the walls 33%

Windows 18%

Floors 8%

Draughts 12%

Doors 3%

▲ Figure 3.60 Sources of heat loss from a house

Insulation must save energy used for heating by reducing heat lost through the various parts of a building. It must also reduce heat entering a building during hot summer weather, as this in turn reduces energy consumption by air conditioning and cooling systems.

▲ Figure 3.61 Air conditioning unit

ACTIVITY

Conduct some research into the term 'passive house'. Discuss with others how the passive house concept can save energy as part of a voluntary arrangement.

VALUES AND BEHAVIOURS

The bricklayer plays an important part in ensuring energy consumption is reduced in buildings. Insulation is installed during the construction of walls and must be correctly selected and positioned in order to function properly.

Work carefully to make sure you're playing your part in supporting sustainable construction. (Methods of installation and materials used for insulation in cavity walls will be discussed in Chapter 5.)

ACTIVITY

Go to www.gov.uk/guidance/developers-get-environmental-advice-on-your-planning-proposals and research the planning requirements for a new development to satisfy sustainability considerations.

Insulation materials

We'll now look at some insulating materials that are available to architects and designers to produce buildings that meet the regulations and official guidance.

▼ Table 3.4 Insulation materials

Type of insulation	Description
Blue jean and lambswool	Lambswool is a natural insulator. Blue jean insulation comes from **recycled** denim.
Fibreglass/mineral wool	This is made from glass, often from old recycled bottles or mineral wool. It holds a lot of air within it and is therefore an excellent insulator that is also cheap to produce. It requires substantial thickness to comply with Building Regulations.
PIR (polyisocyanurate)	Formed as a solid insulation board with foil layers on the faces, it is lightweight, rigid and easy to cut and fit. It has excellent insulation properties. Polystyrene boards are similar to PIR. Although polystyrene is cheaper, its thermal properties are not as good.

→

▼ Table 3.4 Insulation materials (continued)

Type of insulation	Description
Multifoil 	A modern type of insulation made up of many layers of foil and thin flexible insulation material. These work by reflecting heat back into the building. Usually used in conjunction with other types of insulation.
Double glazing and draught proofing measures 	The elimination of draughts and air flows reduces heat loss and improves efficiency.
Loose-fill materials (polystyrene beads) 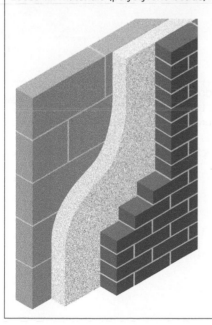	Expanded polystyrene beads (EPS beads) are used as cavity wall insulation. They are injected into the cavity and mixed with an adhesive, which bonds the beads together to prevent them spilling out of the wall. This type of insulation can be used in narrower cavities than mineral wool insulation and can also be used in some stone-built properties.

→

▼ Table 3.4 Insulation materials (continued)

Type of insulation	Description
Expanded polystyrene bead boards 	Graphite-impregnated expanded polystyrene bead boards are lightweight and easy to handle, store and cut on site.
Autoclaved aerated concrete blocks (The photo shows a section of aerated material.) 	Autoclaved aerated concrete blocks are excellent thermal insulators and are frequently used to form the inner leaf of a cavity wall. They can also be used in the outer leaf, where they are usually rendered.
Materials formed on site (expanded foam) 	Spray foam insulation is an alternative to traditional building insulation such as fibreglass. A two-component mixture composed of isocyanurate and polyol resin comes together at the tip of a gun and forms an expanding foam that is sprayed onto areas such as the underside of roof tiles or through holes drilled into a cavity of a finished wall.
Triple glazing 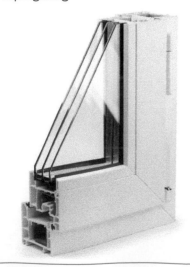	Single glazing has a U-value of 5, older double glazing has a U-value of about 3 and new modern double glazing has a U-value of 1.6, which is mainly due to the fact that the cavity is gas filled to improve the efficiency of the units. Triple glazed units can have a U-value as low as 0.6. An added advantage of triple glazing over double is the improved reduction of external noise. For more on U-values, see below.

Thermal transmittance

Thermal transmittance, also known as U-value, is the rate of transfer of heat through a structure (in **watts**) or more correctly through one square metre of a structure divided by the difference in temperature across the structure. It is expressed in watts per metre squared kelvin, or W/m²K.

KEY TERM

Watt: the unit measurement for power

Well insulated parts of a building have a low thermal transmittance, whereas poorly insulated parts of a building have a high thermal transmittance. The lower the U-value, the better the insulation properties of the structure.

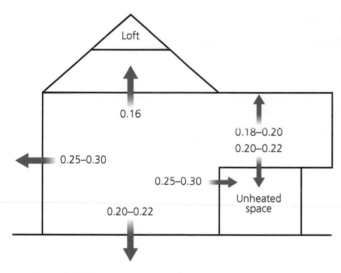

▲ Figure 3.62 Heat loss expressed as U-values in a typical house

Table 3.5 shows the thermal conductivity of different parts of a building's structure expressed as U-values.

▼ Table 3.5 Thermal conductivity of parts of a building

Part of structure	U-value (W/m²K)
Single-glazed windows, allowing for frames	4.5
Double-glazed windows, allowing for frames	3.3
Double-glazed windows with advanced coatings and frames	1.8
Triple-glazed windows, allowing for frames	1.2
Triple-glazed windows, with advanced coatings and frames	0.8

▼ Table 3.5 Thermal conductivity of parts of a building (continued)

Part of structure	U-value (W/m²K)
Well insulated roofs	0.15
Poorly insulated roofs	1.0
Well insulated walls	0.25
Poorly insulated walls	1.5
Well insulated floors	0.2
Poorly insulated floors	1.0

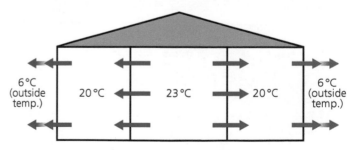

▲ Figure 3.63 Infrared image of heat escaping from a house

ACTIVITY

Go to www.planningportal.gov.uk and look up the minimum U-value required for a roof.

Increasingly, buildings are required to be airtight when completed to limit the movement of air (which conducts heat out of the structure). Controlled ventilation systems feature designs that exchange and conserve heat as air moves into and out of a building.

▲ Figure 3.64 Heat flowing from a building

Every aspect of building design principles is being examined to make structures as energy efficient as possible. Buildings can have their energy efficiency measured by a specialist surveyor who will then produce an Energy Performance Certificate (EPC).

An EPC is needed whenever a property is:

- newly built
- placed on the market for sale
- placed on the market as a rental property.

An EPC gives potential buyers an assessment of how energy efficient a property is, how it can be improved and how much money this could save. It grades the property's energy efficiency from A to G, with A being the highest rating.

If you live in a brand-new home, it's likely to have a high EPC rating. If you live in an older home, it's likely to be around D or E. The EPC also lists ways to improve the rating, such as installing double glazing or loft, floor or wall insulation.

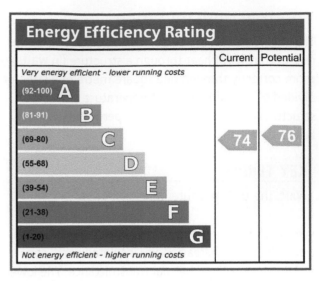

▲ Figure 3.65 EPC rating chart

Thermal performance of different types of structure

Study Table 3.6 to understand the thermal efficiency of different types of structural design.

▼ Table 3.6 Thermal efficiency of structures

Structural element	Description	U-value
Solid wall	Brickwork 215 mm, plaster 15 mm	2.3
Cavity wall	Brickwork 103 mm, clear cavity 50 mm, brickwork 103 mm	1.6

▼ Table 3.6 Thermal efficiency of structures (continued)

Structural element	Description	U-value
Cavity wall	Brickwork 103 mm, insulation 50 mm, lightweight concrete block 100 mm, lightweight plaster	0.48
Cavity wall	Brickwork 103 mm, insulation 100 mm full cavity fill, lightweight concrete block 100 mm, lightweight plaster	0.28
Cavity wall – timber frame	Brickwork 103 mm, clear cavity 140 mm studwork filled with PIR insulation	0.28

→

▼ Table 3.6 Thermal efficiency of structures (continued)

Structural element	Description	U-value
Timber frame and clad Breathable membrane, e.g. Kingspan Nilvent Kingspan Kooltherm K5 EWB Treated softwood counter-batten Cementitious board or expanded metal Render	Tiles, render or cladding on battens, 103 mm masonry, clear cavity 140 mm studwork filled with PIR insulation	0.28
Pitched roof	Tiles on battens, felt, ventilated loft airspace, 100 mm mineral wool between joists, 170 mm mineral wool over joists, plasterboard 13 mm	0.16

▼ Table 3.6 Thermal efficiency of structures (continued)

Structural element	Description	U-value
Warm deck flat roof	150 mm PIR over joists, 13 mm plasterboard	0.18

Choosing and sourcing materials

Energy used in the process of manufacturing and transporting construction materials creates what is referred to as a 'carbon footprint'. For example, the processes involved in manufacturing concrete use a lot of energy. For every ton of concrete produced, some experts state that the energy used to make it and transport it to site produces carbon which combines with oxygen to make around a ton of carbon dioxide. For this reason, many producers are striving to use recycled aggregate in their concrete mix to reduce its carbon footprint.

It is important therefore to choose materials that have the lowest carbon footprint possible. Ask yourself these questions when selecting materials in order to support sustainability and protect the environment:

● How far have the materials been transported? Locally-sourced materials don't have to be transported over long distances, reducing the amount of fuel used. Think about the carbon footprint of slates sourced in the UK compared with slates transported from China or India.

● Are the materials from a sustainable source? For example, timber should come from responsibly managed forests, not from a rainforest which will take generations to recover.

● Have the materials been manufactured and used with minimum waste? Many construction materials, including bricks, can be recycled or reused. Waste management on site is very important. Recycling can be supported if waste is carefully segregated or separated into categories.

KEY POINT

Managed forests, where trees are replanted after harvesting, provide a sustainable source of timber. The Forest Stewardship Council (FSC) is an international non-profit organisation dedicated to promoting responsible forestry. FSC certifies forests all over the world to ensure they meet the highest environmental and social standards.

Energy efficient features

Building designs are increasingly using energy efficient methods of generating their own power and reducing energy use. Solar and wind power are generated by large power companies, but these and other systems can also be applied to buildings individually.

Biomass heating

Biomass heating is gaining in popularity. It uses wood pellets or other sustainable fuels to heat a boiler. This type of heating system can heat water efficiently as well as room spaces that are efficiently insulated.

Ground and air source pumps

Even in seemingly cold ground or air, there is a level of heat that can be extracted using heat exchanger technology. The extracted heat can then be transferred to the interior of a building using a suitable heating system.

By combining efficient and sustainable heating systems with high levels of insulation and technological solutions, building construction and design can support sustainability and protect the environment for future generations.

▲ Figure 3.66 Ground source heat pump system being installed

Summary

Understanding how the construction team is organised will help you understand how and when you can contribute successfully to the building process as a productive team member. Understanding the responsibilities of different personnel on site will enable you to think about the possible career progression routes open to you within the industry.

The information in this chapter will enable you to integrate your work with other skilled personnel on site and to develop a broader understanding of the impact the construction industry has on the environment both now and in the future. You can work as part of an industry that can make a significant contribution to a healthy economy and to the efficient functioning of society in general.

Test your knowledge

1 Which of these descriptions applies to construction work in the public sector?
   a Financed by contractors
   b Financed independently
   c Funded by government
   d Funded by specialists

2 What are residential buildings used as?
   a Businesses
   b Dwellings
   c Offices
   d Shops

3 Which category of personnel describes an architect?
   a Operative
   b Professional
   c Specialist
   d Technician

4 What term describes the method of using individual factory-produced units to build a structure?
   a Managed
   b Modular
   c Monolithic
   d Multiple

5 Which part of a building is the substructure?
   a The section above finished floor level
   b The section below first floor level
   c The section above first floor level
   d The section below finished floor level

6 What type of pile would be used for the foundation of a high-rise building?
   a End-bearing
   b Friction
   c Loaded
   d Pad-bearing

7 In a concrete mix ratio of 1:2:3, what does the '3' represent?
   a Coarse aggregate
   b Cement
   c Fine aggregate
   d Retardant

8 What is an advantage of using modern timber frame construction methods?
   a Better appearance
   b Fast construction
   c Less ventilation needed
   d More moisture resistant

9 What term describes the method of fixing plasterboards with tapered edges to a stud wall?
   a Dry-lashing
   b Dry-leaning
   c Dry-lining
   d Dry-linking

10 What is the angle below which the slope or incline of a flat roof is set?
   a 8°
   b 10°
   c 12°
   d 15°

11 Explain the meaning of the term 'line management' on site.

12 Explain the purpose of a raft foundation.

13 Explain why it is very important to provide adequate ventilation under a suspended timber floor.

14 Describe the advantages of using timber engineered joists instead of solid timber joists.

15 Explain how U-values show the thermal performance of a building.

CHAPTER 4

SETTING OUT MASONRY STRUCTURES

INTRODUCTION

Setting out a new building is a task that requires careful interpretation of drawings and precision in measuring. Accuracy at the setting out stage of the construction process is vital if the finished building is to fully match its design specification.

In this chapter, we will look at how the construction process begins with preparing and clearing of the site for setting out. We will see how the building's outline and position are established in relation to its surroundings and set reference points. We will also consider methods of accurately determining the intended level and height of the structure.

By the end of this chapter, you will have an understanding of:

- preparing the site for setting out procedures
- selecting the right tools, equipment and working drawings for setting out the building
- methods of setting out and levelling a rectangular structure.

The table below shows how the main headings in this chapter cover the learning outcomes for each qualification specification.

Chapter section	Level 2 Technical Certificate in Bricklaying (7905-20) Unit 205	Level 3 Advanced Technical Diploma in Bricklaying (7905-30)	Level 2 Diploma in Bricklaying (6705-23) Unit 205	Level 3 Diploma in Bricklaying (6705-33)	Level 2 Bricklayer Trailblazer Apprenticeship (9077) Module 4
1. Preparing the site for setting out procedures	1.1, 1.2, 1.3	N/A	3.1, 3.2, 3.3, 3.4, 3.5, 4.1, 4.2, 4.3, 4.4, 4.5	N/A	N/A
2. Selecting the right tools, equipment and drawings for setting out the building	2.1, 2.2, 3.3	N/A	2.1, 2.2, 5.1, 5.2, 5.3, 5.4	N/A	**Skills Depth:** 1.1, 1.2 **Knowledge Depth:** 1.1, 1.2
3. Methods of setting out and levelling a rectangular structure	3.1, 3.2, 3.3, 3.4	N/A	1.5, 6.1, 6.2, 6.3, 6.4, 6.5, 7.1, 7.2, 7.3, 7.4, 7.5, 7.6, 7.7, 7.8, 7.9, 8.1, 8.2, 8.3, 8.4, 8.5, 8.6, 8.7, 8.8, 8.9	N/A	**Skills Depth:** 2.1, 2.2, 2.3, 2.4, 2.5, 2.6 **Knowledge Depth:** 1.1, 1.2, 1.3, 1.4, 1.5, 1.6

Note: for 6705-23, Unit 205:
Content for Learning outcome 1 is covered in Chapter 2.
Content for Assessment criteria 1.1 to 1.4 and 1.6 to 1.7 is covered in Chapter 2.

1 PREPARING THE SITE FOR SETTING OUT PROCEDURES

Achieving a successful outcome for a construction project starts with careful preparation of the site. Good preparation will reduce the likelihood of problems arising later in the project. Before a brick is laid, a great deal of work will have been done to ensure that work proceeds safely and efficiently.

Site investigation

If accidental damage were to occur to services such as electricity or gas, the consequences could be dangerous as well as causing delays and unexpected repair costs, so careful investigation is necessary.

▲ Figure 4.1 All service pipes and cables should be located before starting work on a project

> ## VALUES AND BEHAVIOURS
>
> If the project involves work on a customer's property, it is very important that the site investigation is undertaken with care. Give careful thought to how the services will be located so that any excavation work causes minimum inconvenience to the customer and members of the public. Look for ways to avoid damage to drives, pathways or landscaped areas.

> **KEY POINT**
>
> Not all relevant information is available from the local authority. Service providers or utility companies such as electricity or water boards often have their own information database from which locations of cable and pipe runs can be determined.

> **INDUSTRY TIP**
>
> When a construction company's office can reliably confirm the position of services from existing drawings held by the local authority or service provider, this is referred to as 'desktop study'.

Services

Locating services that may be on the site must be undertaken at an early stage of the construction project. The route, extent and type of services adjacent to and crossing the project site must be identified.

Local authorities have drawings on file which identify boundaries, footpaths and roads surrounding the project site where services may be located. Guidance is available from the planning department of the local council through their online planning portal. Information about the location of services that can be confirmed prior to the start of works on site will be included on the working drawings.

Gas and electricity

Because of the potential danger from services such as gas and electricity, special equipment is available to scan the ground to confirm the position of cables and pipe runs.

Often, a detailed physical examination of the site may be conducted on foot by a surveyor who will look for signs of the location of services. This is termed a 'walk over'.

Once the gas and electricity services have been located, their position must be marked and permission to expose the services must be requested from the relevant service provider. It may be necessary for the service runs to be re-routed.

Services that cannot be re-routed or that are to be used on the project site must be protected from potential damage from construction operations.

> ### HEALTH AND SAFETY
> Control measures such as fences and signs must be put in place to protect personnel from potential hazards.

As well as underground services, there may be overhead cables for electricity that must be considered during preparation of the site. Overhead electricity cables can be a hazard to operators of machines such as cranes and excavators operating and moving about the site.

If overhead services cannot be removed or re-routed, they must be clearly marked and measures put in place to signal to machine drivers when they are close to the hazard.

▲ Figure 4.2 Machinery such as mobile cranes can be at risk from overhead cables

Telecoms

Underground and overhead telecoms cables are less of a hazard than electricity cables. However, damage to telecoms services can be costly to repair and can result in significant disruption to communications in nearby businesses and residential areas. Telecoms cables should therefore be identified and protected.

Water and drainage

If water supply pipes cross the project site, they must be carefully located and protected or re-routed. Accidental damage can cause disruption through flooding and may result in contamination of the water supply.

If drainage systems are damaged, they could cause pollution of the environment surrounding the site. Water courses and surrounding natural drainage ditches must be protected from pollution by chemicals used on site during construction operations. Fuel used for construction machinery must be stored in accordance with regulations to avoid spills and leaks.

> ### VALUES AND BEHAVIOURS
> Preventing environmental damage is a major consideration when setting up a construction site.

▲ Figure 4.3 Drainage ditches and water courses near a site must be protected

Site clearance

The process of clearing a site ready for construction work will depend on the type of land being brought into use. Land is classified as a **greenfield site** or **brownfield site** and different approaches to site clearance are needed for each type of land.

> ### INDUSTRY TIP
> Remember, careful planning and thought at the preparation stage will lead to a successful outcome.

Greenfield site

A greenfield site is land that has not previously been used for construction. It could be in a countryside location or an open area such as a school field that has not previously been built on.

To prepare the site, vegetable matter and **topsoil** must be removed from the area of construction activities. This is important because of the low load bearing capacity of this type of material. If the topsoil is clean and of suitable quality, it can be stored for later re-use in landscaping and levelling the site when construction is finished.

KEY TERMS

Brownfield site: this type of land will have been previously used for dwellings or industry

Greenfield site: land that has not previously been used for construction, such as a countryside location or an open area such as a park or school field that has not previously been built on

Topsoil: the upper, outermost layer of soil, usually the top 5–10 inches (13–25 cm)

▲ Figure 4.4 Topsoil includes a lot of vegetable matter

VALUES AND BEHAVIOURS

Storing topsoil in a suitable location on site needs careful consideration. If you have a part in planning removal of topsoil, keep in mind that the material must be stored so as not to interfere with water courses or drainage ditches. Stockpiling in a location that will not mean double-handling the material later is also an important part of efficient preparation of the site.

INDUSTRY TIP

Good quality topsoil is a valuable material and any surplus quantity can be sold on.

▲ Figure 4.5 Topsoil being removed to storage

Calculating volume

Removing the top layers of soil when preparing a construction site will generate a lot of material (sometimes referred to as 'spoil') for removal and possible storage. Soil and similar materials are measured by volume expressed in cubic metres (m³).

When soil is excavated from the ground, something known as **bulking** occurs. This means that the volume of the soil increases by a certain percentage due to the introduction of air. It is important that bulking is considered when calculating excavated quantities of materials like soil that may need to be stored on site or transported elsewhere.

IMPROVE YOUR MATHS

Review the method of calculating percentages (page 69) and then calculate the percentage of bulking in this example.

Several lorries are needed to transport 250 m³ of soil off site. One lorry carries 10 m³ of soil. To allow for bulking, add 30% to the excavated amount and state:

1 the total amount of material to be removed in m³

2 the number of lorries needed to take all the excavated materials off site.

Trees and shrubs

Greenfield sites may contain trees and shrubs that need to be removed as part of the preparation of the site. There may be a requirement to consult with the relevant authorities to obtain permission for tree removal. To protect the environment, trees may be the subject of preservation orders.

VALUES AND BEHAVIOURS

Thought should be given to reducing the environmental impact of any building project by retaining and incorporating existing trees and shrubs into the design layout of the project. Retaining established trees and plants will also make a development more aesthetically pleasing to the intended occupants.

Trees and other vegetation that will be preserved as part of the development will need to be clearly identified at the site preparation stage of the project.

▲ Figure 4.6 Preserving trees on a construction site can be an advantage

Brownfield site

A brownfield site is land that has been previously used for dwellings or industry. As a result of prior use, existing structures may need to be demolished and removed, adding to development costs.

Materials from demolished structures may be able to be recycled. Hardcore, brick, stone or chippings are materials that can be reclaimed and sold for use as recycled aggregate, or they may be processed on site and re-used in the new project.

▲ Figure 4.7 Demolition of existing structures on a brownfield site

▲ Figure 4.8 Some materials can be recycled or re-used

2 SELECTING THE RIGHT TOOLS, EQUIPMENT AND WORKING DRAWINGS FOR SETTING OUT THE BUILDING

Once services have been identified and the site has been cleared, setting out of the structure can commence. A clear understanding of setting out methods and the tools and equipment required is essential if the building is to be constructed accurately to the specification.

Setting out equipment

Traditionally, relatively simple equipment is used by the bricklayer to set out a building. Accurate results can be achieved using string (ranging) lines in conjunction with timber **profiles** as guides for setting out wall positions on the ground, ready for excavation of the foundations.

INDUSTRY TIP

When setting out a building, the string lines used are referred to as 'ranging' lines.

KEY TERM

Profiles: an assembly of timber boards and pegs set up at the corners and other points of a building to allow string lines to be accurately positioned

Profiles consist of timber rails attached to timber pegs and are assembled on site to suit the requirements of the job. (We'll look in detail at the steps used to accurately position the profiles later in this chapter.)

The pegs (approximately 50 mm square) are securely driven into the ground and the positions of corner points and the lines of walls are indicated by nails driven into the top of the cross rails (approximately 150 mm × 30 mm). An alternative method is to use saw cuts in the top of the rail instead of nails to show the specified wall positions.

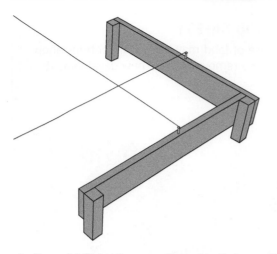

INDUSTRY TIP

Marking the position of ranging lines using nails or a saw cut is preferable since the profiles might be in position for some time. Markings made in pen or pencil can be washed away or fade, leading to mistakes in setting out.

Table 4.1 shows the range of tools and equipment that can be used to establish the position of the profiles and the building outline on the plot.

▲ Figure 4.9 String lines are attached to timber profiles

▼ Table 4.1 Tools and equipment for setting out wall positions

Tools/equipment	Description
Timber pegs and rails 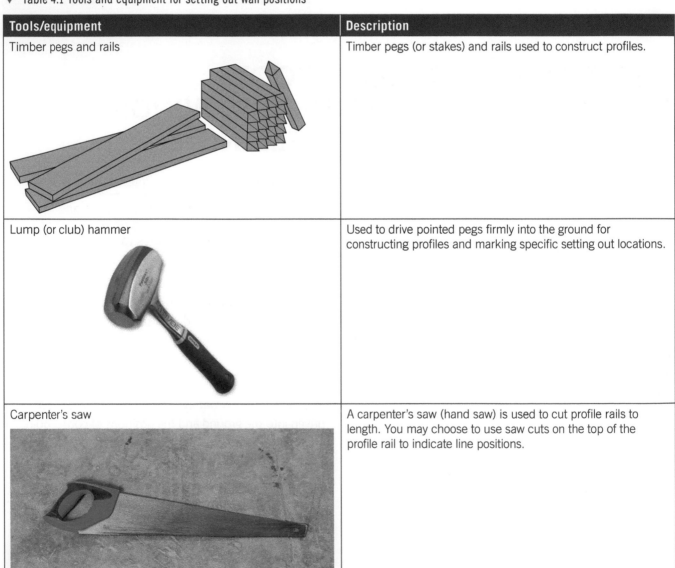	Timber pegs (or stakes) and rails used to construct profiles.
Lump (or club) hammer	Used to drive pointed pegs firmly into the ground for constructing profiles and marking specific setting out locations.
Carpenter's saw	A carpenter's saw (hand saw) is used to cut profile rails to length. You may choose to use saw cuts on the top of the profile rail to indicate line positions.

➡

▼ Table 4.1 Tools and equipment for setting out wall positions (continued)

Tools/equipment	Description
Tape measure/surveyor's tape measure	Used constantly for measuring and marking dimensional details. For small to medium measuring tasks there is a range of lengths from 3 m to 10 m. For measuring long distances, a surveyor's tape measure is available in lengths ranging from 10 m to 100 m.
Mason's line and pins	Using a mason's line and pins makes it easy to attach string lines to reference points. Effective over short to medium distances.
Ranging line	Over longer distances a more substantial ranging line can be used. Made from nylon materials for strength.
Spray paint	Traditionally, fine building sand or lime powder (which is hazardous to health) was used to mark out the line of wall positions. A more modern approach is to use spray paint. When the profiles and ranging lines have been accurately set out and checked, the wall positions can be marked on the ground ready for excavation of the foundation trenches.

INDUSTRY TIP

Surveyor's tapes are made from steel or special fabric. Fabric tapes should be used with care to avoid stretching them and distorting the measurements.

INDUSTRY TIP

When setting out wall positions on profiles, it's helpful to add a carpenter's claw hammer to the list of tools and equipment to make it easier to drive nails when constructing a profile. Remember, you can mark wall positions clearly on a profile by using nails rather than a saw cut if you choose to.

▲ Figure 4.10 A claw hammer makes fixing nails (and removing them) easier

Drawings for setting out

In Chapter 2, we looked at the full range of drawings used to communicate information for a construction project. In this section we'll concentrate on the drawings used to set out the position, outline and internal walls of a structure.

Types of drawings

Two categories of drawing types apply to the process of setting out a building:

- general location drawing
- construction drawing.

We'll review the description and purpose of each type. Make sure you understand each description and analyse how the drawings contribute to the setting out process.

General location drawing

▼ Table 4.2 Types of location drawings

Type of drawing	Description
Block plan Green St Plot 1 Plot 2 Plot 3 Sand St	This shows us a 'bird's eye' view of the whole site in relation to the area around it. Usual scale 1:1250 or 1:2500. A block plan will show individual plots and road layouts on the site as a simple outline with few dimensions. It gives a clearer understanding of access requirements and storage and positioning of materials during preparation.

▼ Table 4.2 Types of location drawings (continued)

Type of drawing	Description
Site plan 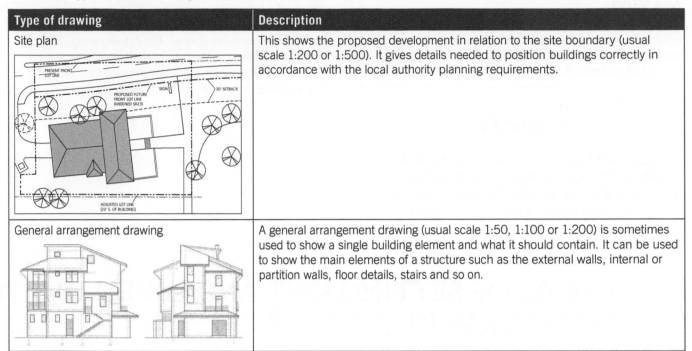	This shows the proposed development in relation to the site boundary (usual scale 1:200 or 1:500). It gives details needed to position buildings correctly in accordance with the local authority planning requirements.
General arrangement drawing	A general arrangement drawing (usual scale 1:50, 1:100 or 1:200) is sometimes used to show a single building element and what it should contain. It can be used to show the main elements of a structure such as the external walls, internal or partition walls, floor details, stairs and so on.

Construction drawing

The main drawing that will be used to set out the building is the floor plan. This view of the planned construction is often the most useful one for a bricklayer and makes it relatively easy to interpret complex information. The drawing shows the dimensions of the rooms and the position and dimensions of key elements such as doors and windows in the structure.

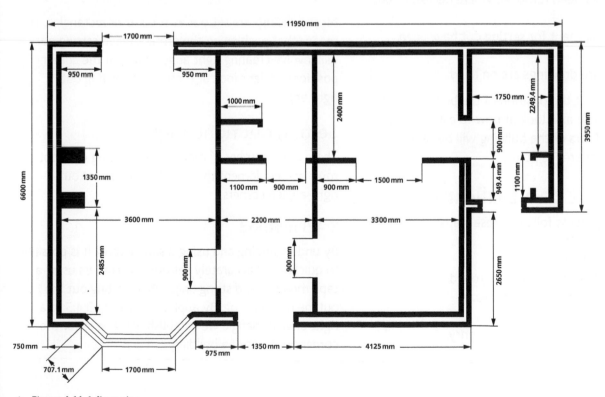

▲ Figure 4.11 A floor plan

③ METHODS OF SETTING OUT AND LEVELLING A RECTANGULAR STRUCTURE

We've seen that a lot of work has been done to get the site ready for setting out the building:

- The services have been located and identified and measures have been taken to ensure that work can continue safely.
- The topsoil has been removed where necessary and the site cleared.
- Tools and equipment for setting out have been assembled.
- The appropriate drawings are on hand.

Now we can set out the position of the building. Later in this chapter, we'll also look at how to establish the levels at which floors of the building will be set in relation to a set reference point.

Setting out the building

The setting out process follows these steps:

1 Position the profiles.
2 Set up ranging lines.
3 Mark the wall positions on the ground.
4 Excavate the foundation trenches.
5 Pour the foundation concrete.
6 Set out and build the footing masonry.

Most buildings are set out as squares or rectangles. This means that positioning the profiles when setting out will involve creating right angles (90°) at the corner positions, so let's look at methods of setting out right angles.

Setting out right angles

There are a number of methods employing different tools and equipment that can be used to set out right-angled corners.

3:4:5 method

By understanding and using a simple ratio, it is possible to quickly and accurately set out 90° corners using a tape measure and string lines when setting out the outline of a building. By applying the ratio 3:4:5 to a right-angled triangle we establish 90° angles for the corner positions of a building.

IMPROVE YOUR MATHS

Look at the example of a right-angled triangle to see how this works.

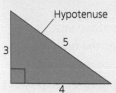

▲ Figure 4.12 A right-angled triangle

If the numbers in this example represent metres, when the height of the triangle is 3 m and the base of the triangle is 4 m, the longest side (the hypotenuse) will always be 5 m when a right angle (90°) is formed.

The ratio 3:4:5 refers to units of measurement. Any unit of measurement (metres, centimetres, millimetres) that is easy to work with and suits the needs of the job can be used, provided the ratio stays the same.

When using the 3:4:5 method to set out a right angle, it's often easier to use two tape measures. You could do it this way:

1 Set out two corner pegs with nails driven into their tops to represent the front of a building. Attach a string line.
2 Measure 3 m along the string line from the first peg and position a third peg with a nail at the exact dimension directly under the line.
3 Now attach one tape measure to the nail on the first peg you set up (tape 1) and attach the other tape measure to the nail on the third peg you set up (tape 2).
4 Forming a triangle with the string line and the two tapes, read 4 m on tape 1 and 5 m on tape 2. Position a fourth peg with a nail at the exact point where the two tapes cross over each other. You've created a right angle!

Builder's square

A builder's square is a large piece of equipment that can be used as a guide to set up string lines at a 90° angle. It can be made on site using available timber to create a right angle using the 3:4:5 method.

Builder's squares manufactured from foldable metal are available; these are more convenient to transport and store.

Optical site square

This is a simple instrument which has either two sighting points set at right angles to each other, or an optical device known as a prism, which allows the user to view two points at right angles to each other at the same time.

The instrument is mounted on a tripod when in use to provide a steady platform. An assistant is directed by the instrument user to mark two positions at right angles to each other. This instrument makes accurately setting out right angles a simple process.

▲ Figure 4.13 Builder's square

▲ Figure 4.14 Optical square

Positioning the profiles

The front wall of the building (**frontage line**) is set out first and must be located on or behind the **building line**. The side walls of the structure are set out at right angles to the frontage line and the rear wall can then be set out parallel to the frontage line. These wall positions are established by positioning pegs at each corner.

The frontage line is often confused with the building line. Remember, the frontage line refers to the front wall of the building which can be moved forwards or backwards. It could be positioned directly on, but never in front of the building line. The building line cannot be moved – it is set by the local authority.

Positioning profiles step by step

To set up profiles ready for work to continue, first consult the location plan and site plan to confirm the position of the building line.

Then set up two pegs along the frontage line corresponding to corner positions of the building. Remember, the frontage line must never project in front of the building line. Study the step by step to see the remaining steps.

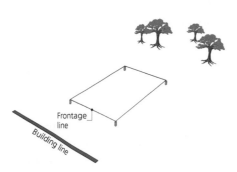

STEP 1 Using one of the methods to create right angles previously discussed, set up pegs at the remaining two corners to create a rectangle.

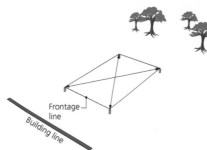

STEP 2 Check the diagonal measurements are the same to confirm the corners of the rectangle are square.

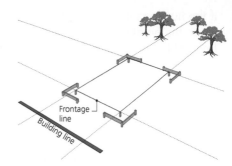

STEP 3 Extend the lines of each side beyond the corner positions to set up corner profiles.

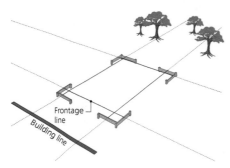

STEP 4 Remove pegs and attach the lines in position on profiles.

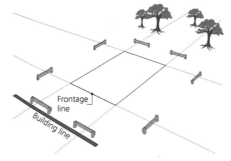

STEP 4A Alternatively, if space is needed for a mechanical excavator to work, position profiles a short distance from the corner positions.

If the diagonal measurements of a rectangle are the same, we can confirm that the corners are 90°. However, this will only be the case if the overall length and width dimensions of the rectangular (or square) building are accurate. Check these for accuracy first.

Most buildings will have load bearing internal partition walls which will require a foundation to be set out for them. This means profile boards must be provided at suitable intermediate points corresponding to the floorplan of the structure.

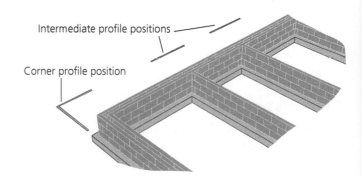

▲ Figure 4.15 Intermediate profiles to set out internal walls

Create a ruler-assisted floor plan drawing of an L-shaped building with two internal loadbearing partition walls. Indicate where the profiles should be positioned to allow ranging lines to be set up for all the walls. The tricky part is where the internal corners are situated.

Preparation for excavation

Once the ranging lines have been set up on the profiles, the markings to guide the excavation of the foundation trenches can be made. Spray paint lines are made directly below the ranging lines for all walls that are to be excavated.

A commonly used method is to mark the centre line of the wall position so that the excavator operator can easily align the centre of the digging bucket when excavating the foundation trench.

The top of the profile rail may have a number of nails showing a range of positions, including the centre line.

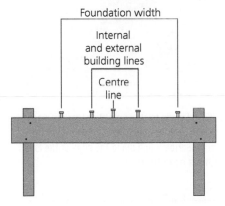

▲ Figure 4.16 Positions marked with nails on the top rail of a profile

INDUSTRY TIP

Remember, the markings on the ground for the excavation to follow are made using spray paint. Traditionally, the marks were made using fine sand or white lime powder (which can be hazardous to health).

When the excavation of the trenches is complete, the concrete foundation can be poured and levelled. (For information on transporting and positioning concrete, see page 225. For information on different types of foundations, see page 83.)

Establishing wall positions

When the foundation concrete has hardened, the bricklayer can set out wall positions ready to construct the footing masonry. This is done by re-attaching the ranging lines to the profiles showing the face line of the building. The bricklayer then plumbs down from these lines to mark the wall position onto the foundation concrete.

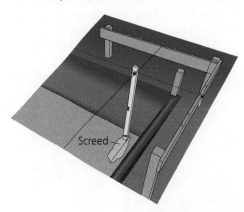

▲ Figure 4.17 Plumbing down from ranging lines to the foundation concrete

INDUSTRY TIP

The traditional method of marking the wall position on the foundation is to spread a thin screed of mortar on the surface of the concrete which can be marked with the tip of a trowel to follow the face line of the wall. Some bricklayers use a line of spray paint to provide a clean surface for marking with a permanent marking pen.

Transferring the face line of the building onto the concrete foundation

Look at the step by step to see the traditional method of transferring the face line of a wall onto the concrete foundation.

STEP 1 Spread a thin screed of mortar on the foundation concrete stretching from the corner, along the line of the wall, about 400–600 mm long. (Judge the position by eye to line it up below the string line.)

STEP 2 Carefully position the spirit level alongside the string line allowing a slight gap. (If you allow the spirit level to touch the line, you risk moving the line and losing accuracy.) Position the level at one end of the screed.

STEP 3 Adjust the spirit level until it is plumb. (You may need to brace the level to stop it moving about. Some bricklayers use their trowel held at an angle to steady the spirit level.)

STEP 4 Carefully make a small mark in the screed at the bottom of the level with the tip of the trowel. (Make sure the mark is the same side of the level as the string line.)

STEP 5 Now position the spirit level at the other end of the screed and repeat steps 3 and 4.

STEP 6 Finally lay the spirit level on the screed using it as a straight edge to join up the two marks. Scribe a line along the edge of the level with the tip of the trowel.

You now have a line of reference alongside which you can lay your first bricks or blocks, which will correspond to the string lines attached to the profiles.

Methods of transferring levels

To establish the correct level of a building, a reference point called a **datum** is set up on site. This often takes the form of a peg which is secured in concrete and is protected from disturbance caused by machinery movements or other construction activities. The datum is known as a temporary bench mark (TBM) and all levels for buildings are transferred from it.

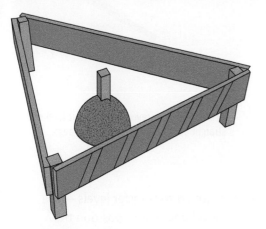

▲ Figure 4.18 Datum point protected from disturbance

KEY TERM

Datum: a fixed point or height from which reference levels can be taken

INDUSTRY TIP

Sometimes a fixed point, such as an inspection chamber cover in a road near to the site, may be used as a datum.

The TBM is taken from a permanent reference point known as an ordnance bench mark (OBM). These are survey marks made by a regulated organisation known as Ordnance Survey. Most commonly, the OBMs are found on buildings or other semi-permanent features.

▲ Figure 4.19 Ordnance bench mark

Although the main network of OBMs is no longer being updated, the record is still in existence and the markers will remain until they are eventually destroyed by redevelopment or erosion. The location of OBMs around the country can now be seen on a licensed database.

ACTIVITY

Visit the Ordnance Survey website at www.ordnancesurvey.co.uk/benchmarks/. Find the UK location all OBMs are referenced to.

Transferring a level from the datum to the location of the building you are working on is usually done to establish the FFL as part of the substructure. (For more on the FFL and substructure see the key points on pages 83 and 87.) This usually corresponds to DPC level.

Levelling equipment

A range of equipment can be used to transfer levels. Whichever equipment type is used, great care is needed to avoid errors and discrepancies in establishing levels.

Straight edge and spirit level

A traditional method of transferring levels makes use of relatively simple equipment; a spirit level and straight edge can be used together to transfer levels over distance. When using these items of equipment together, it's important to 'reverse' them end for end between levelling pegs.

▲ Figure 4.20 Spirit level   ▲ Figure 4.21 Straight edge

Using this method will cancel out any defects in the spirit level or distortions in the straight edge. This is especially important if the straight edge is made from timber which can warp or twist over time.

The method of transferring levels using a spirit level and straight edge is described in the following step by step.

This method needs care when transferring levels over any distance on site and can be time consuming. It is best suited to transferring levels over shorter distances in closer proximity to the building under construction.

STEP 1 Carefully level between the first two pegs, making adjustments to the second peg in the row.

STEP 2 Reverse the spirit level and straight edge end to end and carefully level between the second and third pegs. Only make adjustments to the third peg.

STEP 3 Continue reversing the spirit level and straight edge between subsequent pegs.

Optical level

An optical level is efficient and accurate for transferring levels over long distances on site. Care is needed during setting up to ensure that the 'head' of the instrument is perfectly level and the tripod is stable.

Two operatives are required to transfer levels – one to 'sight' the optical level and one to position the graduated staff. The instrument must be handled and stored carefully and protected against dust and vibration to maintain its accuracy.

▲ Figure 4.22 The optical level must be carefully set up

▲ Figure 4.23 One operative takes a height reading with the optical level

▲ Figure 4.24 The other operative positions the graduated staff

Laser level

Accurate transfer of levels can be achieved easily with an electronic instrument such as a laser level. This instrument is often self-levelling and can be used by one operative working alone. Once the instrument has been set up on its tripod, the operator is free to move anywhere on site carrying a staff mounted receiver.

▲ Figure 4.25 A laser level

The reference level from the TBM can be easily transferred to multiple positions over great distance with accuracy. Awareness of the potential hazard to eyesight posed by laser light should always be kept in mind when operating this equipment.

Setting the level of the masonry

The specified level is transferred to datum pegs at each corner of the building and to any other points where a reference for level is required. The datum peg can be positioned at the same location as the corner profiles. If the corner profiles are at a suitable height, a nail can be positioned on them to serve as a datum.

The bricklayer can then use a spirit level and tape measure to **gauge** down from the datum to the top of the foundation concrete to work out how many **courses** of bricks or blocks will be needed to build the footing masonry to the FFL or DPC level.

KEY TERMS

Courses: continuous rows or layers of bricks or blocks on top of one another

Gauge: the process of establishing measured uniform spacing between brick or block courses including horizontal mortar joints

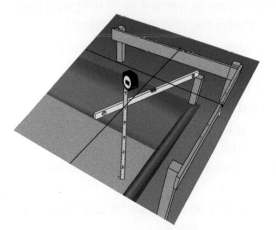

▲ Figure 4.26 Using a tape measure and spirit level to gauge down from the datum peg

INDUSTRY TIP

It is good practice when pouring foundation concrete to set the level at a height that allows for full courses of brick or block between the top of the concrete and the datum. If the level is not set to accommodate gauged masonry, a lot of cutting will be required.

Summary

In this chapter, we've considered the procedures for preparing the site for a construction project and how crucial it is to plan the process carefully.

Investigating the site to establish the position of existing services can be hazardous when dealing with gas or electricity if the work is not done with care, using the right information. Disruption and inconvenience can be caused by carelessness in dealing with water, drainage and telecoms.

Once the site is cleared, the process of setting out the building must also be done with care. Accurate measuring and levelling from given reference points is vital to ensure the building is positioned and orientated correctly in relation to its surroundings.

Test your knowledge

1 For construction purposes, which of these describes a greenfield site?

   a Land in countryside never built on before

   b Land with a large school on it

   c Land with a recycling plant on it

   d Land with houses that have large gardens

2 Why must topsoil be removed from the area on which a building is to be constructed?

   a Because it has high contamination content

   b Because it has high loadbearing capacity

   c Because it has low loadbearing capacity

   d Because it has low vegetation content

3 What term is used to describe the increase in volume that occurs when soil is excavated?

   a Backing

   b Bearing

   c Bulking

   d Bursting

4 When using string lines to set out a building, what term is used to refer to them?

   a Ranging

   b Rating

   c Ringing

   d Routing

5 Which plan drawing will show the proposed development in relation to the boundary of the building plot?

   a Floor

   b Range

   c Section

   d Site

6 Which part of a right-angled triangle is the hypotenuse?

   a Base

   b Height

   c Longest

   d Shortest

7 Who sets the boundary known as the building line?

   a Architect's office

   b Local authority

   c Site engineer

   d Site manager

8 In relation to a datum point, what do the letters 'TBM' stand for?

   a Temporary backing mark

   b Temporary bench mark

   c Temporary bonding mark

   d Temporary building mark

9 Which of these items is used to transfer levels over a long distance?

   a Optical level

   b Optical square

   c Spirit level

   d Straight edge

10 What term is used to describe measured uniform spacing between brick or block courses, including horizontal mortar joints?

   a Gauge

   b Height

   c Level

   d Square

11 Explain why care is needed in dealing with trees on a site during the preparation stage.

12 Describe what is meant by a 'brownfield site' and state one disadvantage of using it.

13 What are timber profiles and how are they constructed?

14 When using a combination of a spirit level and a straight edge to transfer levels, state the reason for 'reversing' them end for end between levelling points.

15 Explain how a laser level can be used by one operative working alone.

BUILDING CAVITY WALLS IN MASONRY

INTRODUCTION

In the past, structures erected in brick were often constructed with solid walls. Beginning in the 1920s and especially from the 1950s onwards, cavity walls were specified as a structural element because of their greater effectiveness in preventing moisture from entering the living and working space of buildings.

In this chapter, you will learn about the important role of the bricklayer in using appropriate methods, specified materials and a range of components when constructing cavity walls. High standards of work are essential in the building process so that the finished product will last a long time and work effectively as a major part of a structure.

By the end of this chapter, you will have an understanding of:
- cavity wall design considerations
- planning ahead and setting up the work area
- positioning and preparing materials for safe and efficient work
- setting out and building cavity walls, including forming openings.

The table below shows how the main headings in this chapter cover the learning outcomes for each qualification specification.

Chapter section	Level 2 Technical Certificate in Bricklaying (7905-20) Unit 202	Level 3 Advanced Technical Diploma in Bricklaying (7905-30)	Level 2 Diploma in Bricklaying (6705-23) Unit 206	Level 3 Diploma in Bricklaying (6705-33)	Level 2 Bricklayer Trailblazer Apprenticeship (9077) Module 5
1. Cavity wall design considerations	4.1	N/A	3.6, 3.7	N/A	**Skills Depth:** 2.2 **Knowledge Depth:** 2.1, 2.2, 2.3
2. Planning ahead and setting up the work area	3.3, 4.1	N/A	3.3, 4.3, 4.5, 6.1	N/A	**Skills Depth:** 2.2, 2.3 **Knowledge Depth:** 1.1, 1.4
3. Positioning and preparing materials for safe and efficient work	3.3, 4.1, 4.4	N/A	3.3, 4.3, 4.5	N/A	**Skills Depth:** N/A **Knowledge Depth:** 1.1, 1.2, 1.4
4. Setting out and building cavity walls, including forming openings	4.1, 4.2, 4.3, 4.4	N/A	3.4, 3.5, 3.8, 3.9, 3.10, 3.11, 3.12, 4.6, 4.7, 4.8, 4.9, 4.10, 4.11, 4.12, 4.13, 5.1, 5.2, 5.3, 5.4, 5.5, 5.6, 5.7, 5.8, 5.9, 6.2, 6.3, 6.4, 6.5, 6.6, 6.7, 6.8, 6.9, 6.10 and Unit 203	N/A	**Skills Depth:** 2.3 **Knowledge Depth:** 1.3, 2.2, 2.3

Note: for 7905-20, Unit 202:
Content for Learning outcomes 1 and 2 is covered in Chapter 6.
Content for Topics 3.1 and 3.2 is covered in Chapter 2.
Note: for 6705-23, Unit 206:
Content for Learning outcomes 1 and 2 is covered in Chapter 2.
Content for Assessment criteria 3.1, 3.2, 4.2 and 4.4 is covered in Chapter 4.

1 CAVITY WALL DESIGN CONSIDERATIONS

First let's look at cavity wall design so that you can understand how the materials and components used work together.

The main reasons why cavity walls are used are:
- to make the building watertight
- to provide thermal insulation
- to allow weatherproof openings to be formed
- to construct a stable and long-lasting structure.

Cavity walls are more complex to build than solid masonry walls. The additional components included in a cavity wall design, such as a damp proof course (DPC), wall ties and insulation, must be installed correctly by the bricklayer to ensure that the wall fulfils its function and purpose.

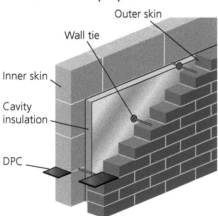

▲ Figure 5.1 Components of a cavity wall

Early cavity wall design allowed ventilation to occur which kept the cavity dry. To meet modern requirements to control **carbon emissions**, the cavity wall (along with other elements of a structure) is now required to be air tight on completion. By controlling the movement of air while at the same time providing controlled ventilation, a building can be made more energy efficient.

The bricklayer plays a part in providing the required ventilation to a structure when installing air bricks in a cavity wall. An air brick must be installed with a **duct** when it passes through the wall to maintain the air tightness of the cavity while allowing ventilation of the occupied area.

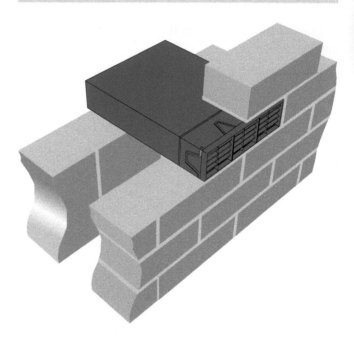

▲ Figure 5.2 A ducted air brick

Damp proof course

A damp proof course (DPC) is an essential part of a cavity wall design. Without a correctly positioned DPC, the ability of the cavity wall to prevent the entry of moisture into the living or working area of a structure will be severely limited. If moisture is allowed to enter a structure, it can cause serious damage to the **fabric** of the building and potentially cause health problems for the occupants.

KEY TERM

Fabric: in this context, the structure or framework of the building

Moisture, or 'damp' as it's more commonly referred to, can enter a cavity wall in two main ways:

- Penetrating damp – moisture that travels through a wall when the masonry materials become saturated. The cavity separating the inner and outer skins of masonry breaks the path of the moisture through the wall.
- Rising damp – moisture that is drawn upwards from ground level and below by a process known as capillary attraction. A horizontal damp proof course installed at a minimum of 150 mm above finished ground level is designed to halt the progress of rising damp. (The position of a horizontal DPC usually corresponds to finished floor level (FFL).)

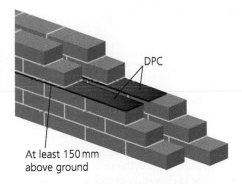

▲ Figure 5.3 Horizontal DPC being laid

A third, less frequent point of entry for moisture is where a cavity wall projects above a roof line in the form of a **parapet**. Moisture can penetrate from above and this requires special DPC arrangements to protect the structure from damage.

KEY TERM

Parapet: a low wall along the edge of a roof or balcony projecting above the roof surface

KEY POINT

A parapet cavity wall is usually in an exposed position and may include a purpose-made coping to close the cavity and 'weather' the top of the wall. The coping will usually be manufactured from pre-cast concrete or stone.

▲ Figure 5.4 Parapet wall with precast concrete coping on top

A cavity wall will have horizontal DPC installed at FFL and vertical DPC installed in the **reveals** of door and window openings.

KEY TERM

Reveals: the masonry forming the side of a window or door

DPC in the form of a tray will also be installed above openings and possibly beneath the sill (sometimes spelled 'cill') of doors and windows. A DPC tray must be installed above an air brick where it passes through a cavity wall. A stepped DPC tray will be installed where a low-level roof meets an adjoining wall as an abutment.

DPC types and materials

There are two main types of damp proof course that can be specified for use as a horizontal DPC in a cavity wall:
- rigid
- flexible.

Examples of materials for rigid DPC are engineering brick and slate, which were used before the introduction of flexible DPCs. Some advantages and disadvantages related to using these materials are described in Table 5.1.

Materials for flexible DPCs are:

- polythene
- pitch polymer
- bitumen felt.

These flexible materials are also suitable for vertical DPC and for use as DPC trays. In the past, copper and lead were used as a DPC but they are very expensive compared to modern materials.

Some advantages and disadvantages related to using these materials are listed in Table 5.2.

▼ Table 5.1 Rigid DPC materials

Material	Advantages	Disadvantages
Slate	● Natural material ● Relatively easy to install	● Heavy to transport in bulk ● Will crack if there is movement in the masonry ● Expensive
Engineering brick	● Very durable and long lasting ● Can be used as a decorative feature ● Good for garden or boundary walls	● Heavy to transport ● Needs space for storage ● Relatively expensive

▼ Table 5.2 Flexible DPC materials

Material	Advantages	Disadvantages
Polythene	● Light and easy to install ● Low cost ● Easy to store ● Suitable for stepped damp proof courses ● Produced in a range of sizes	● Can be punctured
Pitch polymer	● Capable of withstanding high loads (for multistorey structures) ● Easy to install ● Produced in a range of sizes	● Costlier than polythene ● Prone to distort if not stored properly
Bitumen felt	● Relatively easy to install ● Produced in a range of sizes	● Costlier than polythene ● Prone to distort if not stored carefully ● May crack if unrolled in cold weather

ACTIVITY

Look at Tables 5.1 and 5.2 which describe the advantages and disadvantages of each DPC material. Which DPC would you select for use as horizontal DPC in the cavity wall of a four-storey block of flats? Discuss your selection with a colleague to see if they agree.

Wall ties

The individual leaves or skins of a cavity wall are relatively weak on their own. However, when they are properly tied together with suitable wall ties, the stiffness and load-bearing capacity of the combined skins can be almost equal to a similar solid wall. Installing an adequate number of correctly positioned wall ties is the bricklayer's responsibility.

There are several different types of wall tie currently in use which are available in a range of sizes to suit various cavity widths. Ties were previously manufactured from **galvanised** steel, which suffered from long-term failure due to corrosion and rust. Modern ties are commonly made from stainless steel which is much more durable. More rarely, wall ties have been manufactured using polypropylene (a type of plastic).

Wall ties are designed with a feature known as a 'drip'. This serves to shed any moisture that may track across them from the outer skin, preventing it from reaching the inner skin or leaf.

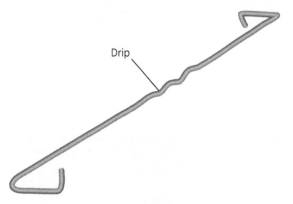

Drip

▲ Figure 5.5 Stainless steel wall tie

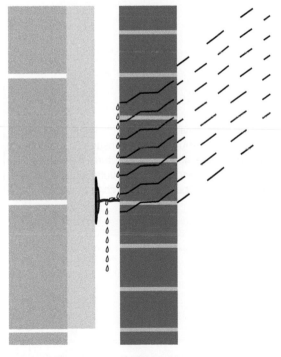

▲ Figure 5.6 Moisture dripping off a wall tie

Insulation

The requirement to improve energy efficiency and reduce carbon emissions in new buildings has led to the introduction of increasing levels of insulation. Cavity walls provide a convenient and efficient way to control heat transfer, because they offer the opportunity to locate insulation materials in a major element of a structure.

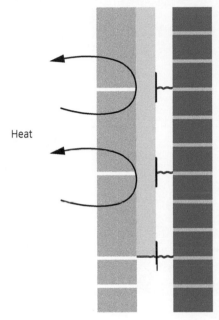

Heat

▲ Figure 5.7 Sectional view of insulation in a cavity wall

There are three main methods of installing insulation in a cavity wall (see Table 5.3).

 Table 5.3 Methods of installing cavity insulation

Method name	Description
Full fill 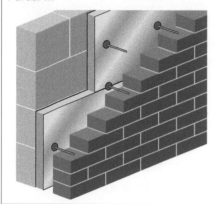	Sometimes called total fill. As the name suggests, this method fully fills the cavity with insulation material so that no air gaps remain. The material used is usually manufactured in the form of flexible slabs or 'batts'. This form of insulation can also be effective in preventing the spread of fire.
Partial fill 	This method allows an air gap of a minimum of 50 mm to be maintained. The insulation material is in the form of rigid sheets of material and is fixed to the inner skin or leaf by special clips attached to the wall ties.
Injection	The method of pumping insulation into the cavity after the building is complete is sometimes referred to as 'retro-fit'. In new-build work the insulation material is injected through holes drilled in the inside skin of the structure. If an older building is insulated by injection, the holes are drilled in the outer skin or leaf of masonry, usually through the mortar joints rather than through face bricks.

The different materials suitable for use as insulation in a cavity wall are described in Table 5.4.

▼ Table 5.4 Materials for cavity insulation

Material	Description	Use
Mineral fibre	Flexible sheets or batts made from glass or rock	Suitable for full fill
Expanded polystyrene (EPS) and polyisocyanurate (PIR) foam	Rigid sheets or boards	Suitable for partial fill
Expanded polystyrene	Beads and chopped strands of mineral fibre	Suitable for injection

The cavity width has gradually been increased over the years since it was first introduced. This has been done to allow the installation of thicker insulation materials to improve energy efficiency.

A newer material used for cavity wall insulation is sheep's wool. Although it is very effective, it is expensive and is not currently widely used. (For a more in-depth discussion of insulation materials, see page 98.)

Other design considerations

A cavity wall is usually wider at the base than a solid wall (unless the solid wall is built to one-and-a-half brick thick or more), so it places lighter loads on the foundation. This means that the design of the foundation for cavity walls can be simplified or reduced in thickness, since it doesn't need to support the greater concentrated weight of a solid wall.

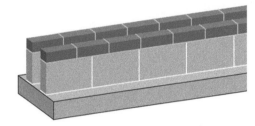

▲ Figure 5.8 A cavity wall is wider at the base

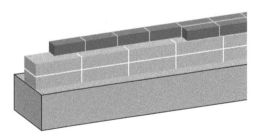

▲ Figure 5.9 A narrower solid wall concentrates loading on the strip foundation

The usual design of foundation for a cavity wall is a concrete strip foundation. This can have measurements that are wide and shallow or narrow and deep, depending on ground conditions and other design considerations such as cost and time-scale requirements.

ACTIVITY

Could a foundation for a cavity wall be affected by trees nearby? Do some research and write down your conclusions.

Experience has shown that since a cavity wall creates lighter loads than a solid wall, it must be designed with provision to prevent cracking caused by **thermal movement** and continuous wetting and drying. To allow for movement, a cavity wall may be designed with vertical movement joints, constructed at intervals along the length of the wall. These are constructed as a vertical 'straight joint' in the bonding arrangement and are filled with a compressible material to allow small amounts of expansion and contraction along the length of the masonry.

KEY TERM

Thermal movement: changes in dimension of masonry or concrete because of fluctuations in temperature over time

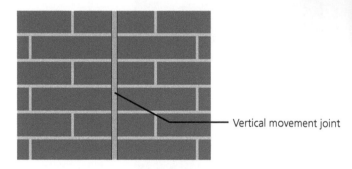

▲ Figure 5.10 Vertical movement joint in a face brick wall

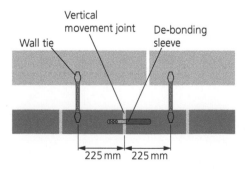

▲ Figure 5.11 A vertical movement joint. Note the de-bonded tie holding the joint together

A bricklayer is often required to use specially designed ties in the bed joints at vertical movement joints. This assists the masonry to resist **lateral movement** at the joint location while still allowing horizontal movement. This is achieved by one end of the tie being firmly bedded in mortar and the other end of the tie having a moveable sleeve bedded in the horizontal mortar joint to allow the tie to move inside it.

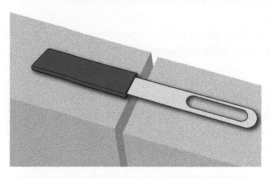

▲ Figure 5.12 Wall tie with de-bonding sleeve

② PLANNING AHEAD AND SETTING UP THE WORK AREA

Whenever you work on site or in a training workshop, your first priority has to be safety. On site you may be called upon to build cavity walls in an excavation or be required to work on a scaffold at height, so you must be familiar with regulations relating to construction health and safety and the Work at Height Regulations 2005.

Always take notice of COSHH statements and the safety guidance on products used in the workplace.

Risk assessments are important documents that help us to recognise hazards and give directions on how to make the workplace safer. (For more on risk assessment, see page 14.)

Working in accordance with risk assessments and maintaining high standards of health and safety is the responsibility of everyone working in construction. You must also make sure that you always use the correct personal protective equipment (PPE). There is more about relevant safety standards and PPE in Chapter 1, and you should review this information often.

Selecting what is needed

When preparing to build cavity walls, the bricklayer will need to carefully refer to working drawings and the specification to find out the exact details and design of the work to be constructed. Documentation and information sources used by the bricklayer when selecting resources for building cavity walls are covered in Chapter 2, which deals with methods of communicating important information about the work required.

Understanding plans, drawings and specifications is vital in making sure that the right materials and other resources are selected for the job during the preparation stage of your work.

Chapter 2 also covers methods of calculating quantities of materials that will be required. Planning ahead and setting up the work area efficiently includes making sure that the right quantities of materials and components are to hand before work commences.

Tools and equipment for the job

It is a good idea to make a list of the tools and equipment that will be required for the job before you start work to ensure things run as smoothly as possible. The tools needed to build any masonry task can be split into three main groups:

- Laying and finishing
- Checking
- Cutting.

Laying and finishing

To build any masonry wall you will need a trowel to lay the bricks or blocks. For face brickwork, you will also need a pointing trowel and a jointer to provide a finish to the joints.

Checking

A tape measure is needed to set out and check dimensions, a spirit level to make sure the work is level and plumb, and a set of line and pins to align the bricks or blocks accurately.

▲ Figure 5.13 Trowel

▲ Figure 5.14 Pointing trowel

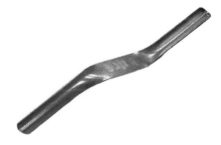

▲ Figure 5.15 Jointer

▲ Figure 5.16 Tape measure

▲ Figure 5.17 Spirit level

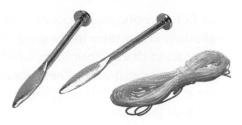

▲ Figure 5.18 Line and pins

Cutting

To cut the bricks or blocks accurately you will need a club (or lump) hammer and a brick bolster. A brick hammer and a scutch hammer (sometimes called a comb hammer) are used to trim and shape bricks or blocks.

▲ Figure 5.19 Club hammer   ▲ Figure 5.20 Brick bolster

▲ Figure 5.21 Brick hammer   ▲ Figure 5.22 Scutch hammer

There is a detailed section on using these cutting tools to cut bricks by hand later in this chapter.

The equipment needed to build cavity walls includes:
- spot boards for mortar
- shovels
- wheelbarrows
- brick tongs for moving bricks.

Characteristics of materials

The materials specified will have particular characteristics that make them suitable for use in cavity wall construction, so it is important that the bricklayer selects the correct materials before work starts. Knowing the characteristics of the materials selected will also help when planning how to store and move them. The key materials you will use are outlined below.

Clay bricks

Clay bricks include the following types.

Facing bricks

These are the bricks that can form the 'face' of the building. They are available in a vast range of colours and textures from many different manufacturers. The clay from which they are manufactured is easily moulded and consists mainly of quartz and clay minerals. The moulded clay is converted into durable bricks by a heating process known as 'firing' which heats the bricks in a **kiln** to temperatures between 900°C and 1250°C.

> **KEY TERM**
>
> **Kiln:** a type of large oven

The firing process may produce bricks with considerable variations in size since the heating process causes the materials to change their form, which can result in shrinkage and distortion. These variations should be kept in mind when selecting, moving and positioning bricks for the work task.

A lot of thought and care can go into choosing the right brick to achieve the desired finish to a building.

▲ Figure 5.23 Brick kiln

▲ Figure 5.24 Well-chosen facing bricks enhance the appearance of a building

ACTIVITY

Take 18 clay bricks and carefully line them up dry, side by side, without any gaps for **perps**. Ensure you use a straight edge to align one end of the bricks. Take out nine of the shortest bricks and set them out dry, end to end, again without any gaps for perps. After doing the same with the remaining bricks, note the difference in the measurement between the two groups. Is the difference surprising?

KEY TERM

Perp: short for 'perpendicular', this is the vertical mortar joint which joins two bricks or blocks together at right angles (or perpendicular) to the bed joint

Engineering bricks

Engineering bricks have a high compressive strength. This means that they can resist the squeezing forces that might be present in a high-rise structure such as a block of flats or in a load-bearing arch.

Engineering bricks are also denser than other clay bricks and do not absorb water, which makes them suitable for cavity walls constructed below ground level (or for use as a DPC).

▲ Figure 5.25 A railway bridge built with engineering bricks

Common bricks

Common bricks are a lower quality brick that is usually used in locations where the finished work will not be on show. For example, they could be used in constructing internal partition walls that adjoin a cavity wall.

▲ Figure 5.26 A common brick

Other types of bricks

A range of other bricks are also used, as outlined below.

Sand/lime bricks

A brick which is relatively easy to cut and shape. These bricks are sometimes referred to as calcium silicate bricks since they are composed of a fine aggregate which is bonded together by **hydrated** calcium silicate.

Pigment can be added during the manufacturing process to produce bricks in a range of colours. The mix is moulded under high pressure in hydraulic presses to produce the shape and then subjected to high-pressure steam in an apparatus called an autoclave for up to twelve hours to **cure** the bricks. This process produces bricks that have very little variation in overall dimensions.

▲ Figure 5.27 A sand/lime brick wall

▲ Figure 5.28 Brick autoclave

Concrete bricks

Like sand/lime bricks, concrete bricks are not fired in a kiln and therefore have more accurate dimensions than clay bricks. They are manufactured from a mix of aggregate, cement and water. They can be coloured during manufacture and given a range of textures if desired. Since they are usually made in solid form without a **frog** in the top of the brick, they are heavier than clay bricks. Concrete bricks are effective in reducing noise transmission and give good fire protection.

▲ Figure 5.29 A concrete brick wall

ACTIVITY

Consider which type of brick would be most suitable for use in the basement of a three-storey house. Write down the description of your chosen brick in your own words and give two reasons for your choice.

Knowing the range of characteristics of the bricks specified for a job means you can make informed decisions during your preparation for building.

You should consider the following points:

- How heavy are they?
- How soft or brittle are they?
- How much moisture will they absorb?
- How much do they vary in size (which will affect how they're stacked)?

Blocks

Blocks include the following types.

Lightweight insulation blocks

In cavity walls, lightweight insulation blocks are usually specified above a DPC for the internal skin (although certain types of lightweight block can be used below a DPC). They reduce heat transmission through the walls of a structure and therefore improve energy efficiency. Because they are lightweight, they are easy to work with and can be readily cut and shaped. However, especially when they're dry, they generate a lot of fine dust when being cut to shape or moved from storage.

> **HEALTH AND SAFETY**
>
> Suitable PPE should always be used to protect against dust contamination which may occur, for example when working with lightweight insulation blocks.

> **INDUSTRY TIP**
>
> To reduce the dust generated by cutting or moving lightweight insulation blocks, lightly spray them with water. Be careful not to saturate them.

Dense concrete blocks

Dense concrete blocks can be specified in cavity walls where the outer skin is not built in brick (to provide a suitable finish, a sand/cement render or other coating would need to be used). They are also used extensively in work below ground such as in the footings of a cavity wall. They are made from a coarser aggregate than concrete bricks.

▲ Figure 5.30 Dense concrete block

Hollow blocks

This type of block is rarely used in a cavity wall. The hollow form of the block allows greater strength to be gained by placing steel surrounded by concrete into the hollow to **reinforce** the wall.

> **KEY TERM**
>
> **Reinforce:** in masonry and concrete, strengthen by adding steel

Lintels

When forming openings in cavity walls, a means of supporting the masonry built above the opening must be provided. Concrete or steel lintels are used for this purpose and are manufactured in a large range of designs and dimensions. Types of lintel are described below.

> **HEALTH AND SAFETY**
>
> Lintels can be heavy items to move and position so prior planning when setting up the work area is necessary to ensure the health and safety of those on site.

Concrete lintels

Concrete lintels must be reinforced with steel to enable them to withstand the forces (tension) placed on them by masonry bearing down on them from above.

Concrete performs well under compression when a load is squeezing it but is not suitable where it is required to withstand tension (in situations where it is stretched or subject to bending).

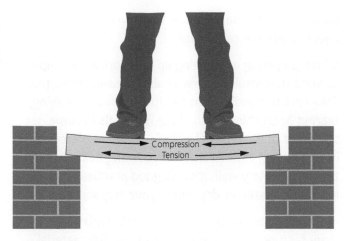

▲ Figure 5.31 A lintel under compression

The reinforcement is placed at the bottom of the lintel when it is cast since this is where the greatest stretching forces will be. The bricklayer must therefore be sure that a concrete lintel is installed the right way up.

Concrete lintels are mostly used where appearance is not so important, such as in bridging openings in partition walls within a structure or for providing openings for service entry points underground.

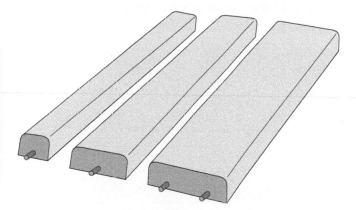

▲ Figure 5.32 Concrete lintels showing steel reinforcement

INDUSTRY TIP

Sometimes, manufacturers label the top of the lintel with the letter 'T' (for 'top') to assist in getting the lintel the right way up.

Steel lintels

Steel lintels are more commonly used in cavity walls which have the outer skin constructed in face brickwork and the inner skin or leaf constructed in lightweight insulation block. This is because a steel lintel is lighter than a concrete lintel of the same span and, unlike a concrete lintel, very little of the steel lintel can be seen when viewed from the outside of the structure. In addition, steel lintels are manufactured to incorporate insulation materials as part of their design so they are more energy efficient than concrete lintels.

VALUES AND BEHAVIOURS

Think about the qualities and characteristics of the materials that have been specified for the job. Are they manufactured from environmentally friendly materials? Do they reduce heat transfer in the structure? Using materials which improve the energy efficiency of the structure is one way you can reduce the environmental impact of your work.

INDUSTRY TIP

Steel lintels are often referred to by the names of the two main manufactures: IG and Catnic.

▲ Figure 5.33 IG lintel

Whichever type of lintel is used, it is the bricklayer's responsibility to make sure that the correct lintel is properly installed and that the **bearing** is as per the specification (a minimum of 150 mm).

Damp proof course

Successful installation of an effective damp proof course (DPC) begins with correct storage of the DPC materials. Flexible DPC is delivered to site in rolls. They must be stored correctly to avoid tearing or puncturing.

When storing any flexible DPC, never stack the rolls horizontally on top of each other. Stack them on end no more than three rolls high. This will avoid the material becoming distorted. If the material is forced out of shape it can make it difficult to lay along the top of a course of brick or block during installation.

▲ Figure 5.34 Flexible DPC materials in stacks

DPC materials containing bitumen and other **thermoplastic** materials should be stored away from direct heat. Bitumen felt rolls can become stiff in cold weather and can crack when the bricklayer attempts to unroll them, so they should be stored in a warm place at a constant temperature if possible.

Insulation

Insulation for cavity walls is commonly manufactured in the form of rigid boards, flexible sheets called bats (or batts) and loose materials that can be poured or injected into the cavity wall.

When preparing and setting up the work area, keep in mind that insulation materials are easily damaged. Take care to handle them carefully to avoid breaking, tearing or crushing them. If the materials become squashed or distorted, they may not be as effective in performing their function of reducing heat transfer through a cavity wall. It's also good practice to keep insulation materials dry during your preparations.

A newer material used for cavity wall insulation is manufactured from many layers of thin foil material.

Mortar materials

Bricks and blocks are, of course, bonded together using mortar to produce the bed and cross (or perp) joints. The materials used must be stored and prepared correctly to produce a mortar that is easy to use and durable when set. A specification for constructing cavity walls may give precise details about the colour of the mortar, the proportions of the different materials in the mortar mix and any additives that are required.

▲ Figure 5.35 Mortar on a spot board

ACTIVITY

Check a builder's merchant's website, such as
Jewson: www.jewson.co.uk

Search the site for two examples of 'mortar
additives' and describe what they are used for in
constructing cavity walls.

Mortar for constructing cavity walls must meet a
number of important requirements throughout the
entire life of the structure. It must possess:

- adequate compressive strength
- durability (resistance to chemical attack and frost
damage)
- strong bonding with masonry components
- a sealed surface to protect against wind-driven rain.

▲ Figure 5.36 Jointing the mortar joints to seal them

To fulfil these demands, the materials from which
mortar is produced need to have certain qualities. The
materials for mortar are outlined below.

Cement

Cement powder is produced from limestone and is
used in mortar as a 'binder'. It effectively fills the voids
between the sand and makes sure that the finished
mortar is hard and durable enough for the job.

The proportions of sand and cement are stated as a
ratio which will be designed to make sure that the
finished mortar has sufficient compressive strength
(resistance to squeezing) while not being overly brittle.

As a general rule of thumb, the mortar should be
slightly weaker than the bricks or blocks being laid. This
allows for any slight movement in the masonry caused
by thermal expansion or by shrinkage occurring during
hardening. A ratio of 1:3 (1 part cement and 3 parts
sand) would be described as a strong mix, whereas a
ratio of 1:8 would be weaker (or leaner).

▲ Figure 5.37 Cement manufacturing plant

A chemical reaction occurs between the cement
powder and water added to the mix. This is called
'hydration' and this reaction makes the mortar set hard.
It is obvious therefore, that cement stored on site must
be kept dry to avoid it going hard prematurely.

There are a number of different types of cement that can
be specified according to the requirements of the job:

- Ordinary Portland Cement (OPC) is the most
commonly used. It is suitable for general masonry
work and, if used correctly, will produce a mortar of
high quality and strength.
- Masonry cement is similar to OPC but contains
25 per cent **inert** filler to make the mix more pliable,
so a mortar using this type of cement requires a
higher proportion of cement powder to provide
adequate bonding characteristics.

▲ Figure 5.38 Masonry cement

- Rapid Hardening Portland Cement (RHPC) can be used where a shorter setting time is required. Whereas the setting and hardening of OPC takes place in seven days or more, RHPC will set and harden to working strength in about four days.
- Sulphate Resisting Portland Cement (SRPC) is better than OPC at resisting **sulphate** attack in damp conditions.

KEY TERMS

Inert: not chemically reactive

Sulphate: a salt of sulphuric acid

INDUSTRY TIP

When storing cement on site, make sure it is stacked so that the older stock can be used first ('stock rotation'). Otherwise there is a risk that cement stored for a long time will harden before you can use it.

Sand

Sand for brick or block laying comes from two main sources:

- Pit sand – dug up from pit deposits, this type of sand produces a mortar that is easy to work with and the finished product has a pleasing colour.
- Dredged sand – sand is increasingly being extracted (referred to as dredged) from the sea, as sand deposits decline and planning permission for removing sand from pit locations becomes more difficult to obtain. Dredged sand must be thoroughly washed to remove salts before it can be used, to avoid affecting the quality of the mortar. It is usually more '**sharp**' than pit sand and may require an additive to make it easier to work with.

▲ Figure 5.39 Dredged sand processing plant

KEY TERM

Sharp: in this context, sand that has pointed or angular grains

All sand used for masonry must be well graded. This means that it must contain a uniform mix of fine, medium and coarse particles.

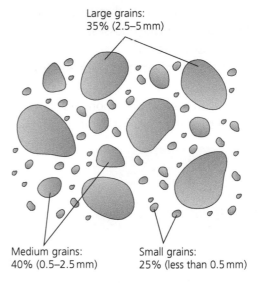

Large grains: 35% (2.5–5 mm)

Medium grains: 40% (0.5–2.5 mm)

Small grains: 25% (less than 0.5 mm)

▲ Figure 5.40 Graded grains of sand

Mortar made from poorly graded sand will be weaker since there will be more tiny spaces that need to be filled. These tiny spaces or voids can cause shrinkage as the mortar hardens, which may result in cracks forming. Rain can then penetrate the cavity wall, which will cause deterioration over time.

Sand used for mortar must be free of mud and silt. A specification may require a 'silt test' to establish whether or not the material is suitable for building masonry walls.

ACTIVITY

Access 'Educational Guide to Aggregates – CEMEX' at: https://bit.ly/2ButMqg

Read about the different sources of sand and how they are graded.

▲ Figure 5.41 Silt test

Water

Water is an essential ingredient and without it mortar will not harden or set. The water used in producing mortar should be **potable.** Using water of drinkable quality means that no chemicals that could interfere with hydration or affect the mortar in undesirable ways will be present.

Plasticiser

If we were simply to mix sand, cement and water together to produce a mortar for brick or block laying, we would find it difficult to use. To make it more 'workable', a **plasticiser** must be added.

KEY TERMS

Plasticiser: an additive that makes a material more pliable

Potable: suitable to drink

Modern plasticisers are chemical additives that are included in the mixing process as a liquid or a powder. They should be used carefully in accordance with the manufacturer's instructions. Chemical plasticisers work by introducing tiny air bubbles which allow the particles of sand to move over each other more freely. This is called 'air entrainment'.

KEY POINT

The air entrainment created by mortar plasticisers produces air bubbles of a controlled and uniform size. Some bricklayers use detergent powder or washing-up liquid as a plasticiser. This creates bubbles of varying sizes in the mix which can weaken it. Always use an established product that conforms to industry standards.

In the past, it was common practice to use hydrated lime as a plasticiser. This increases the powder content of the mix which forms a paste that lubricates the particles of sand. Hydrated lime is still specified on occasions, but it can be hazardous to health and has more health and safety considerations than chemical plasticisers, which are also easier to work with.

▲ Figure 5.42 Hydrated lime

▲ Figure 5.43 Liquid plasticiser

▲ Figure 5.44 Powdered plasticiser

ACTIVITY

Find out about the health and safety considerations when using lime. Start by visiting the Health and Safety Executive website (www.hse.gov.uk) for more information. Write a simple risk assessment for working with hydrated lime.

3 POSITIONING AND PREPARING MATERIALS FOR SAFE AND EFFICIENT WORK

To maintain efficiency and productivity, you should consider these key points:

- how to position materials and components for safe and efficient working
- how to prepare materials and components to reduce waste and costs.

In both of these important areas, it is up to you to be responsible, observant and diligent.

Working efficiently

Good preparation by the bricklayer is important at every stage of constructing cavity walls. Preparation starts when materials are delivered to site and are stored ready for use. Storing materials correctly can have a great impact on how efficiently the job progresses and the quality of the finished work. Stored materials must be positioned so that they are easy to access and are not vulnerable to damage by machinery movements or other operations on site.

Keep your materials dry

When bricks and blocks are delivered to site, they are normally protected from bad weather by polythene wrapping. When breaking open the packs of bricks or blocks to move them from storage into position ready for laying, the bricklayer should make sure that the weather protection is maintained until building work begins.

Wet bricks and blocks are difficult, if not impossible, to lay accurately, since the mortar bed joint becomes unstable as moisture runs out of the brick or block and causes the mortar joint to 'swim'. Mortar can ooze from the joints, making it very difficult to keep the face of the bricks clean.

In addition, if bricks and blocks are allowed to get wet, problems such as **efflorescence** can develop later, spoiling the appearance of the finished work. (Efflorescence can also form if the brickwork is not protected after it is finished. Cover it to protect from bad weather until it is hardened.) Protecting bricks and blocks from bad weather and damage from site operations is simple but important.

KEY TERM

Efflorescence: a powdery white crystal deposit that can form on the face of the masonry

▲ Figure 5.45 Efflorescence

Taking the time to cover brick and block stacks with polythene, or simply placing a spot board on top of the stack, will mean that work can continue immediately after bad weather moves away. When you remove the weather protection from brick or block stacks, keep it nearby. If it starts raining after work has commenced, you can quickly cover up materials to keep them dry.

▲ Figure 5.46 Bricklayer protecting a brick stack with a spot board

Careful positioning

Remember that bricks and blocks are heavy components and need to be moved and handled with care. This is especially the case with dense concrete blocks.

▲ Figure 5.47 Beware of trapping your fingers!

If blocks are handled roughly they can develop cracks in them that are not easy to see. When the block is lifted at the ends it may break into two or more heavy pieces which can fall and injure an operative's legs or feet, so careful handling is very important.

Bricks in a stack can be positioned on edge, with two rows of 6 bricks in each layer and up to 12 layers in each stack. Tools for moving bricks manually are designed to comfortably carry six bricks at a time (on edge) to minimise the risk of damage to the materials and injury to the operative. Other safe methods of stacking bricks can be employed to suit the working space available and the quantity of bricks required for the job.

▲ Figure 5.48 Bricks stacked properly. Note the use of the proper tool for moving bricks

The stacks should be as stable as possible, so it may be necessary to spend some time levelling the area when stacking at ground level. Stacking out on scaffolding is simpler since the deck of the scaffold is level and even.

▲ Figure 5.49 Operatives working with stacks of bricks on a level scaffold

However, preparing the work area and the actual process of building require greater care and awareness when working at height. There may be workers in the area below the scaffold who could be severely injured or even killed if materials were to fall on them from above. Wherever the stacking takes place, at ground level or on scaffolding, never stack too high!

Stacks of blocks are best arranged flat rather than on edge, six to eight layers high, making sure the base on which they are stacked is firm and level. Dense concrete blocks are best moved one at a time if moved by hand.

▲ Figure 5.50 Stacks of blocks and bricks being used to build a cavity wall on site

Remember that the distance at which materials are placed from the wall you're going to build is also important for efficient working. Stacks of materials and spot boards for mortar should be placed around 600 mm from the face line of the wall, but some bricklayers prefer to increase the working space between the materials and the wall to around 900 mm if the space is available.

An important point to keep in mind when stacking out and positioning bricks is to make sure that the bricks are mixed or blended. This is because the colour of bricks can vary from batch to batch due to variations in the manufacturing conditions and the materials used.

Mixing bricks by selecting them from a minimum of three packs while loading out will help to prevent bands of colour showing in the completed work. Such bands would spoil the appearance, especially when building **quoins**.

KEY TERM

Quoins: the vertical external angles (corners) in walling

Positioning lighter components

Bricks and blocks are heavy items that require more care to safely move and position them. By contrast, wall ties and insulation are much lighter and easier to move and position ready for work. However, there's still a need to think about where you position them for convenience, to make sure that you can operate efficiently when working.

Wall ties should be positioned along the length of the wall being built in small bundles, ready for use. A good location for them would be in the space under raised spot boards where they are easy to grasp but won't be stood on and damaged or pressed into soft ground and lost. They may look rather insignificant components compared with stacks of bricks and blocks, but they are made from expensive materials to make them durable, so they need to be looked after.

Insulation materials are light and can be bulky, so they can be blown about by strong winds and damaged or broken. Place them in a convenient position near the cavity wall you are building and make sure that packs or loose sheets of material are weighted down suitably to prevent them being picked up by wind and damaged. A simple way of doing this is to carefully place a dense concrete block or two on top of the insulation sheets.

Fire stops are plastic tubes often referred to as 'socks' which are filled with chopped mineral fibre strands. They are specified at locations such as party walls to prevent the spread of fire through the cavity between individual rooms or occupied units. Since they are fragile and can easily be torn, they should be handled and stored with care.

Other materials and components

Constructing cavity walls will involve handling a range of materials and components other than masonry items. These items will also need protecting from the elements and other causes of damage during storage to contribute to safe and efficient working. Transporting them to the work location should be done with care and forethought, especially in the case of large or heavy items.

Timber components

Although timber components for use on site are treated to resist the effects of moisture, they can distort and warp if saturated and poorly stacked or positioned. Timber frames for doors and windows should be stacked carefully on bearers and covered to protect them from the rain until the time comes to use them. The covering should be done in such a way that ventilation is provided during storage.

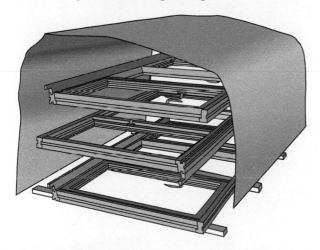

▲ Figure 5.51 Stack frames carefully on bearers and cover them

Timber joists are often long (and can be heavy) and should be stacked neatly on levelled multiple bearers to support them along their length. Solid timber joists are being used less in modern construction. They are being replaced by timber engineered components which, although not as heavy as solid timber joists, can carry loadings over greater spans. These components need

particular care and attention when handling and storing as they can easily distort.

Whichever type of joists are used on the site where you work, they should be treated and handled with care to ensure that the product quality is maintained throughout the building process. This applies when the items are stored after delivery, when they're moved to the work location and if they need to be moved to gain easier access to the cavity wall you're building. Don't be tempted to throw things around if they happen to be in your way!

> **HEALTH AND SAFETY**
>
> If you're moving long joists, don't try to transport them alone. It is easier and safer to have two workers to do the job, one at each end of the joist.

Preparing mortar

Having selected the specified materials and components for building a cavity wall and positioned them for efficient working, it's time to prepare the mortar ready for laying bricks and blocks.

Mortar for brick and block laying must be mixed to conform to the specification and be of a consistency and workability that allows the work to be done efficiently. The pace of work can be slowed down significantly if the mortar mix is either too stiff and dry or too fluid and wet to suit the types of bricks or blocks being used and the conditions in the workplace.

> **KEY POINT**
>
> It takes time and experience to produce a mortar mix that matches all the requirements of the work task. Adjusting the consistency of the mix to suit differing materials and weather conditions is a skill that takes practice.

> **KEY POINT**
>
> Mortar for constructing a cavity wall can be produced in a number of ways:
> - mixed by hand
> - mixed by machine
> - mixed by machine off-site and delivered to site ready to use.

Mixing by hand

Mixing mortar by hand is hard work, but if an operative follows a simple sequence and works on a clean solid base, it can be made a little easier. The common rule when mixing by hand is '3 times dry – 3 times wet'. This means that we move the dry materials from their placed position to the side and then back again three times in order to mix the sand and cement (and lime if specified) thoroughly before adding water.

Once the dry materials are well mixed and positioned in a tidy mound, a large circular dip is formed in the centre and water is added carefully. The water is gradually mixed in, until it is **uniformly** spread through the materials. It can then be turned three times 'wet' in a similar manner to the dry mixing and more water slowly added as required. If a plasticiser is used, this should be added to the water and not to the dry materials.

INDUSTRY TIP

When mixing by hand, don't use a shovel that's too big for you to handle. It is more efficient and much easier if you use a smaller 'taper nose' shovel.

▲ Figure 5.52 Taper nose shovel

The following step by step shows the correct method for mixing by hand.

STEP 1 Add the cement to the sand in accordance with the specified ratio.

STEP 2 Mix the sand and cement three times 'dry'.

STEP 3 Add water to the dry mix (taking care not to add too much).

STEP 4 Mix the sand and cement three times 'wet'. Add water as needed to improve the mortar consistency.

Mixing by machine

Mixing by machine is usually done using a drum mixer. This can be powered by electricity, a diesel engine or a petrol engine. A drum mixer of any size is a powerful piece of machinery and caution must be exercised when using it.

▲ Figure 5.53 Drum mixer

When mixing mortar with a drum mixer, always add an amount of water first. There are advantages to adding the cement next, which will form a paste to which the sand is then added. If the cement is added after the sand, it can form into small balls which prevent thorough mixing and this will affect the final strength of the mortar.

Whichever method of mixing is used, it is vitally important to measure the correct proportions of materials. Measuring materials by the 'shovelful' will not produce a reliably consistent mix. A shovelful of dry powdery cement will have much less volume than a shovelful of damp sand.

Providing accurate amounts of materials for mortar is known as **gauging.** A gauge box is a bottomless steel or timber square or rectangular box which is placed on a clean flat surface and filled with either sand or cement flush with the open top of the box.

KEY TERM

Gauging: the method used to accurately and consistently measure quantities of material for mortar or concrete

If a ratio of 1:4 is required, four boxes of sand will be mixed with one box of cement. A suitably sized strong bucket could also be used. To mix larger amounts, multiples of the gauged amounts will be used.

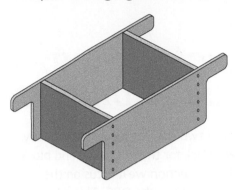

▲ Figure 5.54 Steel gauge box

The carefully measured separate piles of sand and cement can then be either loaded into a drum mixer or mixed by hand as previously described.

By machine off-site

Mortar mixed off-site will usually be mixed by weight batching and not by volume. This is a very accurate method of making sure that the correct proportions are used. The mixed mortar delivered to site may include a 'retarder' which is a chemical additive to slow down the hydration of the cement. This keeps the mortar workable for up to 48 hours and sometimes longer.

▲ Figure 5.55 Mortar batching plant

The mortar may also have additives in the form of pigments to colour the mortar if this is specified for a cavity wall. This method keeps consistency of colour throughout the mortar mix over the course of the project.

▲ Figure 5.56 Pre-mixed mortar being delivered to site

4 SETTING OUT AND BUILDING CAVITY WALLS, INCLUDING FORMING OPENINGS

Cavity walls consist of two skins or leaves of masonry, most frequently face brick for the outer skin and block for the inner skin. In this section we'll focus on the cavity wall construction above the DPC. This is known as the superstructure.

> **KEY POINT**
>
> Remember, the section of a structure *below* the DPC is known as the substructure. The methods used in setting out this section of the building accurately are covered in detail in Chapter 4.
>
> Keep in mind that when 'setting out' is referred to in this section, it relates to cavity wall construction *above* DPC or FFL (finished floor level).

When the outer skin is constructed in face brick, setting out and correctly bonding the brickwork is vital to produce a wall that is visually pleasing and structurally sound. The most common brick bond used in cavity walls is Stretcher bond, which may also be referred to as 'half bond'. This is because each alternate course of brick is laid with an overlap of approximately half the length of a standard brick.

> **ACTIVITY**
>
> Access the Brick Development Association's (BDA) information on brick awards at: www.brick.org.uk/brick-awards/
>
> Look through some of the examples of face brickwork and write down the name of the project and the brick that is used in the example you like most.

Setting out the bond

You should know the standard dimensions of a brick off by heart. This is important information to have in mind when setting out the bond. Figure 5.57 shows how brick dimensions are designed to be 'modular', which means that the dimensions of each face or side of the brick can be connected or combined in different ways.

Keep these important principles in mind when setting out your work:

- Set out the first course carefully and accurately.
- Check the specified dimension for the width of cavity required.
- Make sure that the work is set out to the correct overall dimensions and to the required level.
- Carefully refer to and observe provided datum points.

> **KEY POINT**
>
> Accurately setting out the first course of face brickwork is crucial to help ensure the quality of the rest of the construction process. Taking care to work to high standards at all times means that work will not have to be re-done, which can be costly and time-consuming.

Take some time to establish the brick bond for the face of the cavity wall, setting out the first course of face brickwork 'dry'. This means positioning bricks without mortar to get the spacing right. Using this method will help you to decide whether you need to cut bricks to make things fit within the design dimensions or use **reverse bond**.

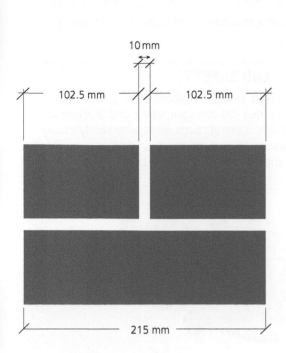

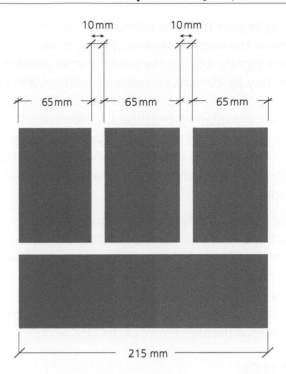

The width of two header faces plus a 10 mm joint is equal to the length of a stretcher face

The width of three bricks on edge plus two 10 mm joints is equal to the length of a stretcher bond

▲ Figure 5.57 How the dimensions of a brick are 'modular'

KEY TERM

Reverse bond: in the same course, starting with a stretcher and ending with a header

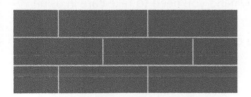

▲ Figure 5.58 Reverse bond

If cut bricks are required to set out the bond correctly, this is known as **broken bond**. Plan where the cuts will be placed, if possible, under a door or window opening. If no openings are to be included in the elevation, then the cuts should be located as close to the centre of the wall as possible.

KEY TERM

Broken bond: the use of part bricks to establish a bonding pattern where full bricks will not fit in

Carefully setting out the bond will ensure the appearance of the cavity wall throughout the rest of the structure will be the best you can achieve. When face brickwork is specified for the outer skin or leaf of a cavity wall, this is the 'finish' of the building and will potentially be on view for a very long time.

▲ Figure 5.59 Broken bond

Methods of work

At the setting out stage, it is good practice to consider the position of door and window openings. The aim should be to position the openings so that full bricks can be laid up to the reveals either side of the opening, avoiding the need for three-quarter cut bricks. Obviously, you will need to cut half bricks for alternate courses of Stretcher bond (or half-bond) to be maintained at the reveals.

This may require the dimensions given for the position of the door or window opening to be adjusted slightly. Moving the position of an opening slightly (say by 20 mm), can save a significant amount of cutting of bricks and blocks, effectively reducing waste and contributing to efficiency. However, beware of moving the position of openings too much. Consider the position of internal partition walls when making your judgement. Ask for advice if you're not confident!

In using good methods of work and maintaining standards, remember these key points:

- Try to avoid introducing cut bricks if possible when setting out the bond.
- Keep perps to 10 mm if possible – although there is an allowable **tolerance** of plus or minus 3 mm, large perps spoil the appearance of the face work.

KEY TERM

Tolerance: allowable variation between the specified measurement and the actual measurement

Cutting bricks and blocks

When masonry materials such as bricks or blocks need to be cut or shaped, you should always make sure that you use the correct PPE:

- Goggles or safety glasses are essential to protect your eyes from flying brick or block chips while cutting by hand or machine.
- Protective gloves are recommended.

- Ear defenders and a dust mask are essential PPE when using mechanical means to cut masonry materials.

HEALTH AND SAFETY

Never attempt to use or handle cutting machinery of any kind unless you are competent and authorised to do so. Proper training, such as how to safely change cutting discs, will reduce the risk of serious injury to yourself and others working nearby.

▲ Figure 5.60 Disc cutter

ACTIVITY

Search 'disc cutter training' on the Health and Safety Executive (HSE) website (www.hse.gov.uk). Read the information and make a note of some hazards that you would need training to deal with.

Cutting by hand

When cutting by hand you need to take care in using and looking after the tools correctly and safely and you must follow the correct work sequence.

STEP 1 Make sure you are using the right PPE (safety glasses or goggles, etc.) and workers nearby are aware that you are cutting masonry materials.

STEP 2 Use a pencil to mark the position of the cut on the face, the opposite side and the bed of the brick. For blocks, it is usually sufficient to mark the position of the cut on the face only, unless you prefer to mark it all around.

STEP 3 Placing the blade of the bolster slightly on the waste side of the pencil mark, strike the brick lightly but firmly with the club hammer. For blocks, you will need to use several more powerful and decisive strikes across the full face of the block.

STEP 4 Now do the same on the opposite side of the brick or block. For blocks, if you've used sufficient strength in your blows the block should break as desired.

STEP 5 For bricks, turn the brick so that the face is uppermost again and strike the last blow. If the strength of the blow is adjusted correctly, this should complete the operation.

STEP 6 If the brick or block doesn't break as desired, repeat from Step 3 until a clean break is achieved.

The aim is to produce a clean, sharp **arris**, especially on the face side of the cut brick or block. This will ensure a satisfactory appearance in the finished wall.

KEY TERM

Arris: any straight sharp edge of a brick formed by the junction of two faces

INDUSTRY TIP

It is a good idea to gently press the edge of the palm of your hand against the surface you're cutting. This will 'damp' the vibrations going through the brick and help to prevent it shattering.

- Never build too high at any one time – each individual skin or leaf of masonry should not be raised too high without support from the other skin. The British Standards official guidance states that a single skin or leaf in a cavity wall design should not be built higher than 6 courses of block or 18 courses of brick in one operation (**BS EN 1996-3:2006**).
- Building too high in one operation can be dangerous because a single skin of unsupported masonry is relatively fragile. It can be easily blown over or pushed over by accident by someone leaning against it, even when the mortar has hardened. (For more on single leaf masonry, see Chapter 6.)

▲ Figure 5.61 Cutting technique

Good practice for cavity wall construction

Employing good practice when building cavity walls will maintain productivity and efficiency along with contributing to a safe working environment at each stage of construction. Take note of these important points:

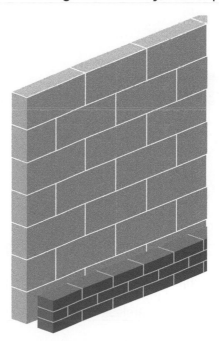

▲ Figure 5.62 A single leaf of blockwork seven courses high, unsupported by the brickwork

- Make sure all bed and especially perp (or cross) joints are full – to speed up the job, some bricklayers have the habit of 'tipping and tailing' their perp joints. This means that instead of completely filling the joint, a small amount of mortar is placed on the front and back edges of the header face, leaving a **void** in the finished joint. The use of cavity walls was adopted in preference to solid walls because they give greater protection against moisture penetration. This protection can only be fully provided if all mortar joints are full.

KEY TERM

Void: an open space or gap

▲ Figure 5.63 Bricklayer incorrectly 'tipping and tailing' the perp joint

▲ Figure 5.64 Bricklayer correctly making a full perp joint

KEY POINT

It may be quicker to 'tip and tail' in the short term while building, but in the long term, expensive problems can be caused by careless or shoddy workmanship.

Installing horizontal DPC

Earlier in this chapter we discussed the different materials used for DPC and the advantages and disadvantages of each one. Now we'll look at the methods you must use for installation of flexible DPC.

Horizontal DPC must be installed in the correct position in relation to finished ground level outside the building. The requirement is that the DPC should be positioned a minimum of 150 mm above the path or ground next to the wall.

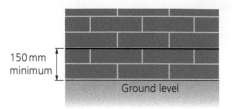

150 mm minimum

Ground level

▲ Figure 5.65 Minimum height of DPC above ground level

When installing a flexible DPC it is always good practice to lay the material on a thin bed of mortar. This protects the DPC material from the risk of being punctured by hardened mortar, which may project from the top of the masonry in the course below.

INDUSTRY TIP

With a thin bed of mortar under the DPC and further mortar above the DPC to lay the bricks on, remember to watch your gauge. Take care not to build up the joint too much.

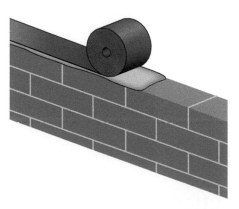

▲ Figure 5.66 DPC being bedded on mortar

If a roll of DPC is too short to run the complete length of the wall, an additional roll should be laid to overlap the first section by a minimum of 100 mm.

An exception to this overlap dimension is when wider DPC is used. Then the overlap should be the same as the width of the DPC being used. So, 200 mm wide DPC will have an overlap of 200 mm.

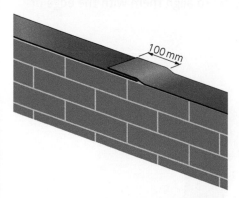

▲ Figure 5.67 DPC being lapped correctly

Building quoins

Once the horizontal DPC is correctly laid, we can start work on the superstructure section of the building.

Having established the bond for the cavity wall face brickwork, the usual method of working is to build quoins at each end of the wall and then fix a string line between the quoins as a guide for building the infill.

When building quoins, remember, it is good practice to set out a corner or quoin as a right-angle or 90°. Laying and levelling away from the corner point maintains accuracy and can speed up your work rate and add to your efficiency. A good sequence of checking for accuracy is:

1 gauge
2 level
3 plumb
4 line.

Techniques that help to maintain quality and standards include gauging – the process of checking the work is built to the correct height. Until you gain greater experience, each course should be checked as the work progresses. With practice it is possible to gauge after several courses have been laid and still maintain

accuracy. Remember, standard gauge is 75 mm for a brick and a bed joint.

KEY POINT

When levelling brick courses in the quoin, remember to keep in mind that the spirit level is a precision instrument.

Never strike the spirit level to horizontally align bricks.

Place the spirit level on the top of the course of bricks to be levelled and gently tap the bricks, NOT the level.

If you need to make excessive adjustments, it could be that you will need to remove bricks in that course and adjust the thickness of the bed joint. Think of the quoin (or corner) brick as a 'control point' to which you refer for the rest of each course.

▲ Figure 5.68 Spirit level being used to level the top course of brickwork

Let's consider some points of good practice when plumbing the work. When only one course is laid, it is virtually impossible to plumb it accurately. Begin careful plumbing from the second course.

To keep the spirit level stable, place your foot against the bottom of the level while holding the top of the level with your free hand. Carefully adjust the bricks at either end of a course and look down the face of the wall to make sure it lines up with the spirit level.

> **KEY POINT**
>
> Never strike the spirit level to plumb a quoin.

Finally, lining the face of bricks between the plumbed ends of each course will help to produce a wall that has an accurate **face plane**. Rest the edge of the spirit level against the face of the bricks in each course. You don't refer to any of the bubbles in the level – you just use the level as a straight edge. Gently tap the bricks into line.

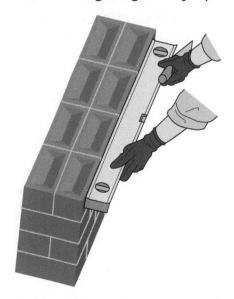

▲ Figure 5.69 Lining in

> **KEY POINT**
>
> Never strike the spirit level to line up a course.

> **KEY TERM**
>
> **Face plane:** the alignment of all the bricks in the face of a wall to give a uniform flat appearance

An accurate face plane will also be achieved if you **range** your work. This again means using your spirit level as a straight edge, but this time you place it in line with the stepped brickwork as it racks back. Gently tap the bricks to align them with the edge of the spirit level.

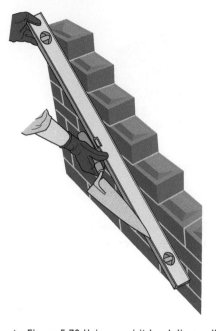

▲ Figure 5.70 Using a spirit level diagonally to range the quoin

Laying to the line

When you are satisfied that your quoins are as accurate as possible, you can use them as a guide to 'run in' the wall between the quoins.

Laying to the line is another process that requires practice to develop speed and accuracy. The main points to keep in mind are described below.

- Make sure the top arris of the brick is level with the line along the full length of the brick.
- Never lay the bricks touching the line. Doing this will cause the line to move away from the face of the wall and will result in a curved face plane. Keep the top arris of the brick about the thickness of the line away from the line.
- 'Eye' down the wall to ensure the face plane is smooth and the perps are lined up vertically.

▲ Figure 5.71 String line attached to quoins

▲ Figure 5.72 Bricklayers laying to the line

Installing wall ties

Wall ties are an essential part of the cavity wall design. They must be correctly installed and positioned if the cavity wall is to remain stable and be durable. Figure 5.73 shows the maximum allowable distance between ties to satisfy Building Regulations.

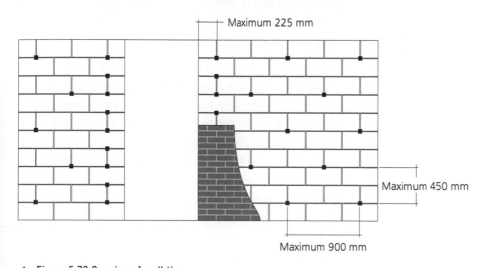

Maximum 225 mm

Maximum 450 mm

Maximum 900 mm

▲ Figure 5.73 Spacing of wall ties

Sometimes the bricklayer will need to make a judgement on positioning additional wall ties to provide enough strength to link the two skins of the cavity wall together. Remember, it's better to have too many ties than too few.

Wall ties are designed with a feature known as a 'drip'. This serves to shed any moisture that may track across them from the outer skin, preventing it from reaching the inner skin or leaf. It's important that the bricklayer makes sure the drip is centrally positioned in the cavity during installation.

▲ Figure 5.74 Wall tie with drip centrally positioned in a cavity wall

The competent and careful bricklayer should always keep the following points in mind when installing wall ties:

- Bed the ends of ties at least 50 mm into the bed joint of each masonry skin.
- Never just push the ties into the bed joint – they will not be effective in tying the two skins or leaves together.
- To achieve maximum strength, press the ties down into the mortar – do not lay the mortar over the ties.
- Make sure the ties are level between the two skins or inclined towards the outer skin. (If ties are inclined towards the inner skin, they can conduct moisture across the cavity.)
- Keep all ties clean – remove mortar droppings to avoid creating a 'bridge' for moisture.

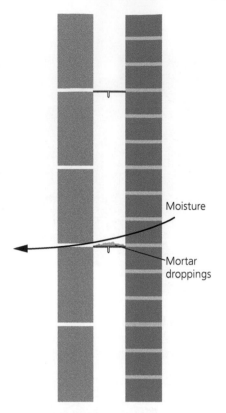

Moisture

Mortar droppings

▲ Figure 5.75 Mortar droppings on wall ties can conduct moisture into the structure

ACTIVITY

What methods could be used to prevent mortar dropping onto the wall ties?

HEALTH AND SAFETY

When installing wall ties into the first skin of masonry, be aware that they project from the wall and can cause injury to the bricklayer, especially those ties at eye level.

Installing insulation

The bricklayer must carefully install insulation in accordance with the manufacturer's instructions in order not to reduce the cavity wall's resistance to moisture entry. When building a cavity wall, the bricklayer must decide whether to construct the inner skin or the outer skin first. The sequence is determined by which method of installing cavity insulation is specified. There are three descriptions of how insulation can be installed in cavity walls:

- full fill
- partial fill
- injected.

Full fill insulation

If full fill (sometimes referred to as 'total fill') insulation is used, as the name suggests, the cavity is filled with insulation material. This means that either skin (or leaf) of masonry can be constructed first. The material used for this type of insulation is usually mineral fibre manufactured in flexible sheets referred to as bats (or batts).

The bricklayer needs to take great care to avoid leaving mortar droppings on top of the installed sheets of insulation as the cavity wall is raised.

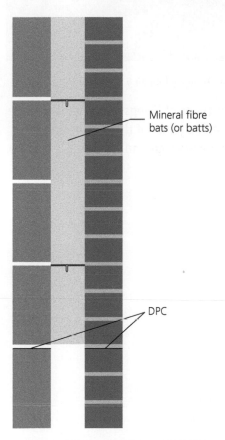

Mineral fibre bats (or batts)

DPC

▲ Figure 5.76 Sectional view of full fill insulation

INDUSTRY TIP

Be aware! Full fill insulation can expand and push the two skins of masonry apart until the masonry has set hard. Monitor your work carefully to maintain accurate plumb.

Partial fill insulation

Partial fill cavity insulation is fixed to the inside of the inner skin using special clips attached to the wall ties in the cavity. This means that the inner skin must be

constructed first and allowed to harden before the rigid insulation boards can be attached to it.

Building the inner skin first will require the bricklayer to take care when setting out the face brick bond to match the position of openings constructed in the inner skin.

The rigid insulation boards are most commonly manufactured from polyisocyanurate foam, more commonly referred to as PIR. This soft material is easy to cut to fit around awkward shapes using a saw.

HEALTH AND SAFETY

When cutting PIR boards to shape, make sure you wear suitable PPE. Use a dust mask and safety goggles. Cutting operations with this material generate a lot of dust and floating debris.

VALUES AND BEHAVIOURS

When doing jobs such as cutting insulation that make a lot of dust, consider how your work might affect others working nearby and let them know what you're going to do before you start.

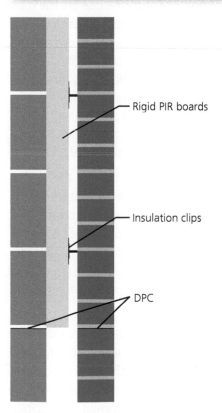

Rigid PIR boards

Insulation clips

DPC

▲ Figure 5.77 Sectional view of partial fill insulation

Because the cavity is only partially filled, a gap between the insulation and the outer skin is maintained. The bricklayer must be careful not to allow mortar joints to project into the cavity as they could carry moisture onto the surface of the insulation.

Injected insulation

Injected insulation is usually installed after the masonry is finished. If this type of insulation is specified it will have no impact on the sequence of work when constructing a cavity wall. The materials commonly used for this method of insulation are chopped strands of mineral fibre or beads of polystyrene. They are injected into the cavity using compressed air.

If the building is being constructed in a location where it will be frequently exposed to harsh weather conditions, the design decision may be to avoid placing any insulation in the cavity. In this case, the internal skin would be constructed in insulation block and the thickness of the block increased to meet energy efficiency requirements.

Specialised methods of work

While building quoins and laying to the line are the most common types of work for the bricklayer when working on cavity wall construction, there are methods and techniques that are a little more specialised in nature that you need to be familiar with. These may require more technical knowledge and greater diligence in developing and using your skills, and are outlined below.

Forming openings in cavity walls

Openings in cavity walls are constructed by the bricklayer to allow for the installation of doors and windows. The methods used to set out and create an opening vary according to the type of door or window frames specified. Openings are also formed to allow for services such as water and electricity below ground, but here we'll concentrate on forming openings for doors and windows.

If softwood timber frames are specified, they can be 'built-in' to the masonry as work proceeds. The frame is set up and temporarily braced in the correct position on the masonry at the door or window sill level and the brickwork or blockwork is carefully formed around it.

▲ Figure 5.78 A timber frame set up on a cavity wall

A range of different types of fixings to secure the frame can be built into the bed joints of the masonry at the required spacing as the work around the frame is built up.

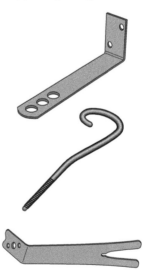

▲ Figure 5.79 Different types of frame fixings

If PVCu frames are to be used, they will need to be installed after the masonry work is completed since they can easily be damaged during construction activities such as bricklaying.

The openings are formed using 'dummy' frames; these form a temporary frame or profile around which the masonry can be built. They are later removed to leave an opening ready for the PVC frames to be inserted into. This method might also be used when expensive hardwood frames are specified which could suffer damage during the building phase.

▲ Figure 5.80 Dummy frame in use

The dimensions of the dummy frame are usually increased by 10 mm or so, to create an opening slightly larger than the finished frame – this allows for ease of fitting later.

INDUSTRY TIP

When constructing openings in masonry, a considerable number of cut bricks and blocks will be required. Some bricklayers prefer to spend time preparing the required cuts before building commences to improve the flow of work.

Closing the cavity at the reveals

If the inner skin is constructed in insulation block and is returned into the outer skin to close the cavity at the reveals, care is needed to make sure that the cutting operation produces accurately sized components. This is because a vertical DPC may be specified at this location, and this will need to be held firmly in place without the risk of it being punctured.

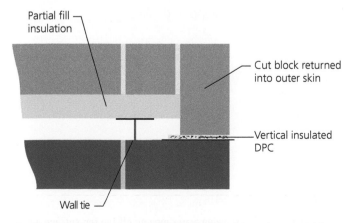

Partial fill insulation

Cut block returned into outer skin

Vertical insulated DPC

Wall tie

▲ Figure 5.81 Plan view of blockwork returned into the outer skin

The outer skin will have cut bricks (or batts) laid to maintain the bond either side of the opening. If the cuts are half batts, the cut end should be placed in the perp or cross joint to make sure that a smooth rear face is presented against the vertical DPC. Again, it is important to ensure there is minimal risk of puncturing the DPC. This also ensures that the position of the DPC follows the correct vertical line.

There are modern flexible vertical DPC materials that have an insulation material attached to them to reduce what is known as the '**cold bridge**' effect around a door or window frame. This material can be easily damaged and must be handled with care. The rolls of insulated DPC can also be bulky, giving rise to storage considerations.

KEY TERM

Cold bridge: low temperatures conducted to the interior of a cavity wall where the masonry of the outer leaf touches the masonry of the inner leaf at a reveal

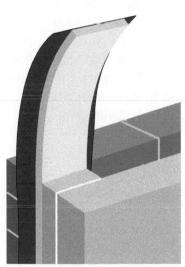

▲ Figure 5.82 Insulated vertical DPC in position

More frequently, no vertical DPC is specified, but instead a **proprietary** PVC cavity closer may be used. This has the advantage of simplifying the cuts required since no masonry closes the cavity at the **jamb** area. The rigid PVC profile closes the cavity, acts as a DPC and can also be designed to include insulation material. In addition, fixings for the door or window frame can be included in the design to make a simple-to-use multi-purpose component.

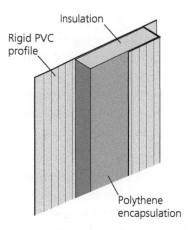

▲ Figure 5.83 Proprietary cavity closer system

Creating an opening in a cavity wall will inevitably cause an area of weakness at that location. To improve the strength at the opening, a greater number of wall ties are built into the jamb area either side of the opening. As already discussed, the vertical spacing of wall ties is 450 mm. This spacing is reduced to 225 mm at the reveals so that the strengthening effect provided by the wall ties is significantly increased.

DPC for lintels and sills

Previously in this chapter we have looked at different types of lintels used to span openings. The position of the lintel above the opening is known as the 'head' of the door or window. This position is vulnerable to moisture penetration and must be protected by a damp proof course tray, which is installed above the lintel.

Study Figure 5.84 and note how any moisture travelling down the inside of the outer skin will be directed out of the structure (through openings called 'weep holes') and will be prevented from penetrating the living or working area of the building.

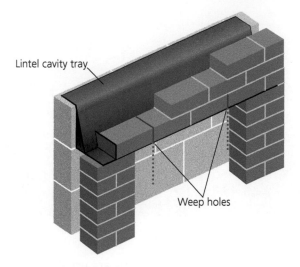

▲ Figure 5.84 The section above a window showing a cavity tray and weep holes

The door or window opening design may include a sill (or cill) detail which must be installed by the bricklayer. These may be manufactured in concrete, stone or slate or can be constructed in brick.

When installing a sill, keep in mind the purpose of this detail; it functions as a means of directing water away from the face of the cavity wall to give protection for the masonry. The bricklayer must make sure that the **fall** that forms part of the design can work properly.

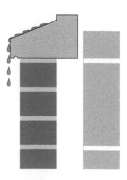

▲ Figure 5.85 Sill showing water moving off the top surface and being thrown clear of the cavity wall

Forming junctions in cavity walls

Depending on the design, a party wall between independent sections of a structure may be constructed as a cavity wall. The point where the party wall meets with the external wall forms a junction which must be correctly bonded to form a strong and stable part of the structure.

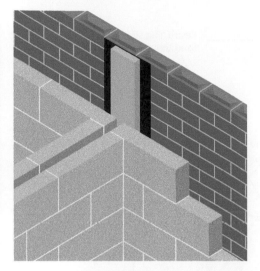

▲ Figure 5.86 Party wall adjoining outside cavity wall

The cavity is maintained from the external wall into the party wall unless the design requires the cavity to be closed, for example, as a fire break. Study the plan views shown below to see how this works in practice.

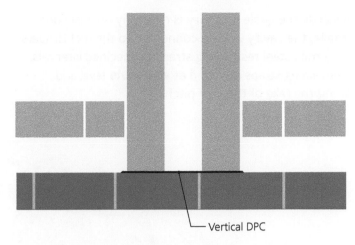

Vertical DPC

▲ Figure 5.88 Plan view of junction with cavity closed

Note the position of the vertical DPC where the cavity has been closed. Great care should be taken to make sure that no mortar squeezed from the bed joints can form a bridge for moisture to travel into the party wall.

It may be necessary to install additional wall ties at the junction to maintain the strength and stiffness of the cavity wall structure. Sometimes a movement joint may be included in the external skin of masonry at the junction location which will require additional wall ties to be installed.

Gables in cavity walls

When constructing gables, the bricklayer should keep these important points in mind:

- Avoid building one skin too high without the support of the other skin; the cavity work can be unstable and prone to collapse in windy conditions, even when the mortar has hardened.
- A lot of cutting will be required to produce the rake angle to match the pitch of the roof; make provision to remove debris safely from the scaffold.
- Keep the perp joints in the gable face brickwork plumb with the perp joints in the face work below. If the perp joints are allowed to 'wander' it will seriously affect the appearance of the finished job.
- Only build small racks at each end of the gable; fix your line and pins to these for building the infill.

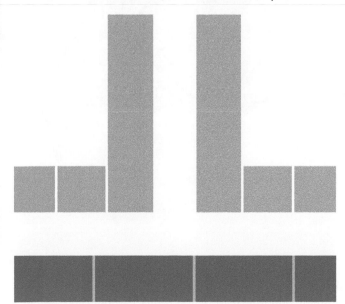

▲ Figure 5.87 Plan view of junction with cavity maintained

Because the gable masonry is relatively unstable and fragile, the cavity work is connected to the roof timbers with mild steel restraining straps at specified intervals. Restraining straps are fixed at **wall plate** level and along the rake of the roof pitch.

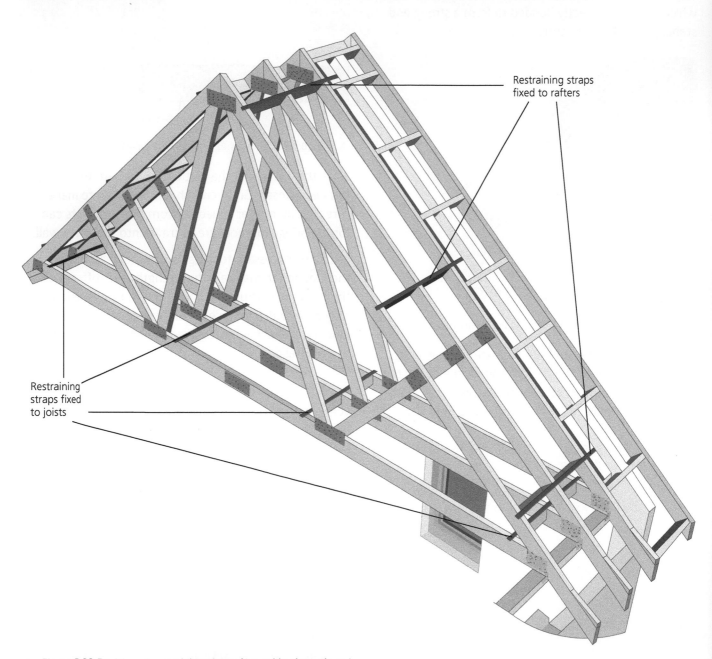

Restraining straps fixed to rafters

Restraining straps fixed to joists

▲ Figure 5.89 Position of restraining straps in a gable shown in red

Timber frame cavity walls

The timber frame method of construction is not new. Traditional buildings using timber framing were made mostly from oak, with various in-fills such as brick or plaster to form the walls. These have proved to be durable and long lasting and many examples are still in use after hundreds of years.

▲ Figure 5.90 Elizabethan oak frame

Modern timber frame homes are generally built from treated softwood, constructed as panels in factories and delivered to site ready to assemble on a prepared floor slab. The design may specify brick or block for the outer skin forming a cavity wall. The cavity prevents moisture entering the structure and must be kept free of mortar droppings. An additional feature is the provision of a vapour barrier within the timber frame panels to enhance moisture resistance.

The brick or block outer skin is secured to the timber frame core by stainless steel frame ties which are anchored using screws or annular ring nails. This gives a secure fixing which is less affected by shrinkage or movement in the timber panels.

Thin-joint masonry cavity walls

Traditional methods of constructing cavity walls in masonry have been supplemented by a relatively new system of building known as 'thin-joint' masonry. This method has several advantages that can make cavity wall construction more efficient. Some features of thin-joint masonry are:

- improved speed and accuracy of work
- improved stability during construction
- improved thermal performance
- reduction of waste.

The thin-joint arrangement refers to the construction of the inner blockwork skin using an adhesive instead of mortar.

The mix is produced from powder using a powered whisk. The mixed adhesive is spread using specialist tools called a scoop and hopper. No trowel is used to form either the bed joint or the perp joint, which is spread to a thickness of 2–3 mm.

▲ Figure 5.92 Whisk

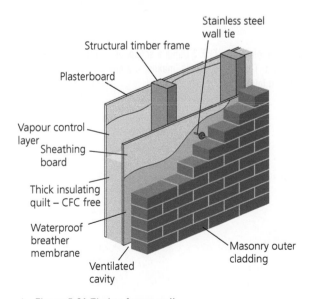

Stainless steel wall tie

Structural timber frame

Plasterboard

Vapour control layer

Sheathing board

Thick insulating quilt – CFC free

Waterproof breather membrane

Ventilated cavity

Masonry outer cladding

▲ Figure 5.91 Timber frame wall

▲ Figure 5.93 Scoop

Since the outer skin or leaf is laid in sand/cement mortar with 10 mm bed joints, and the inner skin is constructed with 2–3 mm bed joints, wall ties cannot be installed in the usual way. The inner skin must be built first and allowed to harden. Then special helical wall ties are driven into the thin-joint blockwork to correspond with the gauge of the outer skin of masonry.

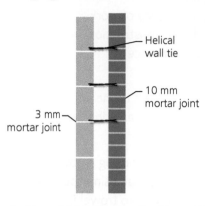

Helical wall tie

10 mm mortar joint

3 mm mortar joint

▲ Figure 5.94 Helical ties fixed in position

Any trimming or adjusting of blocks to maintain accuracy is carried out using a rasp.

▲ Figure 5.95 Rasp

Thin-joint blockwork

Look at the step by step sequence below to see the methods of work used for thin-joint blockwork.

STEP 1 Using the adhesive from the bag, mix the powder in a tub using a hand-held whisk to get the right consistency.

STEP 2 When laying the first course of thin joint blockwork, use traditional sand and cement. This sets a level base on which to start the thin joint system.

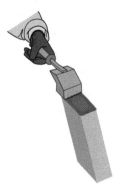

STEP 3 Using a scoop, apply the adhesive to the perp ends, making it 2 mm thick.

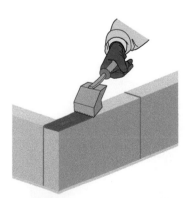

STEP 4 Again using a scoop, apply the adhesive to the bed joint to a thickness of 2 mm. Keep in mind that the adhesive sets more quickly than sand and cement.

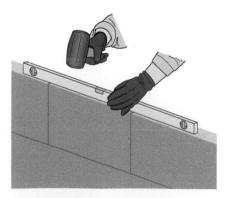

STEP 5 Check your work using a spirit level. Use a rubber hammer to gently knock the blocks into place. Remember to use the hammer on the blocks and not on the level.

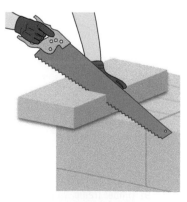

STEP 6 Use a masonry saw to cut any blocks that need to have an accurate cut edge.

STEP 7 After cutting the block, use the rasp to smooth the cut edge before applying the perp joint.

STEP 8 Use traditional blocklaying methods to finish the job. All joints must be full.

Forming joint finishes

Finally, we need to consider the finishing of the wall. Forming the joint finish in a cavity wall requires concentration and patience. With practice, you will become faster and more efficient at producing a finish that's attractive and durable. The effort put forth and the level of skill you employ will determine the quality of the finished job.

The bed and perp joints need to be **tooled or ironed** to satisfy two requirements:

- To produce the desired appearance that is specified.
- To weatherproof the joint to prevent the entry of moisture.

This process, also called 'jointing', forms a finish to the mortar joints as the work proceeds. This is different to 'pointing', which refers to the process of raking out the mortar joints while they're still soft and then re-filling the joints later after the mortar has hardened. This allows for a coloured mortar to be used or for a specialised joint to be used which may be too time consuming to produce at the time of building.

The most commonly used joint is called a half-round joint and uses a specific tool called a jointer to produce a concave finish to the mortar just before it begins to harden. The timing of the operation is very important since if it's performed too soon it will lead to a rough finish and if it's performed too late a black deposit will sometimes form due to what is known as 'lime burn'.

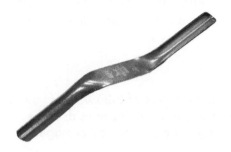

▲ Figure 5.96 Jointer

Because the profile of the half-round joint is concave, the procedure of jointing pushes the mortar tight against the arris of the brick and seals it against moisture penetration.

It is good practice to joint all the perps first, followed by the bed joints. This reduces the number of small projections where the perp joint and the bed joint intersect.

▲ Figure 5.97 'Mouse ears' or 'curtains' on brickwork jointing. These are the small projections which occur when the horizontal and vertical joints intersect

An alternative joint is known as a 'recessed' or 'raked-out' joint. Look at Table 5.5 to see the appearance and description of these two joints. Other joint finishes are detailed on page 189.

Remember, not only does jointing provide a finish to the face of a cavity wall, but it also serves a vital function in the weatherproofing of the brickwork. The bed and perp joints must be fully filled and compressed during the jointing process.

Carelessness in completing the work properly can result in gaps and voids being left which will make the cavity wall vulnerable to the effects of poor weather. When moisture enters the poorly finished mortar joint, it may stay there for some time and may freeze during cold weather.

ACTIVITY

Assess the jointing in examples of brickwork as you travel around. Do you see any holes or voids in the work? Do you see any examples of frost damage in the work?

Since water expands as it freezes, it can potentially destroy the bed joint at that point and will make it likely that further deterioration will take place over time. Not only the mortar joint, but the adjacent bricks can be affected, leading to sections of the face of the brick detaching. This is known as **'spalling'**.

▲ Figure 5.98 Poorly jointed work at the corners of a brick quoin

KEY TERM

Spalling: when the face of a brick or block crumbles away

▼ Table 5.5 Joint finishes

Type of joint	Description
Half-round joint	Often referred to as 'tooled', this is probably the most commonly used joint. It is a joint finish that can be produced quite quickly and has the advantage of disguising irregularities in the arris of the brick. Since the action of ironing or tooling the joint presses the mortar against the arris of each brick, the possibility of gaps being left in the joint is reduced and the masonry has greater resistance to the effects of poor weather.
Recessed joint	Also produced as the work progresses. This type of joint is formed by using a purpose-made tool to remove the mortar from the joints to a specified depth. The tool can be a simple timber block cut to the right shape on site, or it can be a wheeled tool often referred to as a chariot. To form a satisfactory recessed joint, the bricklayer must make sure that all joints are completely full as laying proceeds. This joint should only be used with hard bricks that are frost resistant since moisture can stand in the recess on the horizontal arris of the brick.

The care taken in construction design must be carried through to the building stage by the bricklayer. By consistently employing good trade practices and developing your skills, the cavity wall you build will fulfil its intended purposes of:

- maintaining structural stability
- resisting wind loadings
- preventing moisture penetration into the interior of the structure
- insulating against heat loss.

Summary

Cavity walls in masonry are a major element of many masonry structures that are used as dwellings or workplaces. This chapter has covered why cavity walls are used, the components and materials that are used to construct them and the characteristics of those components and materials.

We've also discussed the importance of the bricklayer working to high standards to produce cavity walls that are durable, long lasting and stable as a structural element. Recall the information we've discussed in this chapter whenever you start work on construction of a cavity wall. It takes skill, diligence and thoroughness to produce work that will stand the test of time.

Test your knowledge

1 What is the process called when bricks are placed in a kiln to harden them?

   a Firing

   b Hydrating

   c Purging

   d Soaking

2 What type of strength do engineering bricks have that makes them suitable for use in a railway bridge?

   a Competitive

   b Compressive

   c Tensional

   d Torsional

3 What method is used to accurately measure proportions of materials for mortar?

   a Gauging

   b Grouping

   c Setting

   d Sorting

4 What additive is used to make a workable mortar mix?

   a Accelerator

   b Plasticiser

   c Retarder

   d Smoother

5 In what conditions would Sulphate Resisting Portland Cement (SRPC) be used?

   a Damp

   b Dry

   c Hot

   d Sealed

6 What term describes how chemical plasticisers affect air to make a mortar mix more workable?

   a Elevation

   b Elongation

   c Entrainment

   d Entrapment

7 What term describes moisture moving across a cavity wall above finished floor level (FFL)?

   a Connecting damp

   b Dripping damp

   c Penetrating damp

   d Travelling damp

8 What term is used to describe the introduction of cut bricks or blocks into a course to create a suitable bonding arrangement?

   a Broken bond

   b Cutting bond

   c Reverse bond

   d Secondary bond

9 What feature is included in the design of a wall tie to prevent moisture from travelling across the cavity?

   a Drip

   b Drop

   c Step

   d Strip

10 What name is given to the method of providing a joint finish as work proceeds?

   a Flashing

   b Jointing

   c Pointing

   d Sealing

11 State one disadvantage to the bricklayer of using mineral fibre full fill insulation in cavity walls.

12 State two things that forming a joint finish in masonry achieves.

13 Explain why a steel combination lintel is a good choice for use in bridging openings in cavity walls.

14 State the reason for installing steel reinforcement in concrete lintels.

15 Explain how a vertical movement joint can be strengthened to resist lateral forces.

CHAPTER 6

BUILDING SOLID WALLS, ISOLATED AND ATTACHED PIERS

INTRODUCTION

The term 'solid walls' applies to masonry that has no cavity in the structural design and often refers to walls which are typically 215 mm (one brick) thick or greater.

However, a half-brick-thick single-leaf wall can also be referred to as a solid wall. A wall of this thickness is relatively weak on its own and is usually strengthened by adding piers, so we will also consider how to construct piers, both those that are bonded into the wall (attached) and those that stand on their own (isolated).

By the end of this chapter, you will have an understanding of:
- where solid walls can be used
- selecting the right materials and damp proof courses for solid walls and piers
- the masonry bonding arrangements used in solid walls and piers
- forming joint finishes and providing weather protection for solid walls and piers.

This table shows how the main headings in this chapter cover the learning outcomes for each qualification specification.

Chapter section	Level 2 Technical Certificate in Bricklaying (7905-20) Unit 203	Level 3 Advanced Technical Diploma in Bricklaying (7905-30)	Level 2 Diploma in Bricklaying (6705-23) Unit 204	Level 3 Diploma in Bricklaying (6705-33)	Level 2 Bricklayer Trailblazer Apprenticeship (9077) Module 5
1. Where solid walls can be used	Introduction	N/A	N/A	N/A	**Skills Depth:** N/A **Knowledge Depth:** N/A
2. Selecting the right materials and DPC for solid walls and piers	2.2, 2.3	N/A	3.3, 3.5, 3.8, 4.2, 4.3, 4.7, 5.4, 6.1, 6.3	N/A	**Skills Depth:** N/A **Knowledge Depth:** 3.2
3. The masonry bonding arrangements used in solid walls and piers	2.1, 2.2, 3.1, 3.2	N/A	3.6, 3.7, 4.4, 4.5, 5.5, 5.6, 5.7, 6.2, 6.6	N/A	**Skills Depth:** 3.1, 3.2, 4.1, 4.2, 4.3, 4.4 **Knowledge Depth:** 3.1, 3.3, 4.1
4. Forming joint finishes and providing weather protection for solid walls and piers	2.2, 4.1, 4.2	N/A	3.9, 3.10, 3.12, 3.13, 4.6, 4.10, 5.8, 5.10, 5.11, 6.7, 6.9	N/A	**Skills Depth:** 3.3, 5.1 **Knowledge Depth:** 3.2, 3.4, 5.1

Note: for 7905-20, Unit 202:
Content for Topics 1.1 and 1.3 is covered in Chapter 2.
Content for Topics 1.3 and 2.4 is covered in Chapter 5.
Note: for 6705-23, Unit 204:
Content for Learning outcomes 1 and 2 is covered in Chapter 2.
Content for Assessment criteria 3.1, 3.2, 3.4, 4.1 and 5.1 is covered in Chapter 4.

1 WHERE SOLID WALLS CAN BE USED

There are several applications and uses for **solid walls** that involve construction in brick, block or a combination of the two. Brick and block may also be combined with other materials and components when constructing solid walls for specific purposes. Let's look at some locations where solid walls could be used.

> **KEY TERM**
>
> **Solid walls:** masonry that has no cavity in the structural design

Free-standing walls

Solid walls will often be specified for use in free-standing walls such as boundary walls. **Stability** is important because a free-standing wall won't be attached to anything else and it must have enough built-in strength to stand independently.

> **KEY TERM**
>
> **Stability:** resistance to movement or pressure

Free-standing walls are typically built in one-brick thick bonds such as English or Flemish bond, which can resist high winds. Don't worry if you're not yet familiar with some of the names of bonds mentioned here, as we'll discuss them all in detail later in this chapter.

Since a thicker wall has more weight, it is not so easy to push over. Using bonds like these can also be a way of introducing decorative patterns in the face of the wall to make large areas of plain masonry more attractive.

> ## VALUES AND BEHAVIOURS
>
> If you are involved in the design and planning stage of a project, keep in mind that your understanding of brick bonds and masonry features will very likely be greater than that of the customer. This means you may be able to offer suggestions on introducing decorative and other features that the customer might not be aware are possible. However, recognise that the client or customer has the final say. Listen carefully to their requirements before making suggestions.

▲ Figure 6.1 Solid wall (Flemish bond)

Retaining walls

A thicker wall with greater weight will be capable of resisting the force of soil behind it when it's required to function as a **retaining wall**. Retaining walls are carefully designed to perform their job and when built in brick can use variations of English and Flemish bonds to increase the thickness if required.

> **KEY TERM**
>
> **Retaining wall:** a wall built to hold back a mass of earth or other material

These days, retaining walls built only in brick are unlikely to exceed a thickness of 440 mm (the length of two bricks). Often, retaining walls are designed using a combination of a skin of brick on the face of the wall, backed by concrete blocks laid flat to achieve the specified thickness.

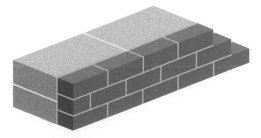

▲ Figure 6.2 Retaining wall in brick and block

To resist strong sideways (or 'lateral') pressures that may be applied to a wall constructed entirely in brick, vertical reinforcement could be included in the design to enable the structure to do its job. One way of installing vertical reinforcement in a wall is to build it using a variation of Flemish bond called Quetta bond. Study Figure 6.3 to see how this works.

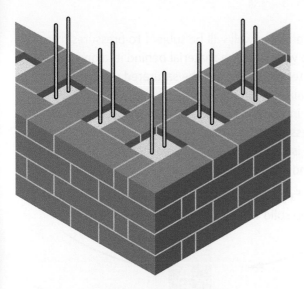

▲ Figure 6.3 Quetta bond wall with reinforcement bars

If a wall needs to be higher and thicker to retain more soil or other material, it is likely to be constructed using reinforced concrete. There are other methods of constructing retaining walls which may be quicker and can be more cost effective.

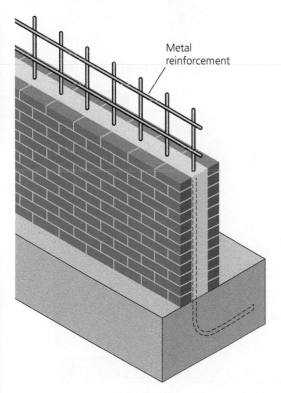

Metal reinforcement

▲ Figure 6.4 Reinforcement in a retaining wall

Inspection chambers

Inspection chambers can also be designed using solid walls to withstand the pressures of material surrounding them. They are often built in English bond since this has the greatest strength of any of the brick bonding arrangements.

KEY TERM

Inspection chamber: a masonry structure that allows inspection of services below ground

Large chambers are now more commonly constructed using pre-formed concrete rings or rectangles, which simplifies and speeds up construction. If a large inspection chamber must be constructed in brick, a variation of English bond can be used to thicken the wall and increase the strength and stability of the structure.

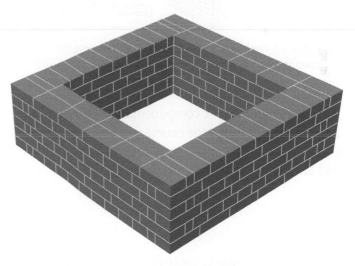

▲ Figure 6.5 Inspection chamber in English bond

2 SELECTING THE RIGHT MATERIALS AND DPC FOR SOLID WALLS AND PIERS

Before you can successfully build any wall or pier, you first need to gather a range of information. As detailed in Chapters 2 and 4, one of the main methods of communicating information for any building activity is by means of drawings. A working drawing makes it possible to provide a great deal of information without having to use lots of writing.

The working drawing may include a specification panel which gives information which can only be shown in writing, such as the type or name of the brick to be used. To provide more detailed information, a separate and more comprehensive specification will be used.

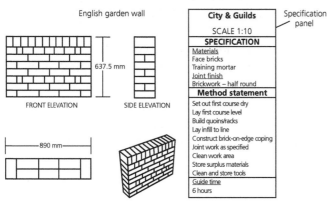

▲ Figure 6.6 Working drawing of a solid wall showing the specification panel

A schedule may also be used where appropriate, to give information about components or materials that are repetitive through the course of a project. For more information on drawings, specifications and schedules, and also bills of quantity, see Chapter 2.

Solid wall and pier design considerations

Whether a solid wall is constructed as a free-standing wall, a retaining wall or an inspection chamber, the materials will need to be resistant to ground pressures and moisture penetration. Consider how these key factors apply to each type of solid wall construction:

- A free-standing wall may be subject to pressures from high winds and driving rain on both faces.

- A retaining wall will be subject to pressures from the weight of the material behind it, along with water pressure from the saturated ground that it retains.
- An inspection chamber will be subject to similar pressures to a retaining wall.
- An isolated pier will be subject to the effects of the wind and weather from all sides.

These factors also apply where piers are incorporated into the solid wall design (attached piers).

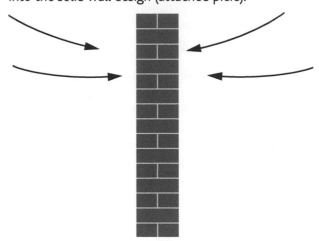

▲ Figure 6.7 A free-standing wall is exposed on both faces

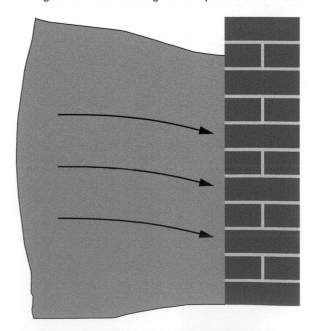

▲ Figure 6.8 A retaining wall must resist sideways (or lateral) forces and pressure from the material held behind the wall

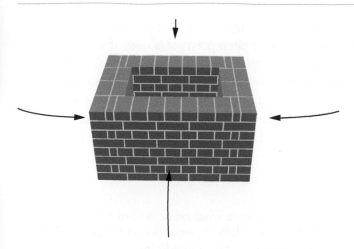

▲ Figure 6.9 An inspection chamber must resist pressures from all sides

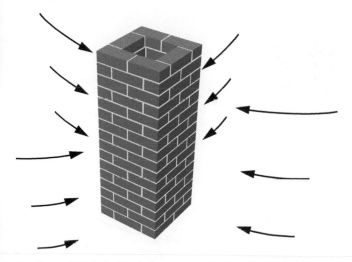

▲ Figure 6.10 An isolated pier must resist the effects of wind and weather from all sides

This means that the materials selected must be suitable to withstand the substantial and persistent pressures the solid wall or pier will be subject to throughout its working life.

Suitable bricks and blocks

In Chapter 5, we looked at the characteristics of several different types of bricks. Remember that an engineering brick is described as having a high compressive strength, which means that it can resist squeezing forces and pressures. It also has low water absorption because of the type of clay and the firing temperature used in manufacture.

Engineering bricks would therefore be highly suitable for use in constructing solid walls. However, there may be cost factors to consider as engineering bricks can be expensive. It may be necessary to select another type

of clay brick that has acceptably low water absorption and high compressive strength and will be more **economical** to use.

> **KEY TERM**
>
> **Economical:** using the minimum of resources necessary for cost effectiveness

> **ACTIVITY**
>
> Find a brick manufacturer's website and use the product selector to check out some of the bricks they produce. Select three different bricks and find the technical data section for each. Fill in the table to show the compressive strength and the water absorption figures for your chosen bricks (an example is given for you). Place a tick against the brick type that you think is most suitable for use in a solid wall.
>
> | Brick type | Compressive strength (N/mm²) | Water absorption (% weight) | Most suitable for solid walls? |
> |---|---|---|---|
> | Ibstock Aldridge Multi Rustic | 35 | 12 | |
> | | | | |
> | | | | |
> | | | | |

If bricks or blocks manufactured in concrete are specified for a solid wall or pier, they will need to satisfy the same requirements as suitable clay bricks. They must have good compressive strength and not deteriorate due to moisture penetration.

> **KEY POINT**
>
> Dense concrete components are graded according to compressive strength in Newtons per square millimetre (N/mm²). The higher the number, the better the compressive strength.

Just as solid walls constructed in brick can be reinforced with steel to increase their strength and resistance to lateral (or sideways) forces, blocks designed with hollow centres can be used to allow steel reinforcement to be installed vertically. The hollows are then filled with concrete.

▲ Figure 6.11 Hollow concrete blocks reinforced with steel and concrete

HEALTH AND SAFETY

The steel reinforcement bars will be installed in the foundation concrete before the bricklayer starts work. The bricklayer will need to 'thread' the hollow blocks over the steel reinforcement to build the wall. Care must be taken when working near cut ends of the steel bars as they can be extremely sharp!

DPC for solid walls and piers

A damp proof course is rarely specified for a pier, free-standing boundary wall or retaining wall. However, the installation of a DPC might be needed if excessive moisture is present and could cause continuous saturation of the masonry from ground level up. A permanently wet wall or pier will be more likely to suffer frost damage when conditions become cold enough. Since water held in the masonry will expand when it freezes, the structure can literally be pushed apart. (For more details on frost damage, see page 198.)

IMPROVE YOUR ENGLISH

Find some images of frost damaged brickwork online. Write a short description of what it looks like and how frost damages the face of saturated brickwork during freezing weather.

Suitable DPC

If a DPC is specified for a free-standing boundary wall or a pier, there is an important structural consideration

to take into account. If one of the flexible DPCs such as polythene or pitch polymer is used, it will create a weak point in the bed joint where the DPC is installed, since mortar will not bond or adhere to any of these flexible materials. Something like a very strong wind putting sustained lateral force on the structure could cause it to fail at this weak point and be pushed over.

ACTIVITY

Discuss with others what other occurrences could put sideways or lateral pressures on a free-standing boundary wall or an isolated pier.

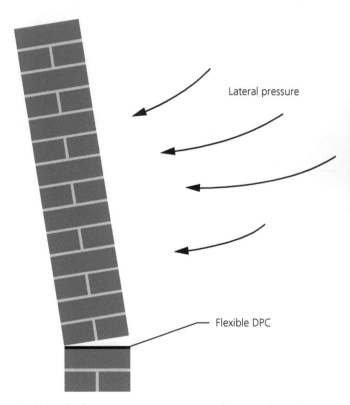

Lateral pressure

Flexible DPC

▲ Figure 6.12 How sideways pressures or forces could fracture the wall at the DPC

However, rigid DPC materials will bond to mortar and, in the case of engineering brick, the DPC can form part of the masonry bonding arrangement to maintain the face bond appearance.

Two courses of engineering brick or two (or more) courses of slate half-bonded will form an effective DPC. This will serve to maintain the strength of the wall or pier while at the same time providing a barrier to rising moisture.

▲ Figure 6.13 Two courses of engineering brick used as a DPC prevent rising damp caused by capillary attraction

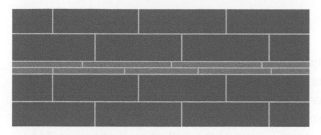

▲ Figure 6.14 Two courses of bonded slate forming a rigid DPC

Movement joints in solid walls

In Chapter 5, we discussed the requirement to introduce movement joints into brick and block masonry to avoid cracking caused by thermal movement or settlement.

Solid walls (such as free-standing boundary walls) which extend over long lengths will also require the construction of movement joints. Review the details on page 134 about how movement joints are correctly tied together to maintain stability while allowing enough movement to protect the wall from damage.

③ THE MASONRY BONDING ARRANGEMENTS USED IN SOLID WALLS AND PIERS

Solid walls can be constructed in a range of bonding patterns, usually one-brick thick or more. If the design of the wall requires it to be constructed as a half-brick thick wall, then the bond will usually be Stretcher bond (sometimes referred to as 'half-bond').

For Stretcher bond, we are required to lap each course of bricks by the dimension of the header face of a brick (102.5 mm). This will result in the perpendicular joints (perps) in any course being lined up with the centre of the stretcher face of the bricks in the courses above and below them.

Similar principles apply to a block wall that is half-bonded. The perps in any course will be lined up with the centre of the face of the blocks in the courses above and below them.

> **KEY POINT**
> Remember, a solid wall in half-brick-thick masonry will rarely be constructed as a free-standing wall. It will require attached piers to be included in the design to be stable and durable.

One-brick-thick bonding arrangements

For a solid wall constructed in one-brick-thick masonry, there are four main bonds that can be used:
- English bond
- Flemish bond
- English garden wall bond
- Flemish garden wall bond.

These are all quarter bond arrangements. In other words, the bricks overlap by ¼ of a brick length.

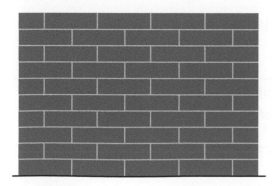

▲ Figure 6.15 Stretcher (or half) bond

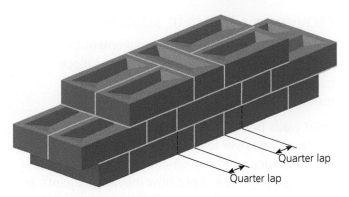

▲ Figure 6.16 Quarter bond overlap

Keep in mind that when setting out quarter bond, Broken bond should be avoided if possible. Reverse bond is more acceptable and is easier to achieve. (To review details about broken and reverse bond, see page 152.)

Let's focus on each of these bonds at a time. In each case, study the illustrations carefully to familiarise yourself with each bonding arrangement.

> # VALUES AND BEHAVIOURS
>
> One-brick-thick walls used to be referred to as '9-inch work'. This goes back to the time when imperial measurements were used, before metric measurements were introduced. You can benefit from the experience of others in the industry – if you get the chance, ask your trainer or supervisor how things have changed in construction over the years. Remember, you never stop learning!

English bond

This is the strongest of all brick bonds. It is arranged as **alternate** courses of stretchers and headers. Since each course of headers is laid across the width of the wall there are no straight joints in the bonding arrangement which could create weak points.

> ### KEY TERM
>
> **Alternate:** interchanging repeatedly

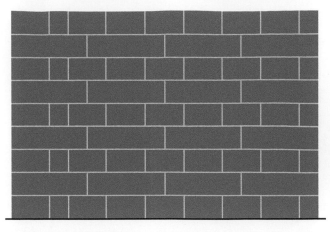

▲ Figure 6.17 English bond

The practice of 'tipping and tailing' when forming the perp or cross joint was mentioned in Chapter 5. If the perp joints for the header bricks in English bond were produced in this manner, a large void would be formed across the width of the wall where moisture could be retained, which could lead to frost damage.

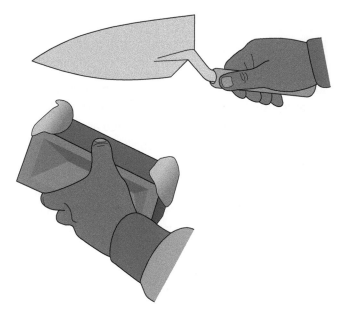

▲ Figure 6.18 Header being badly perped using 'tip and tail' method

> ### KEY POINT
>
> In each of the quarter bond arrangements it is important to ensure that all mortar joints are full.

Flemish bond

Flemish bond is arranged as alternating headers and stretchers in each course. This is considered to be a more decorative bonding arrangement. In each course the centre of the header faces line up with the centre of the stretcher faces in the courses above and below, to give a strong quarter bond and also to produce an interesting pattern.

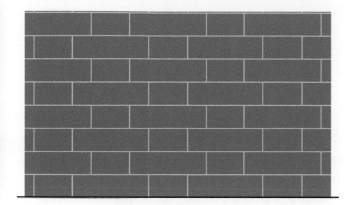

▲ Figure 6.19 Flemish bond

Especially in Flemish bond, care must be taken to make sure that the quarter bond overlap is accurately maintained, otherwise the perps will be out of plumb with each other which will spoil the appearance. Look at the header bricks in Figure 6.20 and note how the uniform appearance of the wall is affected because they are not accurately lined up vertically.

▲ Figure 6.20 Poorly bonded quarter bond brickwork

Flemish bond is not as strong as English bond, so it would not be used to construct an inspection chamber.

English garden wall bond

This bonding arrangement consists of three, five or seven courses of stretchers to one course of headers.

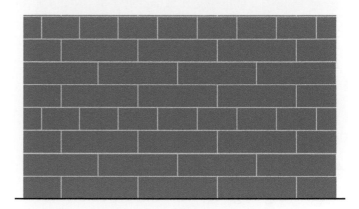

▲ Figure 6.21 English garden wall bond

Although this is not as strong as English bond, it is suitable for garden walls (hence the name) and gives an acceptable appearance on both sides of the wall. Bricks are manufactured to standard dimensions within allowable tolerances. This means that the length of a brick may vary slightly. That being the case, if you lay a header brick to a string line on the face of the wall, while the header face of the brick will be accurately positioned, the back end of the brick may vary in the amount it projects compared to other headers in the wall.

Since English garden wall bond has fewer headers in the bonding arrangement, your wall is more likely to have a better appearance on both sides. The back of this type of wall is often referred to as 'fair-faced'.

Flemish garden wall bond

In Flemish garden wall bond there is a pattern of three or five stretchers followed by one header repeated along the length of each course. The header face in a course should be centred above the middle stretcher of each group of three (or five) stretchers in the course below.

This can be confusing to visualise so take time to study the illustration in Figure 6.22. The header bricks are shown in a darker colour to make them easier to identify.

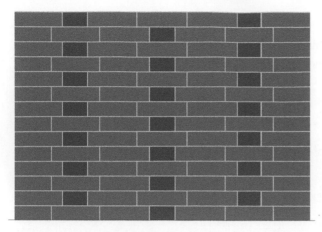

▲ Figure 6.22 Flemish garden wall bond

Just as in English garden wall bond, there are fewer headers in this bonding arrangement, so the rear of the wall is more likely to have an acceptable appearance.

In all of these one-brick-thick walls, it can be a challenge to make sure the rear arris of the bricks is aligned accurately to produce a uniform appearance. To produce high quality work, some bricklayers will decide to use two lines: one on the face and the other on the fair-face (or back line of the wall).

▲ Figure 6.23 Two string lines in use: one on the face side and one on the rear

The face string line is used to line the front arris, but the rear or fair-face line can only be used to establish level and not line. This is because, as already mentioned, the header bricks will vary slightly in length and so cannot be laid accurately to both lines. The usual practice is to lay the header with its rear arris very slightly below the line so as not to push the rear line out of alignment with the run of the wall.

Solid walls in block

When constructing solid walls in block, the selected material will usually be dense concrete block. Thermal insulation blocks are unlikely to be specified as they are lightweight and not suitable for exposure to bad weather.

To produce a one-brick-thick (215 mm) solid wall, the common practice is to lay the blocks 'on flat'. This can speed up production without affecting the strength of the finished job. Look at Figures 6.24 and 6.25 to see two alternative methods of producing half bond.

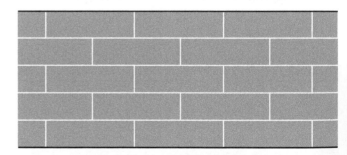

▲ Figure 6.24 Laying blocks 'on flat' method 1

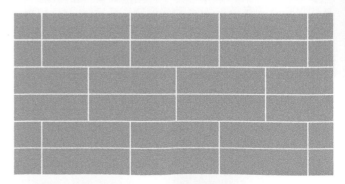

▲ Figure 6.25 Laying blocks 'on flat' method 2

Bonding at returns and junctions

Each of the brick quarter bonds we've discussed requires a **queen closer** to be introduced into the bonding arrangement of each course.

The width of a queen closer is not simply half the width of a header face. It should be cut to a width of 46 mm to create the correct overlap.

Take some time to study these illustrations of English and Flemish bond solid walls (Figures 6.27 and 6.28) and note where the queen closer is placed on each course. This is the arrangement where there is no return or quoin.

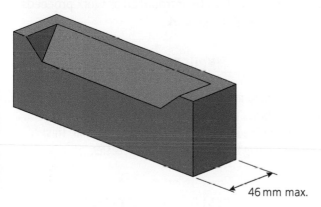

▲ Figure 6.26 Queen closer

▲ Figure 6.27 English bond layout

▲ Figure 6.28 Flemish bond layout

Bonding brick quoins for solid walls

When constructing quoins and returns in any of the quarter bonds already discussed, the queen closer will be placed next to a quoin header. Look at the illustrations of English and Flemish bond quoins (Figures 6.29 and 6.30) and note the positions of the queen closers.

▲ Figure 6.29 English bond quoin

▲ Figure 6.30 Flemish bond quoin

IMPROVE YOUR MATHS

Look carefully at the illustrations of English and Flemish bond solid walls (Figures 6.27 and 6.28). Can you explain why we don't simply split a brick exactly in half along its length to produce a queen closer? (Note: 46 mm is a rounded figure.)

Use the modular dimensions of a brick to calculate the *exact* dimension of a queen closer. Check your results with another student.

Bonding junctions for solid walls

Things get a little more complex when we need to form a junction in one-brick-thick solid walls. The queen closer is now used in conjunction with a 'tie brick'. Study the examples shown in Figures 6.31 and 6.32 to understand the bonding arrangements at junctions and note the position of the queen closer and tie brick in each case.

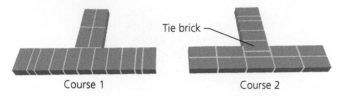

Tie brick

Course 1   Course 2

▲ Figure 6.31 English bond junction (Course 1 and 2)

Tie brick

Course 1   Course 2

▲ Figure 6.32 Flemish bond junction (Course 1 and 2)

A simpler way to form a junction may sometimes be acceptable to speed up productivity and reduce waste generated by cutting bricks. This involves using a special mesh known as expanded metal lathing (EML) which is usually treated to resist corrosion by galvanising (applying a coating of zinc during manufacture). This type of reinforcement is available in a range of widths to suit various thicknesses of solid wall.

VALUES AND BEHAVIOURS

Reducing waste is an important consideration for all construction activities. Always look for ways to be more efficient in your use of materials, so that waste is reduced. Remember that a lot of energy can be used in the manufacture of construction materials and components, resulting in emissions of carbon which can cause damage to the environment.

▲ Figure 6.33 English bond junction with expanded metal lathing

Junctions may also be required where one wall is constructed in brick and the adjoining wall is constructed in block (sometimes called an 'intersection'). In this case the bonding arrangement will involve forming pockets or 'indents' in one wall to allow the other wall to be introduced as work proceeds or at a later stage of the work.

INDUSTRY TIP

If indents are formed for the adjoining wall to be added later, it is always better to make the width of the joints either side of the indent slightly larger than 10 mm. This will make adding the junction wall much easier.

Building isolated piers

Isolated piers are built separate from other masonry structures and are often used in situations such as gated entrances on a drive or pathway. If the pier is built in one-and-a-half- or two-brick thickness (or more), it has the advantage of allowing reinforcement (such as steel rods or bars surrounded by concrete) to be introduced within the hollow centre. The pier can then support the weight of a heavy gate.

▲ Figure 6.34 Isolated piers can bear the weight of a heavy gate when reinforced

Purpose-made hinge support brackets for the gate can be built into the bed joints of the pier as work progresses. The bricklayer will need to take care to install the brackets perfectly plumb, one above the other, so that the gate will operate satisfactorily. The brackets can be spaced to suit the hinge positions on the gate, to allow for later installation.

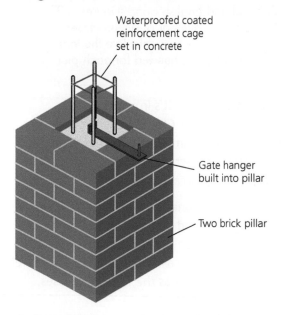

Waterproofed coated reinforcement cage set in concrete

Gate hanger built into pillar

Two brick pillar

▲ Figure 6.35 Gate mounting lugs or brackets

Other uses of isolated piers could be to support an arch spanning a pathway or to support fence panels along a boundary line. If several isolated piers are constructed along a boundary, care is needed to make sure that all the piers are accurately aligned with each other. One method would be to set out and build the first and last piers in the row. Then align the rest of the piers using a string line stretching between the first and last piers as a guide.

Bonding isolated piers in brick

A one-brick isolated pier is difficult to build since the bonding arrangement for each course is simply two stretchers side-by-side. Remember that bricks vary in length, so a brick pier set out this way would be hard to plumb accurately on all sides.

In a two-brick square pier, if we were to use Stretcher bond (half-bond), a course in the front elevation of the pier would show two stretchers side by side and the side or end elevations would show two headers with a stretcher in between.

▲ Figure 6.36 A two-brick Stretcher bond pier

Using the quarter bond arrangements of Flemish or English bond would make it necessary to perform a great deal of cutting in each course to bond an isolated pier correctly.

INDUSTRY TIP

When setting out isolated piers in one of the quarter bonds, it is a good idea to set out the bond 'dry' to confirm the correct position of cut bricks in each course. Straight joints must be avoided to make sure that the pier will have enough strength and durability.

Study Figures 6.37 and 6.38, which show two-brick square isolated piers in Flemish bond and English bond, to see the number of cut bricks required and how they're positioned.

▲ Figure 6.38 Brick pier in English bond

ACTIVITY

After you've studied the illustrations of brick piers in Flemish bond and English bond (Figures 6.37 and 6.38), count the number of queen closers required in each course and calculate the total number of queen closers needed for each pier.

When construction begins, it's important to pay attention to accuracy. Check the diagonal measurements are the same to confirm the pier is 'square'. In a square or rectangular pier or pillar there will be eight plumbing points to control.

Some bricklayers select one corner as a 'control' point, which is carefully plumbed on both faces of the corner. The other corners are then set by measurement from the control corner with occasional use of the spirit level to check vertical alignment as the work progresses.

▲ Figure 6.37 Brick pier in Flemish bond

Working from the control corner

STEP 1 Plumb up both sides of the corner chosen to 'control' the pier.

STEP 2 Level and measure from the 'control' corner in one direction.

STEP 3 Then level and measure in the other direction from the same point.

Whatever method is preferred, great care in plumbing, levelling and gauging is necessary to produce a job with an acceptable appearance that conforms to the specification.

> ### ACTIVITY
>
> Try setting out the bond for some piers 'dry' in a suitable location. See if you can work out the bond for a one-and-a-half-brick square pier in English bond.

Bonding isolated piers in block

Isolated piers can be constructed in block in a similar way to piers in brick. Obviously, a block pier will not have the variety of bonding arrangements that a brick pier can have. Depending on the dimensions of the pier, the blocks can be laid 'on flat' to produce a solid pier or in the usual bonded pattern to produce a hollow pier which will allow for the installation of reinforcement if required.

Building attached piers

Unlike isolated piers, **attached** piers are not free-standing and are used in a structure or wall to add strength or reinforcement.

▲ Figure 6.39 An attached pier

> ### KEY TERM
>
> **Attached:** in this context, bonded into the main section of masonry following the specified bonding arrangement

Some typical situations where attached piers could be used are described below.

- A half-brick-thick garage wall – the attached piers will be incorporated into the wall at intervals of no more than three metres and will usually be included at the end of the wall to provide a solid fixing point for a door. (The dimensions of a pier at this location may be increased to take the weight of a door.)
- Low-level boundary walls which are built in thinner masonry – adding attached piers can make this type of wall more economical to build.
- A boundary wall which needs reinforcement because of its height – even walls built in a one-brick-thick bond may need additional support in the form of attached piers, if built above 1.8 m in height.

When bonding attached piers into face brickwork, the bonding arrangement must be set out to maintain the specified bonded appearance of the face of the wall. Study the examples of one bonding arrangement (Figure 6.40) to see how this works – this shows an intermediate pier.

Plan of course 2

Plan of course 1

▲ Figure 6.40 Bonding arrangement for an attached pier in Flemish bond

Notice the use of half bats and **king closers** to bond the attached pier and maintain the uniform appearance of the face bond.

KEY TERM

King closer: a brick with one corner cut away, making the header at that end half the width of the brick

INDUSTRY TIP

When setting out an attached pier, the bonding pattern should be arranged from the position of the 'tie' brick. In the example shown in Figure 6.40, the tie brick in course one is a king closer and the tie brick in course two is a header.

An attached pier at the end of a solid wall will have a bonding arrangement dependant on the thickness of the wall. A two-brick attached pier constructed at the end of a half-brick-thick wall in Stretcher bond will be set out differently to a similar attached pier built at the end of a one-brick-thick wall built in Flemish bond.

There are many different bonding arrangements possible. Study the illustrations in Figures 6.41 and 6.42 which show a half-brick and one-brick-thick wall with attached piers bonded at the end of each wall.

▲ Figure 6.41 An attached pier at the end of a half-brick-thick wall

▲ Figure 6.42 An attached pier at the end of a one-brick-thick wall

ACTIVITY

Produce some ruler-assisted plan drawings to show the first two courses of bonding arrangements for an attached pier at the end of a wall. Try bonding a two-brick pier to a one-brick-thick wall in English bond. Think of some other examples and draw them.

4 FORMING JOINT FINISHES AND PROVIDING WEATHER PROTECTION FOR SOLID WALLS AND PIERS

In Chapter 5, we discussed how 'jointing' describes the process of producing a finish to the bed and perp joints as the work proceeds. 'Pointing' is a different process which involves raking out the mortar while it is still soft and re-filling the joints later.

Joint finishes

Whatever type of solid wall or pier you may build, if it is classified as 'face work' (or 'best brick', as it's traditionally referred to), you will need to produce a joint finish. There are a number of factors that will affect the choice of joint for your wall:

- the **durability** of the chosen joint
- the time required to produce a satisfactory result
- the appearance required as a design feature.

KEY TERM

Durability: ability to withstand wear and tear from elements that cause decay

Two methods of producing joints were considered in Chapter 5: half-round (sometimes referred to as 'bucket handle') and recessed. Table 6.1 shows some other methods that could be used to provide a joint finish to the mortar joints in solid walls and piers.

Remember, not only does jointing provide a finish to the face of a masonry wall or pier, but it also serves a vital function in the weatherproofing of the brickwork. The bed and perp joints must be fully filled and compressed during the jointing process.

Carelessness in completing the work properly can result in gaps and voids being left, which will make the solid wall or pier vulnerable to the effects of poor weather. When moisture enters the poorly finished mortar joint, it may stay there for some time and eventually cause the masonry to deteriorate. Besides making sure that your joint finish is weatherproof, you need to know about other requirements for providing weather protection for solid walls and piers. Let's look at some details.

▼ Table 6.1 Methods of producing joint finishes

Type of joint	Method
Weather struck joint	Formed by shaping the joint with a trowel (preferably a pointing trowel) so that it slopes to allow rainwater to run off it. The top of the joint is set back from the face slightly and the bottom of the joint remains flush with the brick below. The perps are also angled, with the left side of the joint sloped in slightly. As the name suggests, this joint is used where brickwork is in exposed situations and gives the masonry good protection from the effects of poor weather.
Flush joint	Often used where more character is required. It is produced by smoothing and compacting the mortar with a hardwood timber block as the work progresses. The disadvantage is that it is difficult to achieve a weather-tight finish without giving the joint the appearance of being wider than it actually is.

Weather protection

A free-standing solid wall or pier will need protection from rain penetration through the top of the masonry. When rain protection is installed on a solid wall or pier, this is often referred to as 'weathering' it. Protection from moisture penetration can be provided by what's known as a **coping** (or in the case of a pier, sometimes called a 'capping') which can be constructed in several ways.

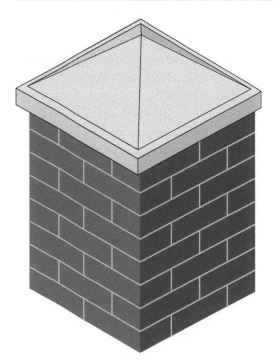

▲ Figure 6.43 Concrete capping

Copings formed in brick

One way to arrange for a coping and a decorative feature at the same time is to lay a course of brick on edge (BOE). The bricks are laid on edge with the stretcher face uppermost. Care is needed to make sure that the bricks are laid accurately to the line to avoid the appearance of **undulations** along the top of the wall.

Some bricklayers use two lines – one at the face of the wall to maintain level and line and another at the rear of the brick on edge – to make sure the rear arris of the brick is accurately levelled.

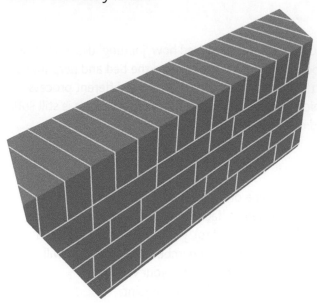

▲ Figure 6.44 Brick-on-edge coping

Brick on edge can be used as a coping on walls greater than one-brick thick by bonding the brick on edge (using Stretcher bond) across the top of the wall.

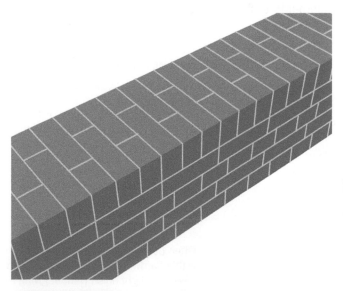

▲ Figure 6.45 Bonded brick-on-edge coping

A protective brick finish to an independent pier can be formed to give a decorative feature. A lot of accurate cutting and skill is required when setting out and constructing brick-on-edge coping (or capping) such as the one shown in Figure 6.46.

▲ Figure 6.46 A decorative brick-on-edge finish to a pier

To form a protective coping and provide a decorative feature, brick 'specials' can be used. One example is to use double **cant bricks** as shown in Figure 6.47.

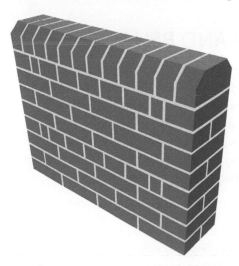

▲ Figure 6.47 Double cant brick specials used as a brick-on-edge coping

Some coping designs require the brick on edge to be laid on a tile creasing. This is formed by laying two courses of concrete or clay tiles, half-bonded and bedded in mortar. Since the tiles will form an overhang on either side of the wall, they allow rainwater to run off the coping away from the face of the wall. This provides both a functional and decorative feature to the wall.

▲ Figure 6.48 A tile creasing

Another method of providing additional weather protection is to install over-sailing courses of brick beneath the brick-on-edge coping. The over-sailing courses must be laid accurately to produce an acceptable appearance. They are commonly set to project from the face of the wall by 18–25 mm to allow rainwater to be directed away from the face of the wall.

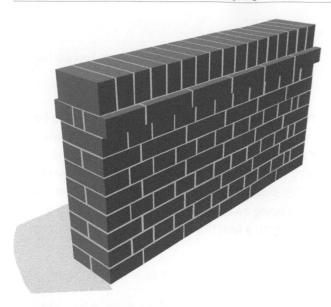

▲ Figure 6.49 Over-sailing course

In a half-brick-thick wall, a coping could be provided by a soldier course, which is formed by laying bricks on end with the stretcher face showing.

▲ Figure 6.50 A soldier course

Other forms of coping

A specified coping may not use any bricks at all. Instead the specification may call for copings (or a capping) made of concrete or stone, designed to suitable dimensions and shapes so that an overhang is formed either side of the wall or pier. This is designed to allow water to be guided away from the face of the main masonry.

Copings can also be manufactured from slate and a type of clay called terracotta. Copings formed from

clay can be moulded into complex decorative shapes that enhance the appearance of a solid wall or pier, as well as providing weather protection.

ACTIVITY

Check the website of a local or national builder's merchant and look for stone and concrete copings. Make a note of the number of different sizes and designs. What size and type of coping would you select to lay on the top of a half-brick-thick wall?

Maintaining quality and productivity

It takes concentration and effort to produce a piece of work that satisfies the relevant industrial standards. Regularly check that your work conforms to the working drawing and that it meets the specification.

It is good practice to check that:
- the specified materials are selected before starting work
- the materials are not damaged before using them
- the correct quantities of materials and components are to hand before construction begins.

KEY POINT

When calculating quantities of bricks, keep in mind that for half-brick-thick walls there are 60 bricks per m², but for one-brick-thick walls we must double this figure to 120 bricks per m². For thicker walls a further adjustment would have to be made in the number of bricks per m².

VALUES AND BEHAVIOURS

Preparing for a work task properly means you will save time and contribute to meeting deadlines set by work programmes. In all aspects of construction activities, thorough and thoughtful preparation is a key to success. An example of good planning would be making lists of tools and equipment that you will need for the job.

As work proceeds, there should be constant checking for accuracy in levelling, plumbing, gauging and ranging the work. When completing your work, make sure that the specified mortar joint is completed to a high standard since this is the 'finish' of the job and will affect the appearance and durability of the solid wall or pier.

Summary

This chapter has demonstrated that, depending on the thickness of the wall or pier to be built, there are a great number of variations in the way the bonding arrangements can be used. This can present a challenge if the drawings you are working from don't provide specific details of a suitable bond, such as how to bond an attached pier to a one-and-a-half-brick-thick wall.

The principles of bonding explained here will assist you to work out the best way of bonding a solid wall, isolated pier or attached pier. If you have difficulty working out the bond for your work task, never be afraid to discuss bonding solutions with someone you are working with. They may see a way of setting out the bond that you have missed.

Test your knowledge

1 What is the strongest brick bond?
   a English
   b Flemish
   c English garden wall
   d Flemish garden wall

2 To establish the best bonding arrangement in a solid wall, how should the bricks be set out?
   a By eye
   b By yourself
   c Dry
   d Slowly

3 In one-brick-thick walls, by how much do the bricks overlap in each course?
   a Half a brick
   b One third of a brick
   c Quarter of a brick
   d Three quarters of a brick

4 What is the dimension of the header face of a brick?
   a 100.5 mm
   b 102.5 mm
   c 105.2 mm
   d 150.0 mm

5 How many stretchers are laid between the headers in each course of Flemish garden wall bond?
   a 2 or 3
   b 3 or 5
   c 4 or 8
   d 6 or 7

6 To what width dimension should a queen closer be cut?
   a 35 mm
   b 46 mm
   c 50 mm
   d 52 mm

7 What is the best material for a damp proof course in a free-standing boundary wall or isolated pier?
   a Bitumen felt
   b Engineering brick
   c Pitch polymer
   d Polythene sheet

8 Which joint finish is it best to use for brickwork in a solid wall or pier that is likely to be exposed to extreme weather conditions?
   a Flush
   b Half-round
   c Recessed
   d Weather struck

9 How many plumbing points are there in a square or rectangular isolated pier?
   a Two
   b Four
   c Eight
   d Ten

10 Which one of the following square isolated piers cannot have reinforcement in the centre?
   a One brick
   b One-and-a-half brick
   c Two brick
   d Two-and-a-half brick

11 State where a queen closer is always positioned in the quoin of a quarter bond solid wall.

12 Which quarter bond would you choose to give the best possible appearance on both sides of a one-brick-thick garden wall and why?

13 State the best type of bricks to be used for a brick-on-edge coping and explain your answer.

14 What three factors should be considered when selecting a suitable joint finish for solid walling?

15 Explain how a tile creasing is formed and state its function.

CHAPTER 7

MAINTAINING AND REPAIRING MASONRY STRUCTURES

INTRODUCTION

A building is constantly subject to many persistent forces and loadings that can cause deterioration and damage. The effects of harsh and changeable weather along with day-to-day wear and tear through use can affect a structure over time, so proper scheduled maintenance and necessary repair work must be carried out.

This chapter will consider the bricklayer's role in carrying out maintenance and dealing with faults in masonry structures and related integrated concrete elements. We will also look at how domestic drainage systems work as part of understanding the maintenance of a building as a whole.

By the end of this chapter, you will have an understanding of:

- the range of faults that can occur in masonry
- repairing masonry faults and replacing components
- producing and placing concrete and repairing faults
- how domestic drainage systems work.

This table shows how the main headings in this chapter cover the learning outcomes for each qualification specification.

Chapter section	Level 2 Technical Certificate in Bricklaying (7905-20)	Level 3 Advanced Technical Diploma in Bricklaying (7905-30) Unit 302	Level 2 Diploma in Bricklaying (6705-23)	Level 3 Diploma in Bricklaying (6705-33)	Level 2 Bricklayer Trailblazer Apprenticeship (9077) Module 6
1. The range of faults that can occur in masonry	N/A	1.2	N/A	1.1, 1.2, 3.1	**Skills Depth:** N/A **Knowledge Depth:** 1.1
2. Repairing masonry faults and replacing components	N/A	1.2, 2.1, 2.2, 2.3	N/A	1.3, 1.4, 1.5, 2.1, 2.2, 2.3, 2.4, 2.5, 2.6, 2.7, 2.8, 3.2, 3.3, 3.4, 4.1, 4.2, 4.3, 4.4	**Skills Depth:** 2.1, 2.2, 2.3, 2.4 **Knowledge Depth:** 1.1, 1.2, 3.1, 3.2, 3.3
3. Producing and placing concrete and repairing faults	N/A	N/A	N/A	N/A	**Skills Depth:** 3.1, 3.2, 3.3 **Knowledge Depth:** 4.1, 4.2, 4.3, 4.4
4. How domestic drainage systems work	N/A	N/A	N/A	N/A	**Skills Depth:** 1.1, 1.2 **Knowledge Depth:** 2.1, 2.2, 2.3, 2.4

Note: for 7905-30, Unit 302:

Content for Topic 1.1 is covered in Chapter 5.

1 THE RANGE OF FAULTS THAT CAN OCCUR IN MASONRY

Understanding the range of possible faults in masonry is essential in identifying the cause of a problem and how to deal with it. The effects of severe weather can greatly affect the fabric of a building, especially in exposed locations. Sometimes buildings suffer damage because they were poorly designed, or the wrong materials were used during construction.

The materials that are likely to be used when maintaining or repairing a building are discussed in Chapters 5 and 6, so in this section we'll concentrate on describing the faults that can occur and what causes them.

Proper maintenance and repair will extend the useful life of a structure and maintain its acceptable appearance.

Identifying faults and their causes

Maintenance and repair of buildings can take two forms:
- preventative – carried out to avoid the building falling into disrepair
- responsive – carried out after a fault has occurred.

Let's look at some faults and damage that can occur in masonry and link them to the two categories. We will also consider whether they require preventative action or responsive action. (We'll look at methods of dealing with each fault later in the chapter).

Cracked bricks and blocks

When building a masonry wall, the bricklayer must take care to select components that are free from faults and damage. Laying cracked bricks or blocks means a weakness is built in to the structure at the start of its working life, which is obviously undesirable. Each brick or block plays a part in evenly distributing the loadings placed on a structure.

▲ Figure 7.1 Cracked brick

We want the structure to last a long time, so a preventative measure such as discarding cracked bricks and blocks as work proceeds is an essential part of the bricklayer's work practice.

VALUES AND BEHAVIOURS

When cracked bricks are discarded, they add to site waste which must be disposed of in an environmentally conscious way. This may involve transporting and processing the waste materials, adding to carbon emissions. If a significant number of bricks must be discarded, alert your trainer or supervisor to the problem. It may be necessary to inform the manufacturer or supplier that there is a quality control issue that needs addressing. Being alert to problems that may seem relatively small can contribute to protecting the environment.

Damp

Damp can cause mould and **fungal** growth inside a building which can affect the health of occupants. If it is severe enough, it can damage the fabric of the structure.

KEY TERM

Fungal: caused by a fungus (a type of organism that obtains food from decaying material)

▲ Figure 7.2 The effects of damp entering a structure

Remember that there are two types of damp that can affect a building:

- rising damp
- penetrating damp.

(For more information on rising damp see page 129.)

If a horizontal DPC fails, rising damp may become apparent as moisture tracks upward (by capillary attraction) through the walls. This can happen in both internal and external walls.

Penetrating damp was more common in solid walls in the past. When cavity walls were introduced as a construction method, the likelihood of damp entering the living or working area was greatly reduced. If internal damp appears in a cavity wall structure, it may be due to poor work practices on the part of the bricklayer which will require responsive measures to deal with it. Mortar build-up in the cavity or on the wall ties is often the problem.

Efflorescence

Efflorescence is white staining on the face of a wall, caused by water passing through the masonry materials carrying soluble salts to the surface. The moisture evaporates to leave a white powder on the surface of the wall as it dries out. While this causes no real harm to the masonry, it spoils the appearance of face brickwork.

▲ Figure 7.3 Efflorescence

If wet bricks or blocks are used or the finished work is not protected from bad weather (by covering with plastic or hessian sheeting), the likelihood of efflorescence occurring is increased.

Lime leaching

Unsightly staining on the face of a wall can be caused by constant water penetration through cement mortar. Water dissolves the calcium hydroxide in the mortar which is then deposited onto the brick face as crystals of calcium carbonate.

▲ Figure 7.4 Staining caused by lime leaching

This is a different chemical reaction to the process that causes efflorescence. Lime leaching is very difficult to remove and will not weather away over time.

The main cause is constant saturation of the masonry. Failure to protect newly completed work from the effects of bad weather will also add to the possibility of lime leaching occurring.

Sulphate attack

Sulphate attack is commonly seen in chimneys or masonry that is in exposed positions. Saturation of the masonry caused by persistent exposure to rain causes soluble salts within the masonry to react with chemicals in the cement mortar. The result is the formation of **ettringite**, a crystal that expands during its formation causing cracking in the joints.

> **KEY TERM**
>
> **Ettringite:** a hydrous calcium aluminium sulphate mineral that causes cracking in joints

The reaction can be reduced by using denser materials that are less absorbent such as engineering bricks. Using sulphate-resistant cement in the mortar mix can also reduce crystalline formation.

Frost damage

Frost damage occurs when saturated masonry is subjected to freezing temperatures. When water freezes it expands, creating pressures and tensions in the brick or block. Over time, the repeated cycle of freezing and thawing can break down the face of the brick or block and cause it to fracture.

▲ Figure 7.5 Frost damaged brickwork

> **KEY POINT**
>
> When the face of a brick or block crumbles away, this is sometimes referred to as 'spalling'. Spalling also occurs in reinforced concrete, where the internal reinforcing steel expands due to rust and **corrosion**, forcing the face of the concrete to break off under pressure.

> **KEY TERM**
>
> **Corrosion:** the gradual destruction of a material by chemical reaction with its environment

▲ Figure 7.6 Brickwork affected by spalling

Taking measures to reduce the amount of moisture entering the masonry by using less absorbent materials or increasing the eaves projection of the roof to shelter a wall can reduce the destructive effect.

Blown render

Render is a sand/cement mix (sometimes including hydrated lime), which can be applied as a finish on masonry walls. It is usually applied as a number of layers or 'coats' depending on the specification.

Cracking in render can occur if the mix ratio has too high a cement content, making it stronger than the backing masonry it is being applied to. The different rates of expansion and contraction in the render and the backing masonry will cause the render to separate from the masonry wall.

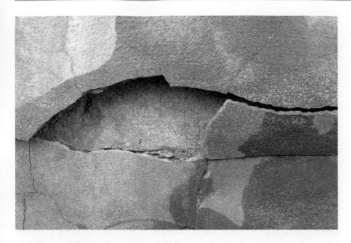

▲ Figure 7.7 Blown render

If the render begins to separate from the backing masonry and water enters the thin gap that forms, the water may freeze and expand, pushing the render completely away from the masonry. This is known as 'blown render'.

Serious faults

Some faults affect more than the appearance of the structure. They can affect the **integrity** of the structure with potentially serious consequences. This sort of fault can lead to the collapse of a structure.

> **KEY TERM**
>
> **Integrity:** in a bricklaying context, being whole and complete

Without proper inspection and maintenance, responsive measures are required which may involve potentially greater expense and hazardous working conditions which could well have been avoided. Let's consider some of the more serious faults and their causes.

Wall tie failure

Wall ties used to be manufactured from galvanised mild steel or iron. The galvanising process (adding a coating of zinc) was intended to prevent corrosion and rust from attacking the metal ties. Experience has shown that this procedure was not sufficient to protect the component from deterioration and wall tie failure has proved to be a significant maintenance and repair challenge.

ACTIVITY

Conduct an internet search of the term, 'galvanised steel'. Find out about the process of galvanisation and write down the sequence of how it is performed. Consider why stainless steel is a better material than galvanised steel to use for cavity wall ties and produce a simple table to show the advantages and disadvantages of each material.

When corrosion affects wall ties, crystals are formed on the metal which expand as they form, with enough force to displace masonry. The mortar joints around the corroded tie can be forced open, resulting in the tie being ineffective.

Wall ties are now manufactured from stainless steel which is much more durable and resistant to corrosion and rust.

> **KEY POINT**
>
> Remember, metal corrosion is also a potential problem in components such as steel reinforced concrete lintels. Other steel reinforced elements such as concrete floor slabs and foundations can be damaged by corrosion, which causes the internal steel to expand and crack the concrete.

Cracking

There is a range of causes that can lead to cracking in masonry walls:

- Large areas of a masonry wall can absorb heat from the sun, resulting in expansion taking place. When the masonry cools down it will contract. This range of movement can be large enough to cause cracking if the design does not include suitable movement joints. (For more details on movement joints, see page 134.)
- As a building is constructed and the weight increases on the foundation, there will often be a slight amount of settlement. This can also lead to cracking in the masonry walls if they cannot absorb the range of movement.
- The roots of nearby trees can cause **subsoil** and clay to shrink or expand, depending on the moisture levels present. This movement will create tensions and pressures (sometimes known as 'heaving')

beneath the foundation of the structure, leading to cracking that can spread through the foundation and into the masonry wall above.

▲ Figure 7.8 Effects of settlement in brickwork

KEY TERM

Subsoil: the soil lying immediately under the surface soil

The direction of cracking in a wall can give an indication of the likely causes:

- Vertical cracking – could occur when mortar is stronger than the individual bricks or blocks; any movement will pass through the bricks or blocks, causing them to crack.
- Horizontal cracking – could be caused by sulphate attack or wall tie failure.
- Diagonal cracking – this follows the brick joints and is usually caused by structural movement such as settlement or **subsidence**.

KEY TERM

Subsidence: the sudden sinking or gradual downward settling of the ground's surface with little or no horizontal motion

ACTIVITY

Use the internet to research images of subsidence. You will see some that are quite shocking! Select some examples to discuss with others. Write down what you think the cause of each example might be and show your written conclusions to your trainer to confirm your ideas.

The seriousness of cracking in masonry can be determined and monitored using a device called a 'tell-tale'. This is fixed over a crack in a wall and shows over time the extent of movement in the masonry each side of the crack.

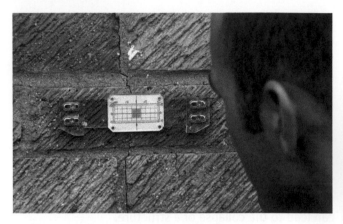

▲ Figure 7.9 Tell-tale applied over a crack in a wall

If the mortar joints are slightly weaker than the bricks or blocks in a wall, this will allow a certain amount of movement in a structure which will reduce the possibility of cracking occurring.

Bulging

Bulging is when the face of a wall is forced out by pressure or loadings from above or behind. Walls distort for a variety of reasons:

- excessive lateral (or sideways) pressure on the wall
- inadequate thickness of the wall
- downward thrust from building elements above (such as floors or the roof)
- loss of bonding between materials (for example, a one-brick-thick garden wall built with two separate skins of Stretcher bond not tied together).

KEY TERM

Bulging: when the face of a wall is forced out by pressure or loadings from above or behind

Buildings are subject to a variety of loadings which can be divided into two types:

- dead load – the load of a structure itself and any other permanent fixtures
- imposed load (or live load) – the load that is applied to the building in use. It could be the weight of major elements of the structure or simply the weight of people using the building. Significant loadings can also be imposed by wind or snow.

Bulging can also be caused by frost damage or vegetation taking root and growing within the masonry structure.

Stone erosion

Stone can be used as a material for door and window sills (or cills) and for window reveals and decorative features, especially in older buildings. Softer types of stone can be seriously damaged by acid rain which erodes the surface. Atmospheric pollution is a major cause of acid rain erosion.

Frost can also damage stone in a similar way to frost damage in brick or block. Spalling of the surface can take place, which spoils the appearance and in serious cases weakens the stone component.

▲ Figure 7.10 Stone coping that has suffered erosion

Since stone features often protrude from a building, softer types of stone can be prone to damage from accidental impact.

Mortar failure

If cement mortar is not mixed correctly or mixed to an unspecified ratio, it can be too weak when hardened and in time it will crumble and fail. Tooling the mortar joint during construction serves to make the joint less prone to water penetration. However, if the mortar is not strong enough, moisture will still have an effect over time and will cause loosening of the masonry components.

Older buildings built using lime mortar are sometimes re-pointed using a cement mortar mix that is too strong. Moisture present in the more absorbent lime mortar behind the cement mortar pointing can freeze and push the stronger material out of the joint, leading to failure.

ACTIVITY

Research mechanical methods of repointing masonry and create a table listing the advantages and disadvantages of using a power tool to repoint masonry, compared with doing it by hand. Discuss your conclusions with others.

Rot and infestation in timber

Rot is decay of wood used for timber components. There are two types of rot that can seriously damage the timber elements of a structure:

- Dry rot – wood decay caused by a species of fungus that digests the part of the timber that gives it strength and stiffness. The wood will shrink, darken and crack. The fungus can travel through building materials other than wood, so it can spread quickly through the building if the conditions allow.
- Wet rot – like dry rot, this problem can severely damage the structural strength of timber. The fungus causes the timber to darken and produce cracks along the length of the timber, usually under a very thin layer of what appears to be sound wood.

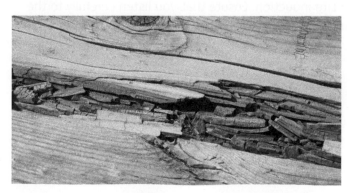

▲ Figure 7.11 Dry rot in timber

▲ Figure 7.12 Wet rot in timber

If left untreated, timber can be prone to attack from many types of insects. This is known as **infestation** and can vary in severity from the appearance of small worm holes to complete structural failure.

Even well-seasoned hardwood timber can be attacked and damaged by certain wood-boring insects.

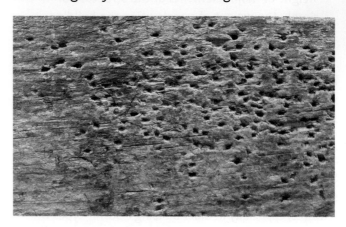

▲ Figure 7.13 Woodworm in timber

② REPAIRING MASONRY FAULTS AND REPLACING COMPONENTS

Before starting a repair job or a maintenance task, carefully consider the health and safety issues that are involved. Review the risk assessments and method statements to inform yourself of the safest and most efficient way to tackle the work task. You should frequently review the health and safety details about how relevant laws affect you and your work practices on site, discussed at length in Chapter 1.

Every site has different hazards that you need to be aware of. These hazards will be explained to you during a site induction. Ensure that you listen carefully to the information given in the induction and act on it.

IMPROVE YOUR ENGLISH

Some hazards that are highlighted during a site induction may have several key points that you need to take note of. The Health and Safety Executive produces many leaflets and guides that communicate in a clear and easy-to-understand way the information and directives that you should pay close attention to.

HEALTH AND SAFETY

Working on repair and maintenance tasks is different to working on new construction. It is likely that the work will involve dismantling or altering masonry and other heavy and durable elements of a structure. This will often result in the generation of large volumes of dust and rubble which can contain hazardous substances. If the repair or replacement job involves working at height, be very aware of the possibility of objects falling onto others below. Think carefully about your own welfare and the effects of your activity on those working nearby.

Selecting the right tools

There is a wide range of tools that can be used for maintenance and repair and correct selection will ensure that the work is carried out efficiently and safely. Tools can be divided into two categories:

- hand tools – unpowered tools requiring the user to employ manual strength and skills
- power tools – actuated by a mechanised power source, requiring the user to be trained and competent.

▲ Figure 7.14 An operative using a masonry hand saw

▲ Figure 7.15 Disc cutter training in progress

Hand tools

Let's look at specific hand tools that might be needed for repair and maintenance tasks.

▼ Table 7.1 Hand tools used for repair and maintenance tasks

Tool	Use
Lump (or club) hammer	This type of hammer is used with a bolster or chisel for cutting operations in masonry.
Bolster	A broad-bladed chisel used for cutting bricks and blocks accurately by hand. It can also be used for trimming openings made in a wall during repair work.
Chisel	Chisels (often referred to as 'cold' chisels) are available in a range of sizes to suit the requirements of the job. They are used with a lump hammer when cutting holes in masonry.
Jointing chisel	Sometimes called a 'plugging chisel', it is used for cutting out mortar joints precisely without damaging the face of the adjoining bricks or blocks.
Brick hammer	Used for roughly cutting or trimming bricks or blocks.
Scutch (or comb) hammer	Used for precisely trimming and shaping bricks or blocks.

Now let's look at some power tools that might be needed to maintain structures and repair faults and damage in masonry.

Power tools

All the power tools below have the potential to be dangerous if used improperly. Always ensure that you follow the manufacturer's safety instructions and make sure you are adequately trained before you start using any power tools.

VALUES AND BEHAVIOURS

At the start of a work task, check power tools for safe condition *every time* you use them. Checking power tools frequently doesn't just protect you from injury; careful checking shows consideration for anyone else who might use the power tool after you and helps to keep everyone on site safe. If there is a fault or damage, label the tool as unserviceable, report the problem to your line manager and put the tool securely out of service.

Always wear suitable PPE, taking care to protect your sight and hearing. Don't forget protection against inhaling dust in addition to using safety boots, safety gloves and a hard hat.

▼ Table 7.2 Power tools used for repair and maintenance tasks

Tool	Use
Disc cutter	Used to cut brick, block, stone and concrete or other dense materials. Can also be used to cut steel with the correct cutting disc. Different types of disc cutter can be powered by petrol, electricity or compressed air.
Masonry bench saw/clipper saw	Works on similar principles to a disc cutter but is bench mounted. Water is fed over the cutting disc to reduce dust in the air around the machine.

▼ Table 7.2 Power tools used for repair and maintenance tasks (continued)

Tool	Use
Hammer drill	Used for drilling holes, this tool has a hammer action to drill more easily through dense materials such as bricks, blocks, stone or concrete.
Tile cutter	This is similar to a masonry bench saw but is set up for cutting tiles and masonry components such as copings.
Grinder	A powerful hand-held tool that is used to grind and cut dense materials. Different types and sizes are available that can be powered by petrol, electricity or compressed air.
Pneumatic breaker	Used for heavy-duty breaking of concrete and other dense materials. The term 'pneumatic' refers to the type of breaker that uses compressed air to power it. Other types use electricity to power them.

Additional information on disc cutters

Never attempt to use or handle a disc cutter unless you are competent and authorised to do so. Being properly trained in using cutting machinery such as disc cutters will reduce the risk of serious injury to yourself and others working nearby.

IMPROVE YOUR ENGLISH

Review the information related to 'competence' on the HSE website (www.hse.gov.uk). Find out if a worker needs the same health and safety competence to work in an office as they would on a construction site. Write a short report defining what 'competence' means.

It is very important to know how to safely change cutting discs in a disc cutter. The disc material must be suitable for the work activity and the correct process used to fit the disc in accordance with the manufacturer's instructions. The disc must be the right size for the machine being used.

Discs or blades made from bonded carborundum have different characteristics to steel discs manufactured with diamond tipped edges.

Make sure that guards are correctly positioned and are secure before starting any cutting or grinding operations.

▲ Figure 7.16 Cleaning stonework with a jet washer

> **INDUSTRY TIP**
>
> Different cutting and grinding discs and blades operate safely at different speeds. This is referred to as the 'peripheral speed' (the speed at the outer edge of the blade). Always consult the manufacturer's instructions and guidance on safe operational speeds for discs or blades. If you're not sure, ask someone more experienced for advice!

Methods of work

Most repair jobs involving masonry will require the removal and replacement of individual components or larger areas of walls, depending on the extent of the damage or fault. Hammering, cutting and drilling will often be involved and selection of the appropriate tools and correct methods of work is a skill that builds with experience.

Let's consider some specific methods of work.

Cleaning masonry as a maintenance task

Cleaning masonry can be done using power/jet washers where appropriate. This type of equipment can cause damage on softer masonry surfaces so consider using other cleaning methods if this is a possibility. Since this type of equipment operates using high pressure water, take care when working with other personnel in your work area.

Efflorescence can be cleaned off and removed from affected masonry by using a stiff brush on the surface. Repeated vigorous brushing on several occasions may be necessary to clean the surface to a satisfactory degree. Avoid excessive wetting of the surface when cleaning it, since the white deposits will dissolve and be drawn back into the masonry, prolonging the problem.

> **ACTIVITY**
>
> Download the 'Brick cleaning' pdf from the Brick Development Association website, accessed at: www.brick.org.uk/admin/resources/cleaning-of-clay-brickwork.pdf
>
> Look at pages 7 and 8 about staining on brickwork. Write a risk assessment to give guidance on safe practice when applying cleaning chemicals.

Applicators and specialist brushes are available for use with masonry cleaning fluids; these products are corrosive and should be used with care. Powerful cleaning fluids are often used to remove lime leaching. Follow the manufacturer's instructions carefully to avoid damage to the masonry and injury to yourself. Wear the correct PPE including mask, goggles, gloves and waterproof clothing such as safety wellington boots and oilskins.

Removing and renewing masonry materials

Careful inspection of the wall can often provide clues to the cause of the fault and this can help you decide on the type of repair that is most suitable.

Cracking in walls

If the crack is following the line of the joints, then it is possible that the ability of the wall to spread loadings will be maintained and a limited repair might be appropriate. You may decide that several bricks in a course or individual bricks should be removed and replaced.

If, however, the crack has formed vertically through the bricks or blocks, then the damage is more serious and the wall has lost its ability to spread loadings evenly. A large section of masonry may have to be replaced. The scale of the repair will depend on the type and extent of the damage.

To minimise the amount of demolition work involved in repairing cracked masonry, specialist methods and materials have been developed that allow the weakened masonry to be reinforced more easily.

In one relatively new process, the mortar joints can be reinforced by a method called 'stitching', which is an effective method used to reduce the amount of disturbance to the damaged masonry.

▲ Figure 7.17 Stitching

Stitching work is carried out by removing sections of bed joint and inserting **helical** stainless steel bars in the gap that is formed. The joints are then filled with cement mortar or epoxy resin grout. This reinforces the wall and improves its ability to spread the loadings more evenly through the masonry.

KEY TERM

Helical: having the shape or form of a helix (a spiral)

If the removal of sections of masonry or individual bricks or blocks is necessary, this should be done with great care. Various methods can be used to remove bricks or blocks without damaging surrounding masonry.

- A disc cutter (Figure 7.18) can be used to carefully cut out the mortar joints. Care is needed to prevent the cutting or grinding disc from touching the face of the surrounding masonry.
- Using the stitch drilling method (Figure 7.19). A series of closely spaced holes are drilled in the mortar joint to allow the masonry components to be loosened.
- A jointing (or plugging) chisel and lump hammer (Figure 7.20) can be used to remove the mortar joints around the damaged masonry components to allow removal and replacement.

Care should be taken to make sure that all the old mortar is removed, so that the new masonry components can be positioned and aligned accurately to the existing work. New mortar should be carefully prepared to a suitable consistency to minimise staining of the surrounding masonry.

The mortar bed can be laid by using a small trowel and the perp joints can be applied to bricks or blocks before the bricks or blocks are placed in position. All the joints should then be carefully filled by pressing mortar into them until compacted.

▲ Figure 7.18 Using a disc cutter

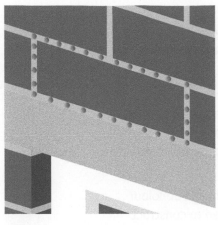

▲ Figure 7.19 Stitch drilling the joint

▲ Figure 7.20 Using a jointing (or plugging) chisel

Study the step by step sequence to see how to use a lump hammer and jointing chisel to remove a damaged brick without spoiling the surrounding masonry.

Removing a brick step by step

STEP 1 First carefully cut out the surrounding mortar joints and remove the damaged brick using a lump hammer and jointing (or plugging) chisel.

STEP 2 Carefully remove the damaged brick.

STEP 3 Take some time to clean out all the old mortar to ensure that the new brick will fit easily and accurately.

STEP 4 Dampen (not saturate) the opening ready to receive the new mortar. This will 'key' the new mortar to the existing masonry.

STEP 5 Lay the mortar bed before positioning the new brick. Form the perps on the new brick before carefully positioning it and making sure all the mortar joints surrounding the brick are full.

Correct selection of materials is very important. Where possible you will want to identify the type, colour and make of bricks that will best match the existing work.

This can be done by visiting a builder's merchant and taking a sample of the brick type that needs to be replaced. You can also look at manufacturers' catalogues to find a match. If you have one of the bricks, it may contain a manufacturer's code which will help to identify its origin.

You will also need to identify the type of sand or colour of ready-mixed mortar that has been used to construct the wall originally, to enable you to use the closest possible match for the replacement mortar.

> ### INDUSTRY TIP
>
> If time allows, it is a good idea to produce a test mortar mix and let it harden to check the colour against the existing mortar joints.

> ### ACTIVITY
>
> Find a website for a coloured mortar manufacturer and check out a colour chart of available colours. How many colours can you find? Write a list of benefits of using a ready mixed coloured mortar on site.

Bulging in walls

Where walls are bulging, it is possible to reinforce them to avoid the need to take them down and rebuild if the **deflection** is not too severe. This sort of work will normally involve consultation with a structural engineer, who will provide calculations and detailed instructions as to how the work should be carried out.

> ### KEY TERM
>
> **Deflection:** the degree to which a structural element is displaced under a load

If the wall cannot be reinforced or restrained, the only alternative will be to demolish and rebuild it.

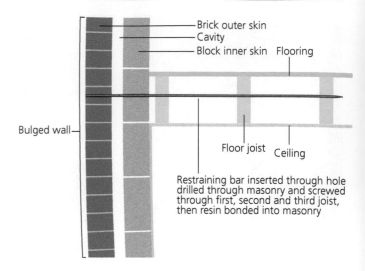

▲ Figure 7.21 Method of reinforcing bulging walls

Frost damage in walls

Often when bricks in a wall are affected by frost damage, the brick faces fall off. Remember, this is known as spalling. This could happen because the brick selected for the job is not suitable for the location. If proper maintenance has not been carried out, a problem could be caused by a missing coping or a leaking gutter or downpipe.

Damaged and unsightly bricks will need to be replaced with new ones using the methods described above. Frost damage develops over time, so the damaged brickwork may be quite old. It may be necessary to obtain reclaimed bricks sourced from older demolished buildings to match the colour, texture and size of the bricks to be replaced.

> ### KEY POINT
>
> Bricks were manufactured to imperial dimensions until metric measurements were introduced in Britain. Reclaimed imperial bricks are larger and heavier than metric bricks. They can also be very expensive compared to new bricks because they are becoming more difficult to obtain. (For a comparison of metric and imperial measurements, see page 64.)

Structural work of a more technical nature

Some repair jobs need a deeper understanding of the structural elements of buildings and the methods available to safely complete the work. Even small areas of masonry weigh a lot and if a wall collapses onto a worker during repair operations, it can cause significant injury or death.

It is essential to be thoroughly familiar with the equipment that can be used and the safe methods of using it.

Removing and replacing a lintel

Sometimes it is necessary to replace a failed lintel over a door or window because the structure has been weakened, as described below.

- A timber lintel (sometimes called a 'Bressumer') that has rotted over time and become structurally unsafe. It is no longer able to safely support the masonry above it.
- A reinforced concrete lintel that has been saturated for long periods so the steel reinforcement has corroded or rusted. This has affected the component's ability to safely support the loadings it is designed to carry.
- A steel lintel has not been properly protected from the effects of moisture in a cavity wall. It has become weakened by rust and corrosion.

VALUES AND BEHAVIOURS

Never work alone on a task like replacing a lintel. Things can get out of hand and become dangerous very quickly if you become over-confident and fail to plan properly. Work alongside more experienced workers and discuss the process both before starting work and continuously during the operation. This will ensure you're using the right methods and equipment and not missing a potential problem.

Replacing a lintel will mean that the masonry above the lintel has to be temporarily supported during the repair work. Each job will need to be treated separately to assess the load that is to be supported and the building's condition; a decision can then be made about how to safely and efficiently achieve the replacement.

For this type of task, it is essential to understand how a load is distributed through a wall by the bonding of the masonry components. Look at Figure 7.22 to see how the load of a wall is transferred through the bonding arrangement of the bricks, blocks or stones.

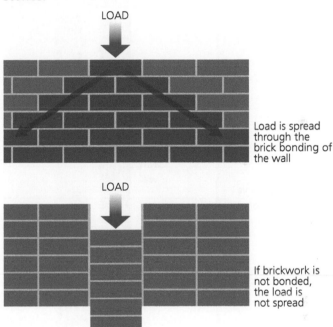

▲ Figure 7.22 How load is distributed through the bonding of brickwork

So how do we provide temporary support to the masonry above the lintel while we remove and replace it?

Using adjustable steel props and 'needles'

One long-established method is to use adjustable steel supports (sometimes referred to by their trade name, 'Acrows') that hold short beams called 'needles' that are placed through prepared openings in the wall at close intervals. The needles can be steel beams or stout timbers.

To use this method there must be access available from both sides of the wall. Look at the step by step to see how this works.

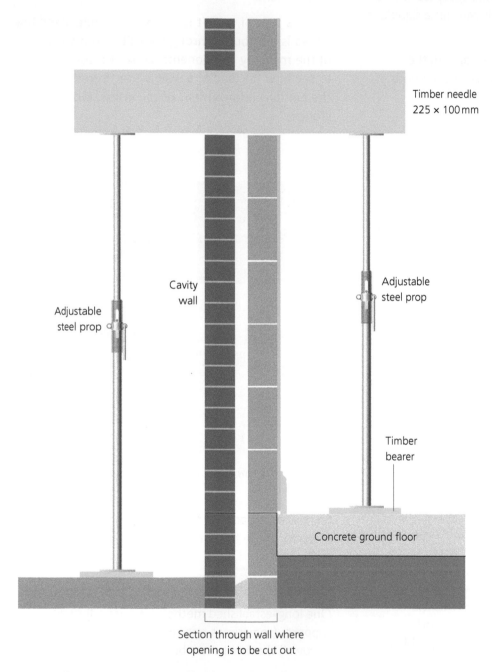

Timber needle
225 × 100 mm

Adjustable steel prop

Cavity wall

Adjustable steel prop

Timber bearer

Concrete ground floor

Section through wall where opening is to be cut out

▲ Figure 7.23 Section through a wall with a needle inserted

Inserting a 'needle' to support a wall

STEP 1 Cut a hole in the wall above the lintel to be replaced, big enough to insert a needle from one side of the wall to the other.

STEP 2 Insert the needle through the wall, making sure there is enough overhang each side to position an adjustable prop under it.

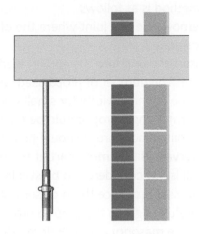

STEP 3 Position the adjustable props to support the needle on either side. Tighten the props under the needle until they are providing support for the area of masonry above the door or window.

INDUSTRY TIP

Be careful not to overtighten the adjustable props; it can cause the masonry joints above the needles to crack and create more damage if you do!

The number of needles used will be determined by the span of the lintel to be replaced and the weight of masonry to be supported above the opening.

HEALTH AND SAFETY

Allow plenty of time to assess the load that will need to be temporarily supported. Many accidents have been caused by a lack of thought and consideration at the planning stage of providing temporary support. Remember – never be afraid to discuss your thoughts or concerns with someone more experienced.

The lintel can now be carefully cut out and replaced. Remember that the type of replacement lintel used will determine the amount of bearing that is needed. Once the new lintel is in place and levelled, the gap between the lintel and the supported brickwork needs to be filled, ensuring that the gap is well filled with mortar. Slate slips can be used to pack the joint above the lintel to ensure that the masonry above is securely supported.

The adjustable props should not be removed until the mortar has fully set. Once the props have been removed, masonry that has been removed to insert the needles can be reinstated. Make sure that all mortar joints are filled completely to finish the job.

Using off-set props to support a wall

A second method of providing temporary support to allow removal and replacement of a defective lintel is to use offset props, sometimes referred to as 'Strongboys'.

INDUSTRY TIP

A Strongboy is a piece of engineered steel equipment that is added to adjustable steel props to form a unit.

▲ Figure 7.24 Strongboy

This method is very useful when you have access to only one side of the wall or when there are separate lintels in each leaf of a cavity wall and only one must be replaced. The sequence of operations when using this method is as follows:

1 Remove the bed joint where the offset prop is to be positioned.

2 Insert the unit into the gap created by removing the joint. The number of props used will vary according to the opening width; for small openings (under 1.5 m) a single prop should be enough. For wider openings more props should be provided at regular intervals of 900 mm apart. If the mortar in the wall is not considered to be stable, then it may be necessary to place them closer together.

3 Carefully tighten the prop to provide support for the masonry above. Before using this type of equipment, always ensure that you refer to the manufacturer's instructions.

4 Once the offset props are safely in position and you are confident that the load above is securely supported, the lintel can be removed and replaced.

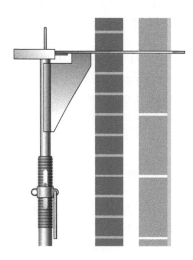

▲ Figure 7.25 Strongboy fitted on the adjustable prop and inserted

HEALTH AND SAFETY

If there is any doubt about the ability of the offset prop unit to support the weight, get advice before proceeding with any work! It may be necessary to obtain the advice of a structural engineer.

Using dead shoring to support a wall

A more traditional method of providing temporary support is to use 'dead shoring'. This consists of stout timbers used to support timber needles, with any necessary adjustment made by using timber wedges.

Dead shoring is not as easy or as versatile to use as the offset prop method or steel prop and needles method of providing temporary support.

ACTIVITY

Dead shoring uses folding wedges to provide any necessary adjustments to the prop. Do some research and then produce a ruler-assisted sketch with labelling showing how folding wedges work.

KEY POINT

A tray DPC above a lintel in a cavity wall may need to be replaced at the same time as the lintel. If this is the case, additional masonry above the opening will need to be removed to allow easier access to the existing tray DPC. Keep in mind that to allow for this, the temporary support will have to be positioned higher up the wall.

Fixing a replacement door or window frame

If a lintel must be replaced, it is often the case that the door or window frame in the associated opening must also be replaced. It is therefore important for the bricklayer to know about methods of fixing replacement frames into an existing opening.

In most cases the frame is fixed on the outer leaf of the cavity and plumbed and levelled in position. Care should be taken to ensure that the vertical DPC is in place and not damaged during the fixing of the new frame and that insulation materials are not damaged or disturbed. (For characteristics of different insulation materials and methods of installation, refer to pages 98 and 160.)

KEY POINT

Older properties that are constructed using solid wall masonry may have no insulation. This leads to energy loss due to transfer of heat through the walls (expressed as a U-value). Insulation can be installed retrospectively, either internally or externally, to reduce the transfer of heat. You may need to consider if this type of insulation should be installed when replacing door or window frames in older buildings.

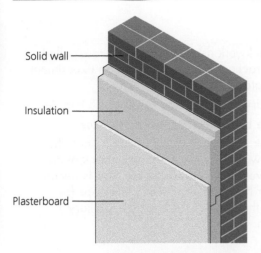

▲ Figure 7.26 Insulation being applied to the inside of a wall

Solid wall

Insulation

Plasterboard

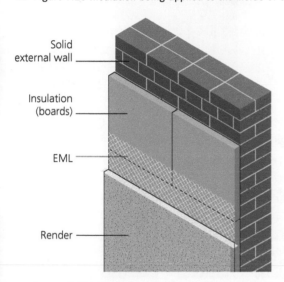

Solid external wall

Insulation (boards)

EML

Render

▲ Figure 7.27 Insulation being applied to the outside of a wall with expanded metal lath (EML) and render

The frame can be fixed in place to the reveal of the opening with purpose-made brackets. Alternatively, holes can be drilled through the frame into the masonry reveal and the frame secured by using window frame fixings which include an expanding plug (fischer type). The two methods are shown in Figures 7.29 and 7.30.

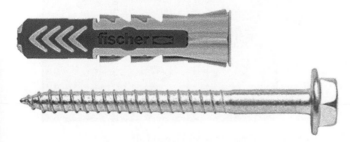

▲ Figure 7.28 Fischer fixing plug

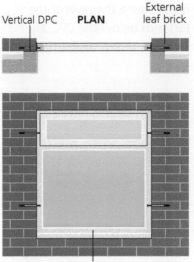

Vertical DPC   **PLAN**   External leaf brick

Vindow frame fixed to outer leaf of cavit with masonry fixings through frame

▲ Figure 7.29 Replacing a window frame: method 1

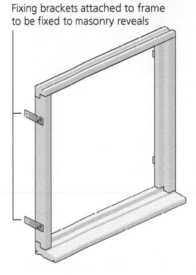

Fixing brackets attached to frame to be fixed to masonry reveals

▲ Figure 7.30 Replacing a window frame: method 2

Replacing a failed DPC

Details about different DPC materials and the pros and cons of each type are covered in Chapters 5 and 6. A DPC may stop working for several reasons. A common problem is shown in Figure 7.31.

Horizontal DPC must be installed at least 150 mm above finished path or ground level to avoid moisture splashing up and tracking above the finished floor level (FFL) into the occupied area of the building. If earth is allowed to build up above the DPC level, it could create a passage for moisture to be drawn into the masonry. If there are mortar droppings in the bottom of the cavity, the problem will be even worse.

The solution in this case is to reduce the level of the earth so that it remains 150 mm below the DPC and remove individual bricks to clean the mortar droppings from the cavity.

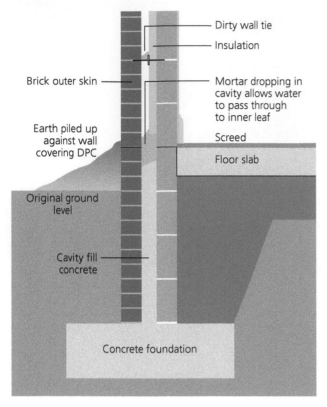

▲ Figure 7.31 Section through a cavity wall showing faults that could cause damp penetration

KEY POINT

Look carefully at Figure 7.31. It clearly demonstrates how important it is for the bricklayer to avoid allowing mortar droppings to fall into the cavity. Good technique must be practised and developed during the construction process to avoid expensive problems cropping up later.

Installing a replacement flexible DPC

STEP 1 Remove four bricks at DPC level using a jointing chisel.

ACTIVITY

Water in brickwork rises through a process called capillary attraction. You can perform an experiment to demonstrate this force:

1 Place a dry brick in a tray of water 25 mm deep.
2 Leave the brick there for several minutes.
3 Now look at the face of the brick and note how high the water has risen. You will note that the brick is wet above the water line. This is because the water has been drawn into tiny gaps (or voids) in the body of the brick by capillary attraction.

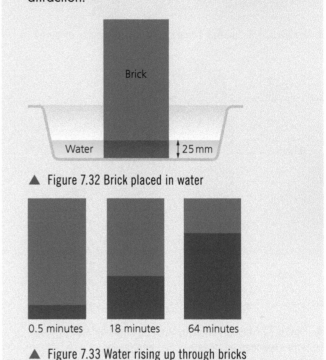

▲ Figure 7.32 Brick placed in water

0.5 minutes   18 minutes   64 minutes

▲ Figure 7.33 Water rising up through bricks

If the DPC material has failed, then it will have to be replaced. Look at the step by step illustrations showing two methods of removing and replacing a failed DPC.

STEP 2 After you've removed the bricks, clean off any remnants of old mortar joints around the opening.

STEP 3 Re-bed a new length of flexible DPC.

STEP 4 Replace the bricks on top of the DPC.

A second method is to take out sections of brickwork and replace them with engineering bricks.

Installing a replacement rigid DPC

STEP 1 Cut out a section of brickwork two courses high using a jointing chisel.

STEP 2 Clean out the mortar joints.

STEP 3 Lay the engineering bricks in the opening and carefully joint the bricks.

When the section of replaced brickwork has hardened sufficiently, a second section can be treated in the same way.

A third method is to inject silicone liquid to penetrate the masonry to provide a water-repellent barrier.

▲ Figure 7.34 Silicone injection to provide a DPC

Replacing defective wall ties

As previously mentioned, in the past wall ties were manufactured from mild steel or iron. The steel was galvanised (coated in zinc) as a protection against moisture damage, but over time this coating deteriorated leaving the wall tie vulnerable to rust and corrosion.

▲ Figure 7.35 Wall tie tying together the inner and outer walls

When wall ties in a cavity wall become rusted or corroded, they can become weakened and unable to perform their function of tying together the outer and inner leaves or skins. The cavity wall will not be able to carry the loadings imposed by floors and the roof and can crack and distort. Wind loadings will also be an issue, when variable air pressure creates suction on the face of the outer masonry leaf.

Remember, steel that is rusting or corroding increases in volume as the deterioration develops and the resulting expansion can cause cracking in the masonry. Cracks running along the bed joints at vertical intervals of 450 mm are a good indication of possible cavity wall tie corrosion.

To replace defective wall ties, follow these steps:

1 Investigate the type of masonry in each leaf of the cavity. Face bricks may have perforations which will affect the choice of method used to remedy the problem.
2 Establish the position and condition of the existing ties.
3 Remove the existing ties if they are found to be rusted or corroded.
4 Install new ties and make good the masonry.

Two methods have been developed to replace defective wall ties, with the minimum of disturbance and damage to the existing masonry:

● Resin grouted tie – helical stainless steel ties are threaded through holes drilled in both leaves of masonry to a specified depth. A chemical resin is then introduced as a bonding agent.
● Mechanical screw-in tie – a percussion drill is used to push a helical stainless steel tie through a small diameter pre-drilled hole in both leaves of masonry. Because the pre-drilled hole is of a small diameter, the helical wall tie mechanically binds with the inside of the hole.

These methods require some training to use effectively. The bricklayer may be called on to use the more straightforward method shown in the following step by step.

Installing replacement wall ties

STEP 1 Cut a hole in the masonry to establish the position of existing wall ties.

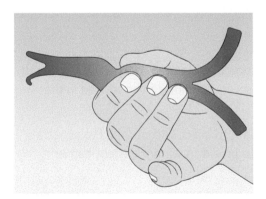

STEP 2 Remove rusted or corroded ties.

STEP 3 Replace the defective ties with new drill-in type ties.

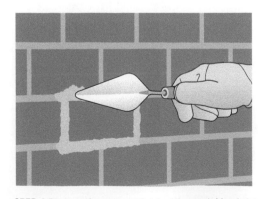

STEP 4 Replace the removed brick with a matching brick and make good.

Repairing a chimney stack

Chimney stacks are usually constructed in an elevated exposed position in a structure, so they can be severely affected by wind, rain and extremes of temperature. A range of faults can occur which will require replacement or repair.

Look at Figure 7.36 and consider Table 7.3 which lists some common problems that may arise.

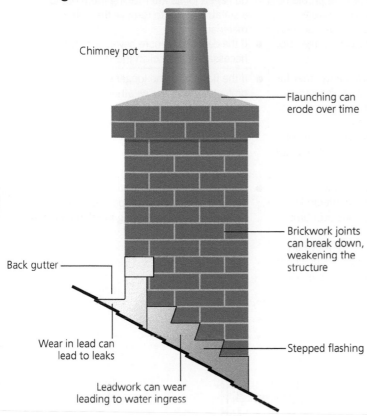

Chimney pot

Flaunching can erode over time

Brickwork joints can break down, weakening the structure

Back gutter

Wear in lead can lead to leaks

Leadwork can wear leading to water ingress

Stepped flashing

▲ Figure 7.36 Parts of a chimney that may need repairing

▼ Table 7.3 Faults in chimneys

Parts and components of a chimney	Things to consider	How to fix the problem
Lead flashings	• These are positioned at the sides of the stack where it meets the roof pitch. Lead is a soft, **non-ferrous** metal that is easy to form into quite complex shapes with practice. • The effect of wind causing small but persistent movement in the soft metal can cause wear and create cracking over time, allowing moisture to enter the structure.	• There are products available that are designed to seal small cracks or tears in the lead. These are rarely successful as a long-term solution. Replacement of the component is a better option.
Lead apron and back gutter	• Positioned at the intersection of the chimney stack with the roof, at the front and rear of the stack. Wind can affect both of these components in a similar way to the lead flashings. • In the case of the back gutter at the rear of the stack, there is greater exposure to water flowing off the pitch of the roof which can cause deterioration over time.	• Replace the component if damaged and leaking. The apron and back gutter are commonly produced by soldering sections of sheet lead together to make a component that suits the size of the chimney and pitch of the roof. • This is not something a bricklayer is usually trained to do.

➡️

▼ Table 7.3 Faults in chimneys (continued)

Parts and components of a chimney	Things to consider	How to fix the problem
Brickwork chimney stack	• Mortar joints in chimney stack masonry are likely to be exposed to continuous wetting and drying which can cause several problems that will lead to deterioration. These were discussed earlier in this chapter in the section 'Identifying faults and their causes', page 196.	• If the mortar joint deterioration is not very advanced, it may be possible to re-point damaged joints after raking them out to a suitable depth to remove the damaged material. • If the damage is more severe it may be necessary to demolish the stack and rebuild it.
Flaunching	• This forms the protection from the weather for the top of the chimney stack and also serves to anchor the chimney pot in position. • Can be manufactured in concrete as a purpose-made unit to suit the dimensions of the chimney stack or formed by hand in sand/cement mortar.	• If the flaunching no longer gives adequate protection from the weather, and the chimney pot is not being effectively secured, the flaunching will need to be replaced.
Chimney pot	• Commonly manufactured from terracotta clay, which, while strong, can be affected by extreme temperature variations over time leading to cracking. • When chimney pots become loose they can be blown off in high winds, causing significant damage to the roof and posing a serious risk to anyone below the chimney stack.	• A cracked chimney pot will need to be replaced. Replacing the pot may not be possible without replacing the flaunching at the same time.

KEY TERM

Non-ferrous: containing little or no iron; resistant to corrosion

Repairing a chimney and replacing components may involve moving and lifting heavy materials to an elevated working platform. Always use the correct equipment to lift and place materials and components. The work needs to be carried out with extreme care and a clear method statement should be produced to outline the methods to be used.

Materials that need to be removed from the elevated working platform should be transported through a properly erected waste chute to a skip at ground level.

▲ Figure 7.37 Waste chute

VALUES AND BEHAVIOURS

Remember that when working on a chimney, small pieces of old mortar and debris are likely to fall down the flue. Protection should be provided inside the property to prevent damage to furniture and carpets. Showing consideration for the customer and respect for their property will go a long way to creating a good reputation for you and your employer.

Using access equipment

Many maintenance or repair tasks are carried out above ground level and will require the use of some form of access equipment. Selecting the right type of equipment is important if the work is to be carried out efficiently and safely.

Working with masonry elements of a structure often involves the use of heavy and bulky components, so working platforms and equipment must be strong and stable enough to carry the loads placed on them and provide enough working space for operatives to work comfortably.

HEALTH AND SAFETY

Working at height multiplies the potential hazards compared to working at ground level. It requires careful planning and careful implementation of measures to protect you and those working near you. Accidents while working at height remain one of the major causes of fatalities and major injuries in the work environment. (For details of the regulations that govern working at height, see page 26.)

Use access equipment wisely. Do not attempt to install or erect access equipment unless you are trained and competent. Independent scaffold should *always* be erected by trained and competent personnel. *Never* alter an independent scaffold yourself unless you have been certified as competent.

▲ Figure 7.38 Scaffolding should be erected by competent scaffolders

Some access equipment for use at lower levels is not so demanding in terms of user competency and can be set up and used after a short period of onsite training.

Table 7.4 shows the range of access equipment that could be selected and some of the features of each type.

ACTIVITY

Look at the HSE website (www.hse.gov.uk) and search for 'Use of ladders'. Make a list of checks you should make before using an extension ladder and a stepladder.

Independent scaffold

The type of access equipment most frequently used by the bricklayer is likely to be independent scaffold, since this is the best equipment for carrying heavy masonry materials safely at greater heights.

To maintain safety, the scaffold is inspected on a weekly basis by a competent worker and also after strong winds or heavy rain. After each inspection, a label is attached recording the inspection date.

Always carry out a visual inspection yourself before starting work. If something doesn't look right, let others who might be affected know and report it to your line manager immediately.

Keep the working platform clear of waste for safety and clean the scaffold deck after the work is finished to remove mortar spills. This will prevent the mortar droppings from staining nearby masonry if it rains after work is completed.

▼ Table 7.4 Types of access equipment

Access equipment	Description
Ladders/extension ladders	Ladders are used for gaining access to working platforms. They can be used as a working platform for light, short-duration work only. Ladders should be placed leaning at an angle of 75° (a ratio of 1:4 or, in other words, one unit of measurement horizontally to every four units of measurement vertically).
Trestles	Trestles are used for gaining access for short-term work activities. They should be properly constructed with handrails and toe boards. A proper access point for a ladder must be provided.
Stepladders	Stepladders are used for short-term work on a solid base. Check all four feet of the stepladder are in firm contact with the ground and the steps are level.

→

▼ Table 7.4 Types of access equipment (continued)

Access equipment	Description
Towers	Towers are used for gaining access for light work only. Heavy materials should not be loaded onto this type of access equipment. Follow the manufacturer's instructions to erect them.
Roof ladders	Like other ladders, a roof ladder is primarily designed for gaining access and should not be used as a working platform for extended periods.
MEWPs	Mobile elevated working platforms (MEWPs) are used for gaining quick access for small repairs and maintenance. They are easily moved from task to task where even firm ground is present. You must be properly trained to use this equipment.

▼ Table 7.4 Types of access equipment (continued)

Access equipment	Description
Independent scaffold 	This is purpose-built scaffolding which is erected to provide a secure working platform. It is formed from steel tubing produced in a range of standardised lengths which are fixed together using purpose-made clips. It can carry heavy materials. (See the health and safety feature on page 219 for guidance on using independent scaffold.)
Stack scaffold	This is a purpose-built scaffold for gaining access to a chimney. A type of independent scaffold using tubes and clips.
Cradles 	Cradles are used for gaining access to the face of buildings. They are specialist pieces of equipment and should be operated by trained and qualified personnel. They are not designed to carry heavy materials.

INDUSTRY TIP

Before leaving site, always remove any ladders to prevent unauthorised access. If they can't be removed, make sure a barrier is secured to stop the ladder being used.

By selecting the right access equipment, using appropriate method statements and risk assessments and being consistently focused on maintaining safety, you can reduce the likelihood of accidents happening to yourself and those you work with. Never take risks!

3 PRODUCING AND PLACING CONCRETE AND REPAIRING FAULTS

Concrete is a versatile and durable material used to form major elements of a structure such as foundations and floor slabs. It can also be used in lintels spanning openings in masonry, as already described earlier in this chapter and in Chapter 5.

The bricklayer's work will often involve interaction with concrete elements and components in a building and may require concrete replacement or repairs to be undertaken in conjunction with masonry construction. Let's look at some useful points about working with concrete.

Producing concrete

Concrete is made from coarse aggregate (crushed stone), fine aggregate (sand) and Ordinary Portland cement mixed with water. Sometimes additives will be used for various reasons such as to speed up or slow down the setting time.

Water reacts with the cement causing it to harden. This process is called hydration and it locks the aggregates together. The water used must be 'potable' which means it is pure enough to drink. Using water of this quality means that chemicals which might affect the hydration process won't be present.

Concrete is very strong under compression (when weight is put upon it) but is weak when it is pulled (put under tension). Tension can be caused in a foundation when it has to bridge softer sections of ground or when ground conditions are unstable.

To prevent cracking and possible failure of the foundation, the concrete may need to be reinforced with steel which is cast into the concrete before it hardens.

▲ Figure 7.39 Concrete with steel reinforcement

Faults in concrete

Some faults that occur in masonry can also occur in concrete:

- spalling – damage causing the concrete to crumble or flake into small pieces
- cracking – caused by the distortion or bending of concrete due to excessive unexpected forces
- attack by chemicals – damage caused by the reaction of certain chemicals with moisture present in the concrete
- frost damage – caused by moisture present in the concrete freezing and expanding
- blowing – deterioration similar to spalling.

▲ Figure 7.40 Damaged concrete column

ACTIVITY

When you're travelling around the area where you live, look out for examples of damaged concrete. Check out bridges and flyovers, concrete structures and paths. When you come across some damaged concrete, take a photograph with the camera on your phone (if you have one) and discuss the example with someone else. See if you can identify the cause of the damage you've seen.

Repair of faults and damage in concrete can be achieved in a number of ways dependant on the type of damage.

- Patching – when a small area of damage can be repaired by introducing a small quantity of concrete mixed to a suitable strength and consistency to the cleaned and prepared area.

- Re-finishing – when the damaged finish of a concrete surface is replaced with new concrete 'keyed' to the old material. Resin-based products may be more appropriate as a durable finishing material.
- Replacement of sections – when larger areas have suffered damage and an entire section must be cut out and replaced.
- Replacement of reinforcement – if reinforcement has corroded seriously it may have to be replaced. This could mean that a section of concrete has to be cut out and recast with new steel reinforcement.

▲ Figure 7.41 Concrete poured over replaced reinforcement

To repair concrete successfully, you will need to give careful consideration to the reasons why damage or failure occurred. The repair should be made in such a way that the possibility of future deterioration is minimised.

Methods of mixing concrete

Concrete can be mixed by hand (which is heavy work), by machine on site or in factory conditions off site, usually located in or near a stone quarry as the source of aggregate. Off-site manufacturing facilities are commonly known as 'batching plants' because the quantities of materials for the concrete are measured in batches by weight or volume. This is a very accurate method of measuring quantities which ensures that the finished product meets the specified standards and maintains consistency between ready-mixed loads delivered to site.

▲ Figure 7.42 Concrete batching plant

Measuring quantities to mix concrete on site by machine or by hand must also be done with consistency and quality in mind. Measuring accurate quantities of materials for smaller amounts of concrete is known as 'gauging' and to measure quantities accurately, a piece of equipment known as a 'gauge box' can be used.

A gauge box is a bottomless steel or timber square or rectangle which is placed on a clean flat surface and filled with coarse aggregate, fine aggregate or cement flush with the open top of the box. If a ratio of 1:3:5 is required for the concrete mix, five boxes of coarse aggregate and three boxes of fine aggregate (sand) will be mixed with one box of cement. (A suitably-sized strong bucket could also be used.) To mix larger amounts, multiples of the gauged amounts will be used.

IMPROVE YOUR MATHS

You are instructed to mix concrete for a new path. The ratio for the mix is 1 part cement, 3 parts fine aggregate and 5 parts coarse aggregate, which will produce $0.5\,m^3$ when using the gauge box you have been given.

If the path needs $4.5\,m^3$ of concrete to complete, how many 1:3:5 batches of gauged materials will you need?

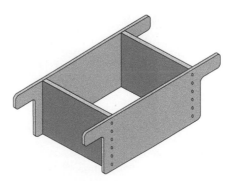

▲ Figure 7.43 Steel gauge box

The carefully measured separate piles of aggregate and cement can then either be mixed by hand or loaded into a drum mixer to produce a consistent product.

▲ Figure 7.44 Drum mixer

Methods of placing concrete

Once concrete is mixed and ready for use, it must be transported to the work area and placed in position ready for finishing in accordance with the specification. Concrete is a heavy material to move and position and may require mechanical means to get the job done efficiently. Moving and placing the materials by wheelbarrow may be too difficult and time consuming and may increase the risk of injury.

There are many pieces of equipment and methods that can be used to position and finish concrete. Table 7.5 describes some of them.

▼ Table 7.5 Equipment and methods used to place and finish concrete

Method	Description
Pouring concrete 	Concrete can be poured directly from the delivery lorry if access is suitable and the ground conditions are firm enough to support this heavy machine. The concrete is poured down a chute at the rear of the lorry, allowing accurate placement if the pour is managed well. Additional sections can be added to the chute to extend the reach of the pour if needed.
Pumping concrete	A concrete pump can be used to transport concrete efficiently from the delivery lorry to site locations that are difficult to access. This method minimises waste and allows precise placement of the concrete with relative ease. The pump can be a large stand-alone piece of equipment that can reach over a long distance or up to high locations. A smaller version can be attached to a concrete delivery lorry to make a compact versatile unit.

→

▼ Table 7.5 Equipment and methods used to place and finish concrete (continued)

Method	Description
Vibrating concrete 	Removing air bubbles from the mix will make concrete stronger, denser and more durable. This can be done by vibrating the concrete immediately after placement. The vibrating 'poker' as it is often referred to, is commonly powered by compressed air, which rapidly vibrates a short heavy steel bar immersed in the freshly poured concrete. This causes air bubbles to come to the surface of the concrete, increasing its density.
Tamping concrete 	'Tamping' is a method of levelling and smoothing concrete with a timber, steel or aluminium beam ready for final finishing. If the process is undertaken vigorously, it can also serve to compact and compress the concrete slightly which will add to its strength and durability. The water content and workability of the concrete when mixed and placed in position will affect the quality of the tamped finish.
Floating concrete 	'Floating' describes a method of providing a finish to the concrete surface. It can be done by hand using a steel (plastering) trowel and timber or polyurethane float to smooth the surface as required. 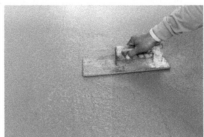 ▲ Floating concrete by hand A faster and more efficient method of finishing the surface to a high standard is to use a power float. Success depends on good judgement in timing when to start the process and adjusting the angle of the machine blades correctly to produce a hardened polished finish.

Underpinning a building

Let's look at a specific use of concrete in a specialised situation. As a bricklayer, you may well be involved in more specialised (and challenging) work at some point in your career.

Previously in this chapter, we discussed some causes of cracking in masonry walls. One cause of masonry cracking can be settlement or subsidence, causing the foundation of a structure to move downwards, affecting the masonry wall bearing on it.

A method of dealing with this fault and the associated damage to masonry is to use a process called **underpinning** of the foundation to prevent or limit further movement. This involves stabilising the failed or damaged foundation by excavating below it a section at a time, and installing an additional specified concrete foundation underneath to support it.

KEY TERM

Underpinning: to prop up or support from below; to replace or strengthen the foundation of a building

To link each section of newly installed concrete foundation, steel reinforcement bars (sometimes referred to as 're-bar') will be installed with the ends of the bars projecting horizontally from the newly installed concrete to tie in to the section to be installed next to it.

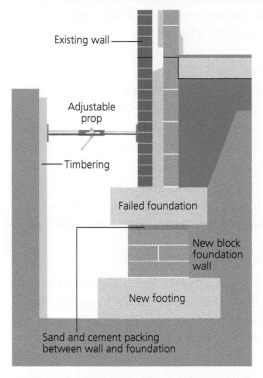

▲ Figure 7.45 Section view of a foundation being underpinned

Safety and efficiency

It is important to undertake the work according to a carefully pre-planned sequence to avoid collapse of the structure during operations. Look at Figure 7.46 to see how the sequencing works.

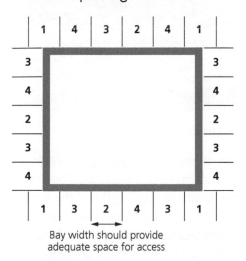

▲ Figure 7.46 Plan view of a foundation and the order of work numbering the bays to be removed and replaced in specific order

VALUES AND BEHAVIOURS

The process of underpinning can be time consuming, expensive and messy. It can involve working in wet, muddy excavations. It is very important to prevent mess and mud from affecting the area surrounding the property being worked on. If excavated material must be taken off site, give consideration to keeping roads and footpaths clean as lorries leave the site.

HEALTH AND SAFETY

There is the potential for collapse of the excavation if it's not properly managed, and consideration must also be given to the location of underground services such as gas, electricity and water, which can create a hazard. As with all construction-related activities, you must be constantly aware of potential threats to your own safety and that of others working with you.

There is now a range of specialist methods to underpin and stabilise a foundation that has sunk downwards causing damage to a structure, including the use of hydraulic jacks, high pressure grout injection and screw anchors. They achieve a result like the one described above but are less time consuming and labour intensive.

ACTIVITY

Go online and research the following types of underpinning.
- Mass concrete
- Screw piles and brackets
- Pile and beam

Select one method and write a short report describing how the selected system of underpinning works. List what you think the advantages might be of using the selected system.

4 HOW DOMESTIC DRAINAGE SYSTEMS WORK

Drainage systems are an essential part of a building's design and layout. Without a well-designed, efficient drainage system there would be no means of disposing of waste liquids generated by the activities of the building's occupants or rainwater falling on the structure.

Domestic drainage systems deal with two main categories of waste liquids:

- **foul water** – sewage from toilets and waste water from sinks, baths and appliances such as washing machines
- storm (or surface) water – rainwater falling on the roof and surrounding grounds of a building.

The bricklayer may be involved in the maintenance and repair of drainage systems and might be called on to repair or rebuild masonry inspection chambers.

KEY TERM

Foul water: includes sewage from toilets and waste water from sinks, baths and appliances such as washing machines

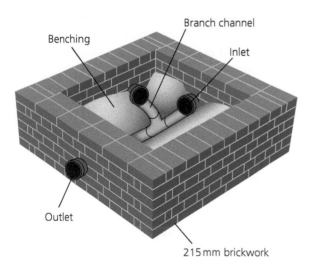

▲ Figure 7.47 Benched drain

When building new structures, such as an individual dwelling or an extension, the bricklayer may be involved in connecting the new drainage system to an existing drainage system. It is therefore important for the bricklayer to understand the types of drainage systems in use and how they work.

Types of drainage systems

There are three main types of domestic drainage systems, each of which has advantages and disadvantages in terms of efficiency and cost effectiveness.

Combined system

Combined drainage essentially means that foul and storm water run in a single pipe together. This was common practice in Victorian times and the advantage was that periodic heavy rainfall would have the effect of flushing the system to prevent the build-up of solid material in the pipework. However, treatment of the waste was minimal and the discharge into the sea created pollution at a level that is unacceptable today.

Separate system

With the determination to reduce pollution in the sea, greater efficiency in sewage treatment was required. However, sewage treatment can be costly in terms of energy use. Treating combined waste water meant that rainwater which could be drained straight into the sea without causing any pollution was going through a sewage treatment plant unnecessarily.

▲ Figure 7.48 Sewage treatment plant

To remove the cost of needlessly treating rainwater, separate drainage systems were introduced. This means that the foul water (containing sewage) and the storm water from rainfall run in separate pipework. The foul water can be directed to a treatment plant and the storm water can be directed to a natural drainage feature such as a river, or direct to the sea, where appropriate, without causing pollution.

Partially combined system

A partially combined system takes part of the storm water drainage and combines it with the foul water drainage to provide a measure of flushing action. The remainder of the storm water is channelled to a separate system. This means that some storm water goes to a sewage treatment plant, reducing the cost effectiveness.

Septic tanks and cesspits

If mains drainage is not near enough to a building to allow cost effective connection, it may be necessary to install a septic tank.

▲ Figure 7.49 Septic tank

This is a treatment system that allows foul water waste to decompose to a degree that allows liquids to discharge into the ground nearby without posing a health threat. It consists of a buried tank constructed of concrete, steel or fibreglass which the foul waste (not the storm water) is drained to.

KEY POINT

'Mains drainage' is the expression used to describe the larger diameter pipes that conduct waste water to treatment plants (for sewage) or to 'outfalls' (for storm water). Outfalls are the drainage openings at a river or the sea.

The internal structure of a septic tank is split into two chambers which allow bacteria to work on the waste in stages to break it down into a sludge. This settles to the bottom and is periodically removed. The remainder of the liquid waste is suitable for draining to a **soakaway** or a system of buried **perforated** pipes that disperse it over a wide area.

KEY TERMS

Perforated: having lines of small holes
Soakaway: a hole dug in the ground, filled with rubble and coarse stone which allows liquid to slowly drain into the surrounding earth. A soakaway can also be used to drain storm water where mains drainage is not available

ACTIVITY

Use the internet to research 'percolation test' and write a short report on how a percolation test is conducted and what its purpose is.

▲ Figure 7.50 An installed septic tank

A cesspit is simply an underground tank that is used to store foul waste and sewage. It does not process the waste product but stores it to await removal by a licenced specialist disposal company when the tank is full.

Materials used in domestic drainage systems

The main material used in the manufacture of drainage pipes and joints for domestic drainage systems is now plastic. It can be formed into a vast range of shapes and diameters to allow any requirement to be met when designing a new drainage system or adding drainage components to an existing system.

▲ Figure 7.51 Plastic drainage pipes

Plastic pipes have the advantage of being slightly flexible, which allows them to accommodate any small ground movements when buried. The type of plastic used is resilient and long lasting and is not affected by chemical cleaning products used in the bathroom or kitchen of the modern home.

▲ Figure 7.52 Plastic pipe straight coupler, bend, junction and reducer

KEY POINT

Plastic pipes are manufactured in orange and grey plastic materials. The orange coloured pipes are intended for use underground since the material from which they're manufactured is resistant to acids and salts often occurring in soil. The grey plastic pipes are designed to be used above ground and are manufactured from plastic material that is resistant to being degraded by ultraviolet light from the sun.

In the past, pipes were more commonly manufactured from clay as a ceramic product. Drainage pipes and other components are still manufactured in clay but are now designed with jointing systems that are easier to use compared with older methods of installation.

INDUSTRY TIP

In the past, clay pipes would be joined and sealed using tarred hemp rope and cement mortar. Clay pipe sections were sometimes referred to as 'salt-glazed' because of the process of firing and glazing in a kiln to produce a durable product.

Lengths of pipe and a range of bends and junctions are available in different diameters manufactured in both plastic and clay.

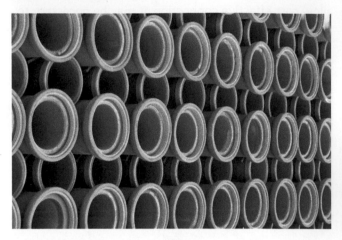

▲ Figure 7.53 Modern ceramic/clay pipes

ACTIVITY

Visit Wavin's online shop at: www.wavin.com/en-GB Select '110–160 mm drain pipes' from the 'Rainwater and stormwater' tab and look at the range of fittings and joints available. Pick one of the following and write a short report on what you think the chosen item is for:

- Covers and frames
- Couplers
- Rodding point

(If the item you select is not on the webpage, use the search facility to find it.) Compare what you've written with someone else, then pick another item and repeat the process.

Types of inspection chambers

Before plastic came on the scene, most shallow inspection chambers were constructed in brickwork and this is still the case where it is decided that masonry is more appropriate. The brickwork is constructed in English bond since this is the strongest brick bond and is more able to resist the ground pressures that develop around an inspection chamber.

Benching is formed either side of a half-channel to maintain the efficient flow of foul or storm water. Step irons are built into the masonry to make access into the chamber easier and safer. Look at Figure 7.54 which shows a brick inspection chamber to see how it is constructed.

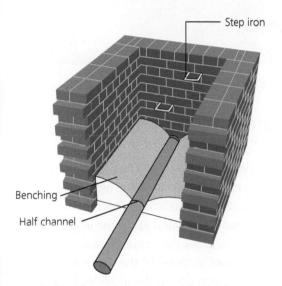

Step iron

Benching

Half channel

▲ Figure 7.54 Inspection chamber in English bond

KEY POINT

If a work task involves connection to or repair of an existing inspection chamber, it may involve cutting through the side of the inspection chamber and the benching to introduce a new channel. This is sometimes known as a 'slipper'. The slipper bend will be arranged to conduct the waste water into the half-channel (or invert) smoothly to maintain an uninterrupted flow.

Larger inspection chambers could also be constructed in brick with the thickness of the masonry increased to deal with greater depth. However, it may be decided that this is too labour intensive and expensive. In addition, the bricklayer could spend long periods working in deep excavations which could have health and safety considerations.

HEALTH AND SAFETY

Working in deep excavations exposes workers to the risk of collapse of the surrounding soil. The soil or other surrounding material in deep excavations should be supported by shoring installed by trained and competent personnel or, alternatively, the sides of the excavation should be angled or battered if appropriate, to minimise the risk of collapse. Your supervisor or line manager must ensure the safe condition of the excavation before anyone is allowed to enter. As a bricklayer, keep the following important points in mind when working in excavations:

● Never work in an unsupported trench or excavation where there is a risk of collapse.
● Keep machinery a safe distance from the excavation edge.
● Never work underneath an excavator.
● Be alert to risks from underground services or undermining adjacent structures.
● Maintain fencing and other safety measures to protect other workers nearby.

A quicker method of constructing larger or deeper inspection chambers is to use prefabricated concrete sections.

Sections of a concrete inspection chamber can be manufactured in a range of sizes and shapes to suit different requirements. The chamber can be assembled in a short time and the excavation quickly filled in around the chamber to restore a safe working area.

▲ Figure 7.55 An inspection chamber being constructed from concrete sections

---

INDUSTRY TIP

Inspection chambers used to be referred to as 'manholes'. They are still sometimes called that.

In a domestic drainage system, the pipework and associated inspection chambers are likely to be installed at a relatively shallow depth. Inspection chambers manufactured in plastic are easy to install in shallow excavations since they are light to work with and versatile in the way that they can accommodate multiple pipe junctions.

▲ Figure 7.56 A small plastic inspection chamber at a pipework junction

The type of inspection chamber cover will be selected according to the level and frequency of traffic crossing it. If vehicles cross the manhole cover frequently, it will need to be stronger than if just pedestrians are expected to be travelling above the inspection chamber location.

Covers and the frames that they sit in can be made from various thicknesses of steel or cast iron to cope with the expected load they will carry. Smaller inspection chambers can have covers manufactured entirely from heavy duty plastic.

▲ Figure 7.57 Storm drain inspection chamber cover

Falls or gradients for domestic drainage systems

For a drainage system to efficiently transport waste liquids and any associated solids to the desired destination, it must be laid to a 'fall'. This means that a steady gradient must be created along the length of the drainage pipework to allow the contents to be moved by gravity. If this is not achievable, the contents of the drain will have to be pumped to the intended location, adding to costs.

The fall is expressed as a specified ratio to indicate the gradient:
- 1:40
- 1:60
- 1:100.

The smaller the second number in the ratio, the steeper the gradient or fall. The correct fall will be determined by the diameter or bore of the pipework and the type of waste to be carried by the drainage system.

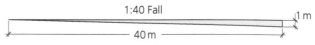

▲ Figure 7.58 How the fall ratio is applied

If a fall or gradient is too steep, any solid materials in the waste could be left standing in the pipe as liquids move quickly onwards. If the fall is too shallow, the risk of blockage is higher since the movement of the waste will be slower.

Setting the fall can be achieved in several ways. An optical level can be used after a period of training. This is an instrument which needs to be used and stored carefully to maintain accuracy.

A less technical method is to use boning rods. These are simple T-shaped pieces of equipment which are relatively easy to construct on site from available timber. Sighting rails are set up at intervals along the drain run and positioned at heights to suit the specified fall or gradient. The boning rods are constructed to a length that corresponds to the set distance between the sighting rail and the top of the drainpipes being laid.

The principle is that once sighting rails with fixed levels are established, any level along the gradient between them can be confirmed by a line of sight lining up the top rail of the boning rod with two sighting rails. Study Figure 7.59 to see how this works.

Use of an optical level will require two personnel to set accurate levels. One operative 'sights' through the instrument to take readings from a staff with graduated markings, held by a second operative. To work out the required fall between two points will need some knowledge of surveying techniques and principles.

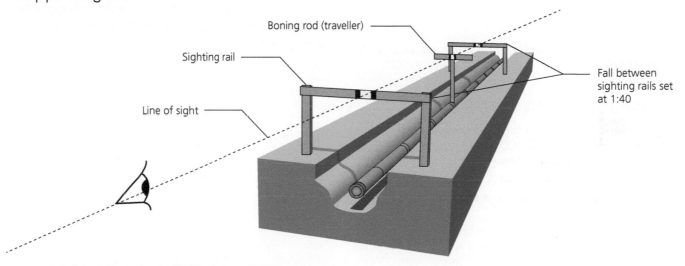

▲ Figure 7.59 Using the line of sight between sighting rails and a boning rod

▲ Figure 7.60 An optical level set up ready for use

▲ Figure 7.61 Using an optical level to check the fall of a drainpipe

A third method for laying pipework to the correct fall is to use a tapered length of timber. The timber is tapered to suit the required fall and is placed on top of the pipe being laid. When the spirit level is placed on the edge of the timber and indicates accurate level has been achieved, the pipe will be sitting at the correct angle of fall.

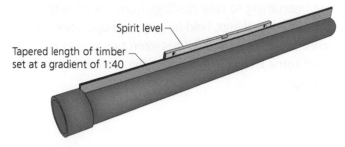

Spirit level

Tapered length of timber set at a gradient of 1:40

▲ Figure 7.62 Using a tapered length of timber and spirit level to set the fall

IMPROVE YOUR MATHS

If you needed to make a tapered straight edge from timber 20 mm thick and 100 mm deep, what length of timber would you need to create a fall of 1:40?

Next time you're on site, find a suitable piece of timber and draw the taper on it to check your conclusion.

Summary

The range of potential faults that can occur in masonry, concrete and drainage systems is extensive. To be effective in tackling repairs and maintenance, you need to develop the ability to analyse the cause of faults and carefully plan your work to avoid causing further damage to a structure or its surroundings. Keep safety matters to the fore when planning repairs and maintenance; many of the tasks detailed can be hazardous if not performed properly.

A great deal of information is contained in this chapter, much of which is specialised in nature. Take time to review and absorb it frequently to safely equip yourself for work tasks that can be both challenging and interesting.

Test your knowledge

1 Which of the following can cause spalling in brickwork?

   a Laying the bricks the wrong way up

   b Laying the bricks with oversize bed joints

   c Rainwater soaking through the brick

   d Water freezing in the body of the brick

2 What type of crack travels vertically through both the brickwork and the bed joints?

   a Movement

   b Non-structural

   c Settlement

   d Structural

3 What is sulphate attack caused by?

   a A chemical reaction between the sand and the bricks

   b A chemical reaction caused by dirty water

   c A chemical reaction in concrete blocks

   d A chemical reaction with the cement in the mortar

4 What is a 'tell-tale' used for?

   a To find the overall height of a wall

   b To determine the maximum thickness of masonry in a wall

   c To measure movement either side of a crack in a wall

   d To measure the strength of a wall

5 What name is given to the process of water rising through microscopic gaps in building materials?

   a Capillary attraction

   b Chemical reaction

   c Moisture transfer

   d Water penetration

6 From what material must wall ties be manufactured?

   a Aluminium alloy

   b Copper

   c Iron bar

   d Stainless steel

7 What is the correct PPE to use when using a grinder to cut out mortar joints at ground level?

   a Gloves, ear defenders, safety boots, dust mask and a combined helmet/visor

   b Gloves, safety boots, dust mask and a combined helmet/visor

   c Goggles, safety boots, dust mask and a combined helmet/visor

   d Safety boots, gloves, ear defenders, safety harness and a fall arrest system

8 What is the name of the component inserted through a wall which provides temporary support when removing a lintel?

   a Dead shore   b Needle

   c Steel prop   d Wall plate

9 What name is given to the process of adding support beneath a failed or damaged foundation?

   a Dead shoring   b Oversailing

   c Reinforcing   d Underpinning

10 What component is built into an inspection chamber to allow easier access?

   a Step beam

   b Step frame

   c Step iron

   d Step rung

11 Explain what is meant by 'preventative' and 'responsive' in the context of maintenance and repair of masonry structures.

12 Describe one cause of spalling in reinforced concrete.

13 State one cause of the following three types of cracking in masonry:
   • Vertical cracking
   • Horizontal cracking
   • Diagonal cracking

14 State three methods that can be used to support the masonry above a window opening when removing a defective lintel.

15 Explain the difference between a combined drainage system and a separate drainage system.

235

CHAPTER 8

CONSTRUCTING DECORATIVE AND REINFORCED BRICKWORK

INTRODUCTION

Many modern masonry buildings incorporate decorative design features such as ornamental brick panels, oversailing and string courses and brick specials to add interesting architectural detail and visual appeal to their appearance. There has also been a growing interest in conservation work, with decorative brick features in many older buildings being repaired and restored to their original specifications.

In this chapter we will look at how we set out and build decorative brick features. To produce decorative work to a high standard, the bricklayer must develop an eye for detail when preparing, setting out and constructing the work. In some cases, the bonding arrangement of decorative brickwork may mean that reinforcement of the masonry must be introduced to make it stable and durable, so we will also look at methods of achieving this.

By the end of this chapter, you will have an understanding of:
- methods and techniques used to build decorative brickwork features
- setting out and building obtuse- and acute-angle quoins
- when to reinforce brickwork and how to do it.

This table shows how the main headings in this chapter cover the learning outcomes for each qualification specification.

Chapter section	Level 2 Technical Certificate in Bricklaying (7905-20)	Level 3 Advanced Technical Diploma in Bricklaying (7905-30) Unit 304	Level 2 Diploma in Bricklaying (6705-23)	Level 3 Diploma in Bricklaying (6705-33) Unit 304	Level 2 Bricklayer Trailblazer Apprenticeship (9077) Module 7
1. Methods and techniques used to build decorative brickwork features	N/A	1.1, 1.2	N/A	1.1, 2.1, 2.2, 2.3, 2.4, 2.5	**Skills Depth:** 1.1, 2.1 **Knowledge Depth:** 1.1, 2.1
2. Setting out and building obtuse- and acute-angle quoins	N/A	2.1, 2.2	N/A	3.1, 3.2, 4.1, 4.2, 4.3, 4.4, 4.5, 4.6	**Skills Depth:** N/A **Knowledge Depth:** 4.1, 4.2
3. When to reinforce brickwork and how to do it	N/A	3.1, 3.2	N/A	5.1, 5.2, 6.1, 6.4, 6.5	**Skills Depth:** N/A **Knowledge Depth:** N/A

1 METHODS AND TECHNIQUES USED TO BUILD DECORATIVE BRICKWORK FEATURES

Brick is one of the most versatile and durable construction materials available. The different colours, shapes and textures obtainable allow bricks to be used in an almost endless number of imaginative ways. The decorative bonds and patterns discussed in this chapter have been in use for centuries and are still very much in use today.

Decorative brickwork is carried out to improve the **aesthetics** of plain masonry and to make it look more interesting. That is why decorative brickwork must be constructed to high standards so that the intended visual effects are created.

> ### KEY TERM
>
> **Aesthetics:** related to attractiveness and appeal

Planning and preparation

Before any work is carried out on site, careful planning and preparation are essential to achieve a successful outcome. By planning ahead and giving careful thought to organising the sequence of work, potential problems can be identified and avoided.

Careful preparation of the work area, materials and tools before starting work will make it easier to produce a piece of decorative brickwork you will be proud of. Having the right 'mind-set' about the job is an important part of the preparation and will help you to focus on getting the detail right. Rushed or sloppy work is never acceptable when bricklaying; this is especially true when constructing decorative work where investing time and care will ultimately produce a better finish.

> ### ACTIVITY
>
> You will come across many examples of decorative brickwork as you travel around. (You may be surprised at how many there are when you start to take notice.) Stop and examine examples you come across and note down what you think are faults that affect the visual appearance of the work. Take a photograph and discuss the quality of the work with others.

Information sources

To ensure the work is carried out correctly, you must pay close attention to the relevant drawings to establish the correct location and type of work required. When reading drawings, study the whole drawing first to establish an overall picture of the work that needs to be carried out. Then consider these points:

● Do you need to order any special bricks or components in advance?
● Is there any part of the drawing that needs further clarification before work starts?

Review the information about drawings on page 38 paying particular attention to the description of detail drawings.

> ### INDUSTRY TIP
>
> If you are in any doubt about the details of the decorative feature you're going to build, get further clarification from your supervisor straight away. Don't wait until after you have started the work to ask questions.

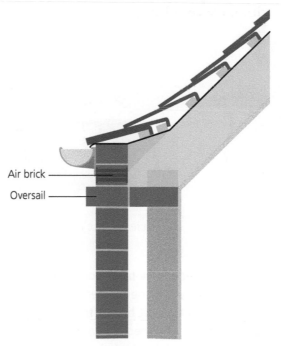

Air brick
Oversail

▲ Figure 8.1 Detail drawing showing a section through an oversailing detail

With any construction activity, it is always important to check the following before commencing work:

- Risk assessments – these help to organise a working area that's safe for you and those working nearby. Decorative brickwork can often require a lot of brick cutting, which can create a number of hazards.
- Method statements – these are helpful in planning the sequence of work for decorative brickwork, since you may not come across the task you are working on very frequently. It is therefore helpful to spend some time getting familiar with the methods and techniques required before starting work.

Let's look at some points about cutting bricks to form decorative features.

Cutting bricks for decorative work

To reduce the risk of accidents, any cutting of bricks should be carried out at ground level, if possible. Cutting bricks on an elevated scaffold platform increases the risk of waste materials falling on workers below. Accumulated waste materials can also form a trip hazard for those working on the scaffold platform.

If a large number of bricks must be cut, it can be more efficient and safer to cut all the bricks in one operation. This can also improve productivity levels by maintaining a good flow of work.

Cutting bricks by machine

▲ Figure 8.2 Masonry bench saw/clipper saw

▲ Figure 8.3 Disc cutter

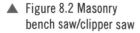

It is often the case that decorative brickwork will require cuts formed to an angle. If engineering bricks or perforated wire-cut bricks must be cut to an angle, it may be easier to cut them by mechanical means. Carefully consider the following points about using machinery to cut bricks.

- You must be fully trained and competent to use disc cutters and table-top cutting equipment. Under no circumstances are you allowed to change a blade on a cutter without specific training.
- When using any mechanical cutting equipment, it is important that you follow the manufacturer's instructions on the safe way to use it.
- Use the correct PPE:
  - Hard hats are to be worn where there is a risk of falling objects.
  - Safety glasses must be worn to prevent any flying brick fragments or debris causing eye injury.
  - Wear a suitable dust mask when dust is generated. If fumes and smoke are present, make sure that suitable respiratory protection equipment (RPE) is worn.
  - Use hearing protection. Brick cutting can be noisy, especially when disc cutters are being used.
- Clothing should be sturdy and snug fitting: any loose clothing could get caught up in the machinery. For the same reason, long hair should be tied up and scarves and jewellery should not be worn.

> **HEALTH AND SAFETY**
>
> Always make use of the correct ear protection when cutting bricks – hearing loss can be permanent.
>
> Cutting bricks with a disc cutter often gives off an enormous amount of dust. Materials such as bricks, concrete and stone can contain large amounts of crystalline silica. Cutting these materials produces airborne dust particles that can penetrate deep into the lungs, causing serious health effects such as lung cancer or silicosis.

Under COSHH (2002), employers not only have a duty of care to protect workers but also the general public who may be nearby. **Water suppression systems** and **local exhaust ventilation (LEV)** systems might need to be put in place to control dust.

> **KEY TERMS**
>
> **Water suppression system:** a system where water is sprayed onto the cutting disc to keep dust to a minimum when cutting bricks, blocks and stone
>
> **Local exhaust ventilation (LEV):** an engineered control system that reduces exposure to airborne contaminants by sucking the dust and fumes away from the workplace

It is the responsibility of the employer to ensure that employees are correctly trained in the use of powered equipment used to cut masonry and that supervision is given to new trainees. All items of equipment must be properly maintained and safe to use. Even once you have been trained, take your responsibilities seriously, as mistakes could prove fatal.

Cutting bricks by hand

When cutting bricks by hand, using the correct PPE is just as important as when using a machine. There are still many hazards that can cause injury. Try to have a separate cutting area away from your work area, to keep things clear of debris and to prevent any dust and flying brick fragments affecting other workers.

Look at the following step by step to see the sequence of operations when cutting bricks to an angle by hand. Notice that a string line has already been set to the required angle as a guide.

Cutting a brick to an angle

STEP 1 Place the brick to be cut, without a mortar bed, in position on the wall. Set it 20 mm (two cross joint thicknesses) away from the brick already laid in that course.

STEP 2 Put a pencil mark where the guide string line crosses the top and bottom arris of the brick.

STEP 3 Draw a straight line between the two pencil marks to show the line of the cut.

STEP 4 Extend the marking onto the rear of the brick to ensure that the brick is cut accurately across its width.

STEP 5 Mark the part of the brick that is wastage with an 'x' or a 'w', then place it onto a cutting mat.

STEP 6 Before cutting the angle, cut the brick 'square' with a hammer and bolster, placing the bolster edge slightly onto the waste side of the brick.

STEP 7 With the bolster chisel carefully placed on the line of the angle, give a gentle tap with the lump hammer to 'bed' the blade of the bolster in position.

STEP 8 Without moving the bolster, give a sharp blow with the lump hammer to produce the angled cut. Produce the angle on the rear face of the brick with a separate blow if needed.

STEP 9 With a scutch hammer, trim any excess to the required angled line as needed.

STEP 10 Lay the brick in position, aligning it to the guide string line.

ACTIVITY

In your workplace or training centre, try cutting some bricks to different angles and lengths. Compare your results with your peers to see how accurate you are. Don't forget to use safety glasses!

Decorative brick panels

Decorative panels can be creatively designed to enhance the appearance of an area of plain masonry. They can be constructed in the half-brick outer leaf of a cavity wall or incorporated into the face of a solid wall.

The most commonly constructed types of decorative panel are herringbone and basket-weave patterns. Let's look at how to set out and build these two decorative designs and some of their possible variations.

Setting out herringbone panels

The herringbone pattern gets its name because it resembles the shape of fish bones. A 'zigzag' pattern is created by setting the bricks at 90° to each other.

This can be carried out vertically, horizontally or diagonally.

Herringbone can be achieved using single or double stretcher bricks laid at angles to each other. Look at the variations in Figures 8.4, 8.5 and 8.6.

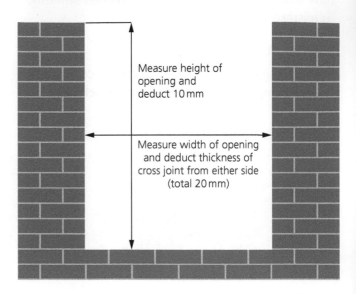

Measure height of opening and deduct 10 mm

Measure width of opening and deduct thickness of cross joint from either side (total 20 mm)

▲ Figure 8.7 Wall with opening ready to receive the decorative brickwork panel

To set out a decorative herringbone pattern, first establish the dimensions of the opening the panel will fit into from side to side and from top to bottom. Deduct 20 mm from the width to allow for a 10 mm joint at either side edge of the panel and 10 mm from the height of the panel to allow for the bottom bed joint.

INDUSTRY TIP

You only need to deduct 10 mm from the height dimension of the panel since the brickwork at the top of the panel should finish flush with the top of the surrounding masonry. Study Figure 8.7 to see how this works.

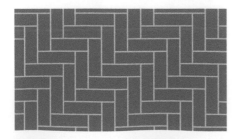

▲ Figure 8.4 Diagonal herringbone pattern

▲ Figure 8.5 Horizontal herringbone pattern

▲ Figure 8.6 Double diagonal herringbone pattern

Setting out methods

A common method employed to accurately set out decorative panels is to position the bricks on a sheet of plywood to establish which ones will be cut. The plywood sheet needs to be large enough to extend well beyond the edge lines of the panel.

The dimensions already established can now be used to draw the square or rectangular panel onto the plywood sheet. Extend each of the edge lines beyond the outline of the panel; this will be useful at a later stage of setting out.

Setting out lines are then drawn on the panel, diagonally from corner to corner and horizontally and vertically across the centres. This gives you the starting position of the first bricks and ensures that they are set out to the correct 45° angle.

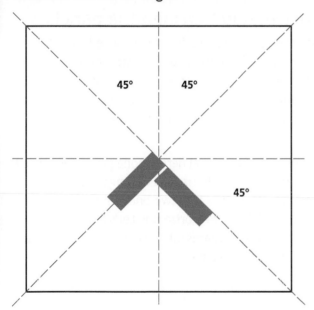

▲ Figure 8.8 Diagonal, vertical and horizontal setting out lines. Note the position of the first two bricks

The rest of the bricks can now be laid out dry on the board in the herringbone pattern, working outwards from the starting bricks. The setting out shown here is for horizontal herringbone decorative brickwork.

INDUSTRY TIP

A vertical herringbone pattern is very similar to a horizontal herringbone pattern. To set it out, simply follow the sequence for a horizontal herringbone pattern and rotate the starting bricks through 90°.

KEY POINT

Make sure that you position your first bricks at the central point, as shown in Figure 8.8. This will help you to get an even cut along the edges of the panel. Work your way carefully outwards, positioning the bricks so that they will have even joints on all sides.

Extend the pattern of bricks beyond the perimeter lines of your panel. Keep checking that the joints in the panel are even and visually line up with the vertical and horizontal lines on the plywood sheet. A lack of accuracy will lead to cut bricks being out of line when they are laid into the wall, spoiling the appearance of the panel.

Next, we need to transfer the edge lines drawn onto the plywood sheet onto the positioned bricks. The pencil lines we extended earlier from each edge line of the perimeter will help us to do this. In Figure 8.9, you can see how a spirit level is lined up with the extended lines and used as a straight edge to mark the outline of the panel onto the positioned bricks.

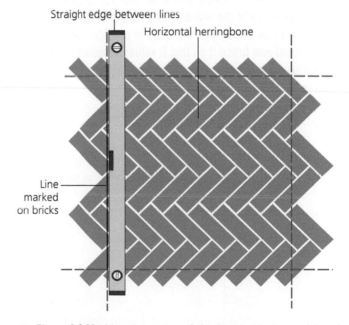

▲ Figure 8.9 Marking the outline of the panel onto the positioned bricks. Note the use of the extended lines

Cutting the marked bricks

The marked bricks can now be cut. As bricks are cut, position them back where they were located on the plywood sheet, ensuring that the edges form a neat line; if they don't, then re-cut them. Remember, an uneven joint line around the edges of the panels will spoil the decorative effect.

ACTIVITY

Using your marked sheet of plywood, try setting out (without cutting any bricks) a diagonal and double diagonal herringbone pattern.

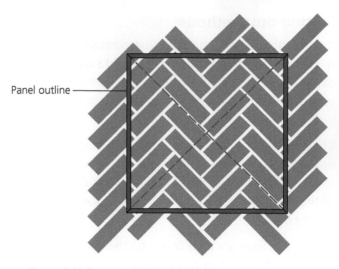

▲ Figure 8.11 Construct a lightweight timber frame to match the outside measurements of the panel and place it on the positioned bricks to mark them

Constructing herringbone panels

If the setting out process described above has been carefully carried out, the actual construction of the herringbone decorative panel will be simpler. Good preparation allows you to confidently approach the job with a clear understanding of the sequence of work in mind.

Remember that a vertical herringbone pattern is set out in the same way as a horizontal herringbone pattern; you simply set out the first two bricks at the centre of the panel at 90° to the horizontal. Let's look at how the preparation translates into a successful result for a vertical herringbone panel.

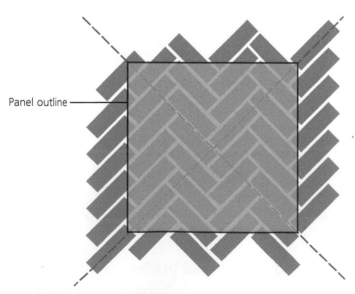

▲ Figure 8.10 Accurately position the bricks to cover an area larger than the proposed panel

Vertical herringbone panel

Study the step by step sequence to see how the vertical herringbone panel is constructed using carefully positioned string lines, gauge rods and templates.

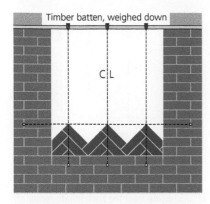

STEP 1 Set up the string lines as guides, making sure that they intersect at appropriate brick corner points in the pattern. The measurement for the height of the horizontal line can be taken from the bricks that were set out dry on the plywood panel or floor. Vertical lines must be positioned to ensure that the bricks line up accurately as the panel rises. Use a spirit level to check the lines are plumb.

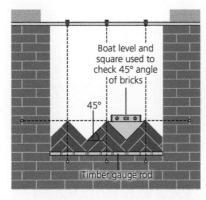

STEP 2 From the bricks previously set out dry, mark a gauge rod with the starting positions of the bricks to be laid for the first course. Use the rod as a template to lay and accurately position the first layer of the pattern, ensuring the top corner of each relevant brick follows the vertical line. Lay the bricks starting from the bottom corner working along the wall from corner to corner. Notice the small template used with a boat level to establish an accurate 45° angle.

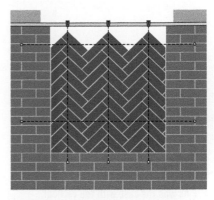

STEP 3 Continue with each course until the panel has been built up to the point where the last layer needs to be laid. Remember to tie in your panel to the masonry behind to ensure a secure fixing. (We'll look at methods of tying-in and reinforcing decorative work later in this chapter.) Step back from the job frequently to visually check your work, making sure that the joints line up with one another.

STEP 4 Position the string line across the top of the panel in line with the surrounding brickwork. To ensure complete accuracy on the top edge of the panel, you may find that you have to re-mark and cut these bricks to ensure an even bed joint is achieved.

STEP 5 Take time to joint your work neatly, being careful to make sure that you form a uniform and continuous joint around the border line of the panel. High quality of jointing is crucial to producing a herringbone panel that is aesthetically pleasing.

Diagonal herringbone panel

Next let's look at a variation of the herringbone pattern. This time study the step by step sequence to produce a diagonal herringbone decorative panel.

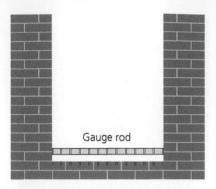

STEP 1 As was the case with the vertical herringbone panel, a timber gauge rod can be made from bricks positioned 'dry' during the setting out stage. Mark the gauge rod to show the intended brick positions along the bottom edge of the panel. The markings can be transferred onto the wall to make accurate laying easier.

STEP 2 Set up vertical and horizontal lines across the panel as shown here.

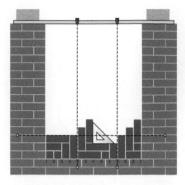

STEP 3 Starting from the bottom left corner of the panel, lay the bricks using the gauge rod to ensure the correct joint spacings between the bricks. Move the vertical line as necessary to follow the gauge rod. A small square can also be used as shown to make sure that the bricks are accurately laid at 90° to each other.

STEP 4 Work your way up the wall, ensuring that the soldier bricks line up with one another and are vertically plumbed up through the panel.

IMPROVE YOUR MATHS

Obtain a sheet of plywood and create a template with an accurate 90° angle to be used to maintain square. Set out the right angle for your template using the 3:4:5 method.

Setting out basket-weave panels

Basket-weave is a more straightforward pattern to construct than herringbone. To maintain an acceptable decorative appearance, the panel should be designed to suit standard brick dimensions. If cut bricks must be introduced because the panel measurements are not to standard brick dimensions, it will spoil the decorative effect.

Horizontal basket-weave

Horizontal basket-weave is created by laying three bricks vertically and three bricks horizontally in alternate groups throughout the panel. When setting out the panel, it's good practice to make sure that the bricks are carefully selected to match each other for size. Bricks that vary noticeably in length will be difficult to align accurately and will stand out when laid, spoiling the appearance.

▲ Figure 8.12 Horizontal basket-weave pattern

Sometimes contrasting bricks are specified for the panel to add to the decorative effect. The bricklayer must select bricks that are consistent in colour to maintain the desired contrast. Bricks tend to have multiple colours within them, and one colour can merge into another, so the bricklayer must develop an eye for selection of bricks that are suitable.

INDUSTRY TIP

Developing an eye for quality and accuracy in bricklaying is something that improves with experience. Lining things up for level and plumb should become a habit that will give you the ability to instinctively know when something needs adjusting.

To set out a horizontal basket-weave panel, you can position the bricks dry in a similar manner to setting out a herringbone panel, to make sure that uniformly-spaced bricks will fit the prepared opening in a wall. This method also allows you to view the grouped bricks from a distance to assess if the colour of bricks selected is uniform enough.

Alternatively, since the groups of three bricks laid in alternate directions will match standard brick dimensions, the panel can be constructed at the same time as the wall is raised if you are confident that the prepared wall opening has been built accurately to suitable measurements.

Diagonal basket-weave

A variation of the basket-weave decorative pattern is diagonal basket-weave. Bricks are laid at 45° to the horizontal line of the main wall to form the basket-weave pattern following diagonal lines. Like horizontal herringbone panels, diagonal basket-weave panels require a great deal of cutting to prepare the bricks that will form the edge of the panel.

Diagonal basket-weave panels are set out in a similar way to horizontal herringbone panels. The bricks are laid in a slightly different arrangement within the opening before being cut. How you want your panel to look and the types of cuts that it leaves at the edges of the panel will determine whether the first brick sits centrally in the opening or not.

Look at these examples of alternative setting out arrangements. Subtle differences at the start of the setting out process will produce a different appearance in the finished result.

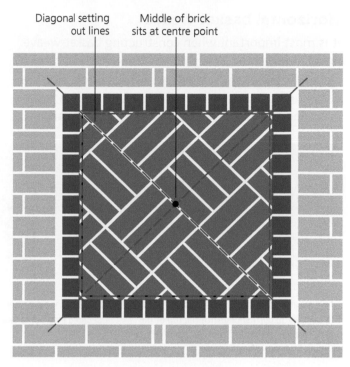

Diagonal setting out lines   Middle of brick sits at centre point

▲ Figure 8.13 Diagonal basket-weave panel

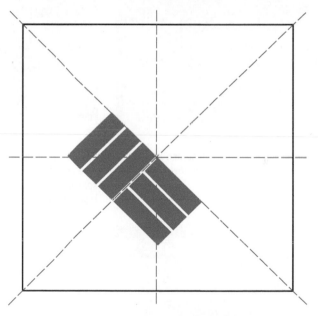

▲ Figure 8.14 Setting out for an alternative diagonal basket-weave

Constructing basket-weave panels

Remember, good preparation is a key part of producing a quality product when constructing decorative panels. Having planned carefully for constructing a basket-weave panel, let's look at the sequence of building for horizontal and diagonal basket-weave patterns.

Horizontal basket-weave

It is most important when constructing basket-weave panels to maintain straight lines through the face of the panel between each group of three bricks. You are essentially creating extended cross joints which must be aligned for the panel to look neat, accurate and symmetrical. To assist you in controlling the uniformity of the construction process, use a timber gauge rod with markings showing where each brick is to be laid on the first course of the panel.

The following step by step shows a horizontal basket-weave panel being constructed in a prepared opening.

STEP 1 Set up a string line horizontally across the opening to correspond with the top of the first groups of three bricks in the panel (three courses high in the wall opening).

STEP 2 Place a timber batten on top of the panel secured by placing blocks at either end.

STEP 3 Set up a vertical line from the batten at the top of the wall, held by a brick placed at the bottom edge of the panel. Make sure that your line sits at the end of the first full brick and check it for plumb using a spirit level.

STEP 4 Starting at the bottom corner, lay the first three bricks, ensuring each brick is plumb level and to gauge.

STEP 5 Continue laying the first course, making sure all the bricks are plumb with even joint spacings. If the joint spacings become uneven, take out the affected bricks and start again.

STEP 6 Progressively move the line as required for each course, ensuring that bricks line up vertically as well as horizontally throughout the panel.

STEP 7 Continue until you reach the top, working from left to right. Check the face plane of the panel. Joint up the brickwork.

As with any decorative panel you're working on, throughout the construction process, frequently step back from the work and check that it is consistent and uniform. Be diligent in controlling the build and adjusting where necessary.

Diagonal basket-weave

For a diagonal basket-weave panel, bricks are positioned accurately by using a boat level resting on a small timber square. This will ensure that a true 90° angle has been maintained between the bricks. A level can be used as a straight edge to check that the brick joints line up diagonally.

If the joint line is located centrally within the panel, a line can be attached diagonally from corner to corner to ensure correct alignment of the joints. It is important also to check the face-plane of your work with a straight edge as the work proceeds or alternatively place a line horizontally running through the brick face as the work proceeds.

If the panel is to be stepped back from the main wall, then a timber template can be made to fit across and into the opening which will maintain a consistent setback across the panel.

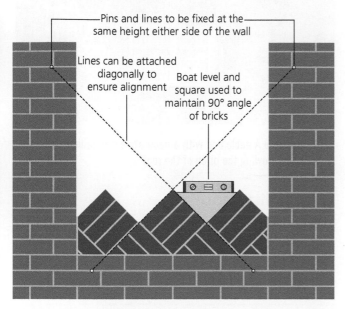

▲ Figure 8.15 Constructing a diagonal basket-weave panel using string lines

ACTIVITY

Carry out internet research into 'basket-weave brick panel'. Find an example of a stepped-back panel. After examining the arrangement of the panel, produce a sketch of a template that could be created to aid accuracy when building the recessed panel. Discuss your idea with others to decide if your idea for a template would work.

Setting out and building panel surrounds

A panel surround can be added as a decorative border with the panel built inside it. Panel surrounds are generally built from bricks laid as headers and can project from the face of the wall to further enhance the decorative effect.

If the face of the panel surround is to project from the face line of the main wall, it is important to plumb the outer vertical edges of the panel surround since this will be a prominent line of sight. Attach lines to the top

and bottom edges of the surround to make sure that a straight line of sight is produced horizontally.

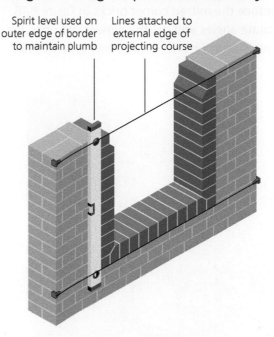

▲ Figure 8.16 Panel with projecting surround

It is essential when building the surround to ensure that the designed width of the opening remains the same throughout. This can be achieved by using a tape measure or, for ease of use and consistency, a **pinch rod** can be made.

KEY TERM

Pinch rod: a piece of timber cut to the width of an opening and used to check that the opening size stays the same width throughout the construction of the panel

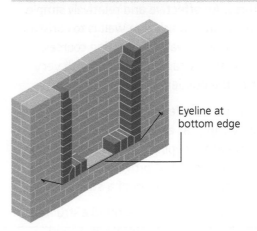

▲ Figure 8.17 Prominent lines of sight must be accurate and straight to produce work that looks visually acceptable

Carefully control the construction of the panel surround, frequently checking for plumb reveals and square of the opening. Notice the mitred corner bricks in Figure 8.18. To cut accurate angles and get the dimensions right, it's helpful to produce your own detail drawing at 1:1 scale.

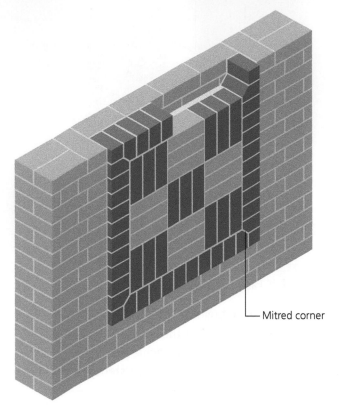

Mitred corner

▲ Figure 8.18 Panel surrounds can be constructed using mitred corners to improve decorative appearance

String courses

Bricks can be used in many imaginative ways to form decorative features. An effective and relatively simple way to form a decorative finish to a wall is to construct a string course. By using a variety of string courses, many decorative features can be achieved with very little extra cost to the builder.

String courses can be included at several positions in a structure. For example, at eaves level, at the top of a boundary wall beneath a coping or located in the face of a building at storey height to provide a visual break in the elevation between the floors of a structure.

Where the string course continues around a structure on all elevations, it is commonly known as a band or banding course. Let's look at different types of string course.

▲ Figure 8.19 A gable end with a decorative oversailing string course following the pitch of the roof

Oversailing courses

Oversailing courses are brick courses that extend beyond the face line of the wall. They can be used to increase the thickness of a wall as well as being used as a decorative feature.

KEY TERM

Oversailing: projecting from the main face of a wall

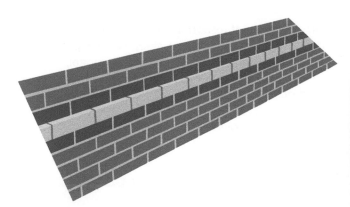

▲ Figure 8.20 An oversailing string course in the elevation of a wall

If several courses are stepped out from the face of the wall to increase its thickness, this is known as corbelling.

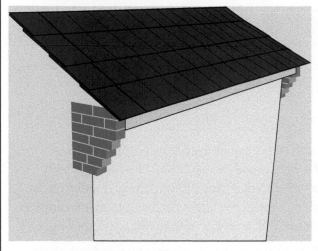

▲ Figure 8.21 Corbelling in a gable end

If an oversailing course is constructed above the eye line, it is best to position the string line on the bottom arris of the bricks when laying them, to make sure that the projecting course appears accurately aligned from the viewing position below it.

When constructing oversailing courses, pay attention to the stability of the wall, especially if a number of courses are stepped out as a corbel.

Dentil courses

Dentil courses are a tooth-like feature in the brickwork produced by laying alternate projecting and stepped-back headers within a course of bricks. Dentil courses should have at least one course of bricks laid over the

top of them to protect the projecting brick from the elements – an oversailing course.

When laying dentil courses, take care to make sure that the correct bonding arrangements are maintained so that straight joints are avoided. The headers must sit centrally above and below the stretcher brick face or cross joint in alternate courses. As with oversailing courses, position the string line on the bottom arris of the bricks if the work is above the eyeline.

▲ Figure 8.22 A dentil course

Dog-tooth course

A **dog-tooth** decorative feature can be a single course or multiple courses formed by laying individual bricks at a 45° angle to the face of the wall. It can be stepped back or projecting to provide additional visual interest.

KEY TERM

Dog-tooth: type of string course with bricks laid diagonally to the face line of the wall to give a serrated effect

INDUSTRY TIP

A template can be made to check that each brick is laid to the correct angle.

▲ Figure 8.23 A dog-tooth template used to ensure that the bricks are laid at 45° to the face of the main wall

As with dentil courses, an oversailing brick course should be laid directly over the dog-tooth course to protect it from the weather.

▲ Figure 8.24 Dog-tooth with brick course above for protection

It may be necessary to cut bricks at the ends of the dog-tooth course or courses to produce a pattern that is even across its length. It is good practice to build the masonry at either end of the feature first and set out the dog-tooth course 'dry' between the ends to establish an even pattern.

▲ Figure 8.25 Plan view of a dog-tooth course. Notice the cut bricks at either end to establish a uniform pattern

Plinth courses

Plinth bricks can be used to reduce the thickness of a wall, usually above DPC. They can be used to form decorative panels in garden walls and piers. In some cases, plinth bricks can be used to increase the thickness of walls in the same way as corbelling.

Plinth bricks have a 45° sloping face and reduce the thickness of a wall by 46 mm (or a quarter of a brick) per course. When setting out walls using plinth bricks, it is good practice to first set out the arrangement of the brickwork that will be above the plinth course 'dry', to make sure that half-bond is maintained throughout the elevation. Any broken bond requirement that is identified can then be arranged in the courses below the plinth course.

Once plinth bricks have been laid, it is important to cover them over with polythene sheeting built into the first bed joint above the plinth work so that any mortar droppings from the work carried out above do not stain them. Once the work is complete the sheet can be cut out with a knife and the joint filled as needed.

Plinth bricks are manufactured in a range of shapes and configurations to allow construction of plinth work in different shaped buildings.

ACTIVITY

Search 'plinth bricks' on Ibstock's website: www.ibstockbrick.co.uk. Look at the plinth external return and produce your own isometric sketch of the brick, complete with labelled dimensions. Write a short report explaining why you think there are two options for dimensions available.

▲ Figure 8.26 An internal plinth return

▲ Figure 8.28 Cant bricks used to form a sill

▲ Figure 8.27 An external plinth return

▲ Figure 8.29 A chamfered corner using single cant bricks laid flat

Using cant bricks

Cant bricks have a 45° angle on the header face. They are made with either one angled face known as a single cant, or with both header faces angled known as a double cant. They can be used to form a sill (or cill) for a window or a brick-on-edge coping for a boundary wall.

When laid flat, single cant bricks can be used to form chamfered corners in right-angled quoins or piers.

INDUSTRY TIP

When constructing a quoin using cant bricks to form a chamfer, be careful to maintain plumb on all faces, including the face of the chamfer.

Constructing circular and straight ramps

Ramped work is constructed from brickwork that has a sloping edge to the face, such as the line following the pitch of a gable end. A ramp can also be constructed to allow a smooth, attractive change of height in a boundary wall.

Ramps can be constructed to form a curved shape as a feature in a garden wall.

Straight ramps

Setting out for a straight ramp such as a gable is carried out after the main wall below it has been constructed. A string line is set up as a guide using timber profiles accurately plumbed in position. The masonry can then be raised by building small racks of brickwork either end of the wall, cutting the bricks as needed to the angle of the ramp (following the guide string line) and laying the infill masonry as the work progresses.

▲ Figure 8.30 A gable end with decorative oversailing courses

For a straight ramp as part of a boundary or garden wall, timber profiles can also be used to position a string line at the required angle of the ramp. Figure 8.31 shows a one-brick-thick boundary wall with a brick-on-edge coping constructed as a straight ramp.

▲ Figure 8.31 A straight ramp providing a change of height in a boundary wall

Concave ramps

Concave ramps can be used as a decorative feature in garden or boundary walls as a more attractive method of providing for a change of height. The wall needs to be constructed with the masonry racked up to the height of the **striking point**.

A timber batten or beam with a trammel attached is then positioned on the wall and weighted down and secured with bricks or blocks. The trammel is positioned to describe an arc from the striking point. Study Figure 8.32 to see how this works.

The dimensions for the striking point will be found on the working drawing. Alternatively, you can work out the position of the striking point by using the dimension for the standard brick gauge that corresponds to the number of courses between the lower and upper height of the wall.

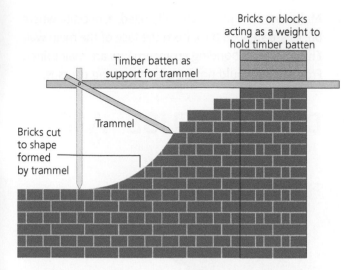

Figure 8.32 A timber batten is securely positioned on the striking line. A trammel is attached at the striking point to mark the bricks for cutting along the concave curve of the ramp

The trammel will have a small notch in the pointed end to allow a pencil to be held accurately when marking the bricks along the radius. The cut bricks that form the curve can be laid using the trammel to accurately position them.

IMPROVE YOUR MATHS

A boundary wall is constructed with a concave curved ramp to transition between two levels.

If the boundary wall is 18 brick courses high overall, and the height difference between the lower and upper sections of the wall is 9 brick courses, what will be the dimension from the edge of the curve to the centre of the radius in millimetres?

Convex ramps

Convex ramps can also be used as a feature at the change of height in a boundary or garden wall.

KEY TERM

Convex: curving or bulging outwards

The striking point for a convex ramp will be in the main section of brickwork below the curve. A trammel will be used for a convex ramp in a similar way to the

method for a concave ramp, but this time it will be fixed directly to the masonry and not supported on a timber batten or beam.

If the mortar joints have not fully hardened in time for work on the ramp to commence, a timber wedge can be carefully driven into the bed joint at the location of the striking point and the trammel fixed to this. The trammel must be allowed to swing freely to mark the shape of the curve ready for the bricks to be cut to the required radius.

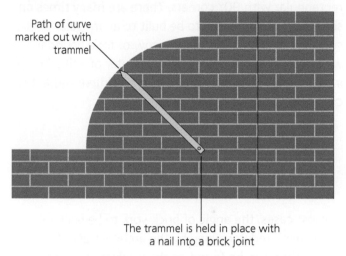

Figure 8.33 Trammel set up at the striking point marking the radius for a convex curved ramp

Good practice when building decorative features

The tips below will ensure that you achieve the best possible result when constructing decorative features.

- Always make sure that projecting bricks are lined in along the 'eye line'. The line must be positioned along the lower arris of the bricks when the work is above the eye line.
- The rear of dentil and dog-toothed bricks need to be cut accurately so as not to protrude into the cavity.
- Pay close attention to plumb and level. Remember this type of work catches your eye so it needs to appear accurately 'in line'.
- Use templates to maintain accuracy in the work.
- Always select bricks that are uniform in size and have neat square arises with no chips.

- Any work that has cuts involved should be laid out dry to check for accuracy before laying commences.
- Projecting bricks are vulnerable to weather damage so ensure that they are hard-wearing.

- Make sure that joints are fully filled, especially where bricks are stepped back from the face of the main wall.
- Ensure correct bonding arrangements are maintained.
- Frog bricks should not be used for string courses.

② SETTING OUT AND BUILDING OBTUSE- AND ACUTE-ANGLE QUOINS

Not all walls and buildings are built square or rectangular with 90° corners. There are many times on site where a wall needs to be built to an irregular angle, such as under a squint bay window or for a boundary wall that follows the irregular perimeter of a site. These irregular-shaped angles are known as **obtuse** and **acute** corners.

KEY TERMS

Obtuse: an angle between 90° and 180°
Acute: an angle less than 90°

In most cases, the angle of brickwork to be built will have been predetermined by the architect and this information can be found on the working drawings. However, there will be times when you need to set out and build walls with angles greater or less than 90° where you will be required to find the angle yourself.

▲ Figure 8.35 Acute-angled brickwork

INDUSTRY TIP

Wall features that are formed using obtuse and acute angles are sometimes referred to as 'splayed' walls.

Using geometry

Having a basic knowledge of geometry will help you to work out acute and obtuse angles. Angles can be established by **bisection**. Many obtuse and acute angles can be worked out by bisecting other angles. Let's look at some examples.

▲ Figure 8.34 Obtuse-angled brickwork

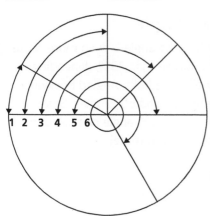

1. Acute angle: less than 90°
2. Right angle: exactly 90°
3. Obtuse angle: between 90° and 180°
4. Straight angle: exactly 180°
5. Reflex angle: greater than 180° but less than 360°
6. Full angle: exactly 360°

▲ Figure 8.36 Circle with angle descriptions

KEY TERM

Bisection: in geometry, the division of something into two equal parts, usually by a line, which is then called a bisector

KEY POINT

Most irregular angles are set out on site using a builder's square, a tape measure and a set of trammel rods or a large set of compasses. When setting out obtuse or acute angles, it is best to start with a 90° angle.

IMPROVE YOUR MATHS

Draw a straight line on a sheet of paper and mark a centre point on the line. From this centre point, set out and draw lines at 45°, 90° and 135° using a protractor.

Bisecting a 90° angle

To find an angle of 45°, you can use a builder's square to mark out a 90° angle on a flat surface and bisect it. Follow the step by step to see how this works.

Bisecting a 90° angle step by step

STEP 1 Establish a base line using a straight edge. (You choose the length of the line.)

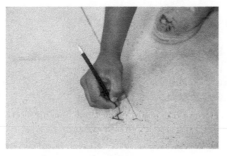

STEP 2 Mark point A on the far left.

STEP 3 Using a builder's square or bisection, draw a perpendicular line from point A; this will give you a 90° angle.

STEP 4 Using a set of compasses or trammel rod, mark two points from the corner (point A) at equal lengths along the vertical and horizontal lines to create points B and C.

STEP 5 Keep the compasses or trammel rod at the same width opening and from point B, scribe a small arc. Repeat the process from point C crossing the two arcs to form point D.

STEP 6 Draw a line joining point A to point D forming a 45° angle.

Setting out using this method gives a 45° angle or an obtuse angle of 135°. Look at Figure 8.37 to see the relationship between the two angles.

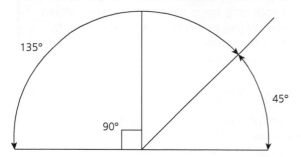

▲ Figure 8.37 The relationship between 45° and 135° angles

Setting out a 30° angle

There may be times when the angle is even more acute than 45° or even more obtuse than 135°, depending on the design of the wall to be built. Angled brickwork can also be set out using 30° and 60° angles. Study the step by step to see how a 30° angle can be set out using a set of compasses or a trammel.

Setting out a 30° angle step by step

STEP 1 Draw a line that will be the frontage line of the wall and mark the two ends A and B.

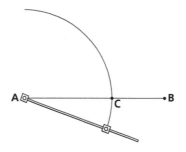

STEP 2 Set up the trammel or compasses at point A, place the pencil end just past the midway point and scribe an arc; where the arc strikes the straight line is point C.

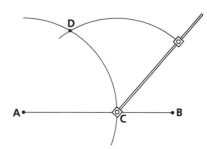

STEP 3 Keep the trammel or compasses at the same spacing and move the point of the instrument to sit on point C. Scribe an arc to form point D.

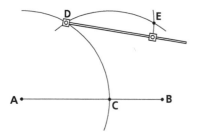

STEP 4 Keep the trammel or compasses at the same spacing and move the point of the instrument to point D. Scribe an arc along line D springing to create point E.

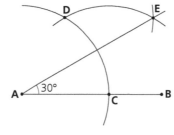

STEP 5 Join point A with point E to give you a line at a 30° angle.

Bonding angled quoins

Maintaining correct bond in 90° quoins is straightforward since in a half-brick-thick wall the overlap creating the bond corresponds to half the length of the brick. In a half-brick-thick wall with angled quoins however, creating an overlap to maintain half bond is not so easy.

Obtuse-angle quoins

To facilitate construction of angled quoins, special shaped bricks are manufactured. For obtuse angles, these are referred to as 'squint' bricks when the faces correspond to a three-quarter and a quarter brick, and 'doglegs' when the faces correspond to a header face and a stretcher face of a standard brick.

Obtuse angle bonding arrangements using brick specials

These special bricks are generally used for walls that are set at an angle of 135°. Many manufacturers will produce specially made angled bricks to order, which match the design angle of the quoins specified for the project.

▲ Figure 8.38 Squint bricks used to bond a half-brick-thick wall. The bonding arrangement is set out from the angled quoin

▲ Figure 8.39 A dogleg brick in an angled quoin allows the stretcher bond arrangement to continue throughout the wall

An option that simplifies the method of maintaining correct bond in half-brick-thick obtuse quoins is to use a universal (or 'easy angle') brick at the change of direction. This allows any angle to be easily constructed and adds visual interest to form a decorative feature.

▲ Figure 8.40 Universal special brick used to form an obtuse-angle quoin

In a one-brick-thick wall, a closer will be placed next to the squint special brick to maintain the correct bond. Look at Figure 8.41 to see how each course is arranged to prevent straight joints between courses.

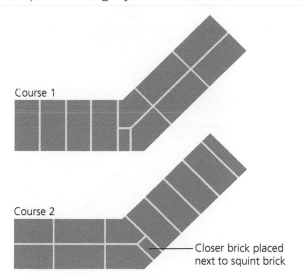

Course 1

Course 2

Closer brick placed next to squint brick

▲ Figure 8.41 A squint brick and closer incorporated into an English bond obtuse quoin

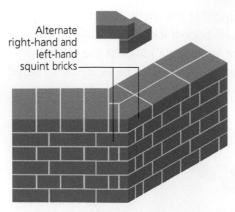

Alternate right-hand and left-hand squint bricks

▲ Figure 8.42 One-brick-thick obtuse quoin showing how the bond changes with the change of direction in each course

When building obtuse quoins in a one-brick-thick solid wall, it is preferable to use dogleg bricks in the internal angles as the solid brick will provide a strong internal bond at the change of direction.

Obtuse angle bonding arrangements using standard bricks

Special bricks give a neater, more attractive result, but they can be very costly, so it may be more appropriate to utilise standard bricks to form an obtuse quoin. There are a variety of methods that can be used to achieve this. The illustrations show two methods of bonding an obtuse quoin using standard bricks.

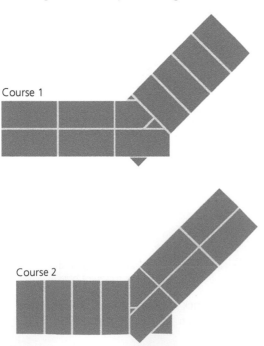

Course 1

Course 2

▲ Figure 8.43 Obtuse quoin in a one-brick-thick wall using standard bricks with a projection

▲ Figure 8.44 Using standard bricks with this set-back method creates an indent at the change of direction

Notice that, in one of the methods above, the bricks are projecting from the corner point, leaving them prone to damage and weathering. However, with this method the angle of the wall can be altered in any way to suit the building lines required for the job.

In the other method, an obtuse quoin is formed with bricks butted together to form the angle and gaps left at the corners. This method has the disadvantage that there is minimal strength at the corner of the wall due to poor overlapping between the bricks and the indent left at the corner is prone to weathering.

Acute angle quoins

Acute angles are formed by bringing the corner to a sharp point. There are no special bricks available for this, so standard bricks must be cut to shape, preferably using a mechanical saw. Cutting bricks with a hammer and bolster will leave a jagged face whereas a table saw will produce a more accurate cut with a smooth finish to the face of the brick.

Study Figures 8.45 and 8.46, which show one-brick-thick and one-and-a-half-brick-thick walls and how they're bonded at an acute angle.

Notice that the one-and-a-half brick thick example has what is termed a 'birdsmouth' arrangement to achieve the correct bond at the acute angle. This arrangement can also be achieved in a one-brick-thick acute quoin and has the advantage of removing the point at the angle which may be vulnerable to accidental damage.

There are tools available to assist in setting out the required angle such as a digital angle finder (Figure 8.47).

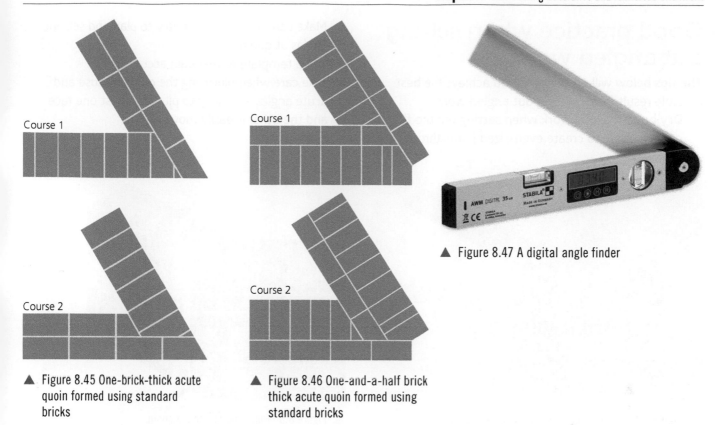

Course 1

Course 2

▲ Figure 8.45 One-brick-thick acute quoin formed using standard bricks

Course 1

Course 2

▲ Figure 8.46 One-and-a-half brick thick acute quoin formed using standard bricks

▲ Figure 8.47 A digital angle finder

Using templates to form angled quoins

To maintain the correct angle in an obtuse-or acute-angle quoin when setting out, it can be helpful to make a timber template. This is especially useful if a design has a repetitive element such as a number of squint bay windows.

Templates can be made on site by a carpenter to fit the exact shape of the angled feature to be built. In the case of bay windows, a profile can be laid into the opening to mark around it. During the construction of the bay, the profile can be laid on top of the brickwork to check that the angles are still correct.

Using a purpose-made template in this way ensures that accuracy is maintained, and windows manufactured off site to given measurements will fit the finished masonry.

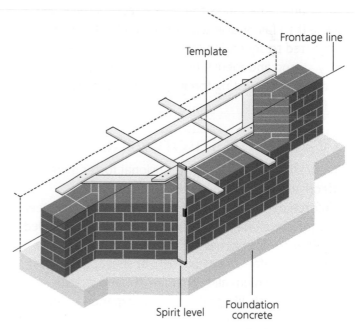

Template

Frontage line

Spirit level

Foundation concrete

▲ Figure 8.48 A timber template used to ensure that masonry is constructed accurately to the correct angles

Good practice when setting out angled work

The tips below will ensure that you achieve the best possible result when setting out angled work.

- Dry bond the brickwork when setting out the first course. Strive to create even-sized joints throughout the work.

- Make careful use of geometry to plan and set out angles at quoins.
- Use a template to maintain accuracy.
- Take care when plumbing the quoin; obtuse and acute angles are tricky to plumb. Adjust one face and the other is easily moved.

▲ Figure 8.49 Dry bonded brickwork on the first course

▲ Figure 8.50 Plumbing an obtuse quoin

③ WHEN TO REINFORCE BRICKWORK AND HOW TO DO IT

There are many forces that can affect masonry and cause it to fail or disintegrate. When brickwork is subjected to squeezing forces (compression) it is very strong and robust, but when it is subjected to stretching or twisting forces (tension) it is more likely to fail.

Types of reinforcement

To enable masonry to resist tensional forces more effectively, it can be designed to incorporate reinforcement. This commonly takes the form of steel bars (often referred to as 're-bar') or steel rods welded together to form a mesh. Another material used for masonry reinforcement consists of sheets of stainless steel or aluminium alloy that are cut and stretched to form a mesh-like grid known as expanded metal lath (EML).

Horizontal reinforcement

EML should be stretched out horizontally before laying in the mortar bed joint; this is to take out any bends in the mesh that might cause it to protrude from the bed joint. When positioning horizontal reinforcement, it is important to keep the reinforcement a minimum of 15 mm back from the face of the brickwork. This is to reduce the risk of any moisture that penetrates the bed joint causing the mesh to corrode. This type of mesh reinforcement has excellent bonding qualities as the gaps in the mesh provide a good key with the mortar.

> **HEALTH AND SAFETY**
> Be careful when handling EML. It is very sharp, especially along cut edges, so it's a good idea to wear gloves to avoid injury.

In decorative brickwork, EML is an effective material to strengthen straight joints that are part of the design such as the sectional patterns of a basket-weave panel or the brickwork that forms the surround of a feature panel.

Another type of horizontal reinforcement is known as 'welded fabric'. This type of reinforcement consists of two 3 mm-diameter stainless steel rods joined together with connecting rods spaced at intervals. The function is similar to expanded metal lath, providing a good key with the mortar bed.

Reinforcement can also be provided by installing simple stainless steel rods in the mortar bed joint. The bond between the reinforcement and the mortar is not as strong with this simple system as with other methods of reinforcement.

▲ Figure 8.52 Hoop iron reinforcement used in a one-and-a-half brick thick wall

▲ Figure 8.51 Welded fabric reinforcement in the facework bed joint and steel rod reinforcement in the blockwork bed joint

In older structures a type of horizontal reinforcement known as hoop irons was used. The purpose of the reinforcement was to strengthen the brick bond, especially in areas prone to settlement. The iron reinforcement banding was covered in tar to prevent corrosion and then coated in sand to act as a key for the mortar.

The iron bands were joined together at the corners with welt hooks to ensure a continuous band was formed around the wall. However, if the metal strips were insufficiently coated or laid in a damp wall they could deteriorate due to corrosion.

Vertical reinforcement

There are instances where vertical reinforcement is required to provide resistance to tensional forces that the masonry structure may need to deal with. Let's look at some applications.

Quetta bond

Brickwork constructed in Flemish bond is sometimes used as a decorative bond, perhaps with the use of contrasting header bricks to add to the decorative effect. A variation of Flemish bond known as Quetta bond (see page 175) can be used to include vertical reinforcement where added strength is required, such as in a retaining wall which must also be visually pleasing.

▲ Figure 8.53 Quetta bond incorporating vertical reinforcement

Hollow piers

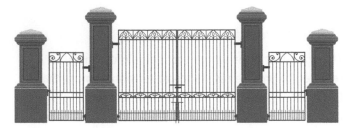

▲ Figure 8.54 Gate piers are under tension from the heavy gates pulling on them sideways

Hollow piers can be reinforced by adding steel bars in the internal space or core and filling the hollow with concrete when completed. The reinforcement can be incorporated into the concrete foundation before brickwork begins. Wait for the pier to fully harden before pouring the concrete into the pier, as compacting the concrete too early may cause the pier to expand.

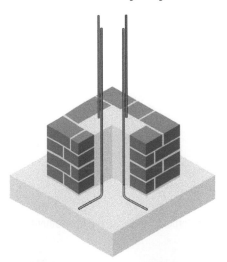

▲ Figure 8.55 Vertical reinforcement built into the concrete foundation of a hollow pier

Summary

Using bricks to produce complex decorative features is skilled work that can give you a great deal of job satisfaction when successfully achieved. All the decorative features covered in this chapter require careful planning before work begins, followed by skilled observation, monitoring and control of the work as it proceeds.

Many of the activities suggested in this chapter involve observing brickwork features you come across or sketching details of specific decorative features. These activities are intended to improve your ability to analyse how decorative features are constructed in the context of real projects. Continue to be observant and analyse what you see and you'll learn a lot about decorative brickwork and how it's constructed.

Test your knowledge

1 What type of brick can be used to form a chamfered corner in a half-brick-thick wall?

  a Bullnose   b Cant

  c Dogleg   d Plinth

2 What are quoins built with angles set at less than 90° known as?

  a Acute   b Obtuse

  c Reflex   d Reverse

3 What type of brick can be used to form an obtuse-angled quoin?

  a Double cant   b Radius

  c Squint   d Voussoir

4 Which brick bond can be used to incorporate vertical steel reinforcement?

  a English

  b Header

  c Quetta

  d Stretcher

5 What aid can be used to check the correct width of a decorative panel opening during construction?

  a Gun template

  b Pinch rod

  c Straight edge

  d Timber profile

6 What is the name of the herringbone pattern shown in this decorative panel?

  a Diagonal

  b Double diagonal

  c Horizontal

  d Vertical

7 What name is given to a string course consisting of alternating set-back and projecting headers?

  a Corbelled   b Dentil

  c Oversailing   d Plinth

8 Which section of brickwork should first be set out 'dry', when bonding in a wall that incorporates plinth courses?

  a Above datum level

  b Above the plinth course

  c Below ground level

  d Below the plinth course

9 At what angle to the main wall are the bricks in a dog-tooth course laid?

  a 30°   b 45°

  c 60°   d 90°

10 What type of quoin can be formed using a birdsmouth?

  a An acute-angled quoin using standard bricks

  b An obtuse-angled quoin using brick specials

  c An obtuse-angled quoin using standard bricks

  d A right-angled quoin using brick specials

11 List the essential PPE that must be used when cutting bricks by machine and describe the protection given by each item.

12 Explain the possible health consequences of cutting bricks by machine without using the correct RPE.

13 a State two factors that the bricklayer must consider when selecting bricks to construct a decorative brick panel.

  b Explain why the factors you have identified are important.

14 Explain when it is good practice to position the string line along the lower arris of a decorative course of brick.

15 a Name three types of ramped brickwork that can be constructed as a decorative feature.

  b State the function of ramped brickwork in a free-standing wall.

CHAPTER 9

CONSTRUCTING RADIAL AND BATTERED BRICKWORK

INTRODUCTION

In your career as a bricklayer, you will need to understand different methods of constructing radial and battered brickwork. Radial brickwork refers to brick features and masonry structures that are curved or rounded, such as arches and curved or radiused walls. Battered brickwork refers to masonry that is deliberately built leaning inwards from its base. This type of construction could be used in retaining walls or piers constructed to support other masonry.

Arches and radiused brickwork are becoming more common in new buildings that use brick as a construction material. Inspiration taken from the past is resulting in a desire for period brick designs and features in modern structures. To be able to construct these designs and features effectively, the bricklayer needs a knowledge of basic geometry, an eye for detail and to be familiar with more complex setting out methods.

By the end of this chapter, you will have an understanding of:

- the different types of brick arches and how they are set out
- methods and techniques used to build brick arches
- setting out and building brickwork that is curved on plan
- methods and techniques used to set out and build battered brickwork.

This table shows how the main headings in this chapter cover the learning outcomes for each qualification specification.

Chapter section	Level 2 Technical Certificate in Bricklaying (7905-20) Unit 204	Level 3 Advanced Technical Diploma in Bricklaying (7905-30) Unit 303	Level 2 Diploma in Bricklaying (6705-23) Unit 206	Level 3 Diploma in Bricklaying (6705-33)	Level 2 Bricklayer Trailblazer Apprenticeship (9077) Module 7
1. The different types of brick arches and how they are set out	1.1, 1.2	1.1, 1.2	5.4, 5.6	1.1, 1.2, 2.1	**Skills Depth:** 3.1 **Knowledge Depth:** 1.1, 3.1, 3.2
2. Methods and techniques used to build brick arches	2.1, 2.2, 2.3, 2.4	1.3	6.5	1.3, 1.4, 2.4, 2.5, 2.6, 2.7, 2.8, 2.9	**Skills Depth:** 3.1, 3.2 **Knowledge Depth:** 3.1, 3.2
3. Setting out and building brickwork that is curved on plan	N/A	2.1, 2.2	N/A	3.1, 4.1, 4.4, 4.5, 4.6, 4.7	**Skills Depth:** 4.1, 4.2 **Knowledge Depth:** N/A
4. Methods and techniques used to set out and build battered brickwork	N/A	3.1, 3.2	N/A	7.1, 8.1, 8.4, 8.5, 8.6	**Skills Depth:** 5.1, 5.2, 5.3 **Knowledge Depth:** N/A

Note: for 6705-33, Unit 303:
Content for Learning outcomes 5 and 6 is covered in Chapter 9.

1 THE DIFFERENT TYPES OF BRICK ARCHES AND HOW THEY ARE SET OUT

Arches have been used in construction for thousands of years. The Romans developed and used arches extensively in their structural designs to build bridges, sewers, aqueducts and other structural elements that have lasted for centuries. The main arch used by the Romans was the semi-circular arch, which we'll look at in more detail later in this chapter.

ACTIVITY

Search online for images of a Roman arch, for example the Arch of Constantine in Rome. Select five examples of arched structures that are still standing today. Which is the oldest example?

Since Roman times, different periods of architecture have developed different forms of arch with distinct characteristics. Some are elaborate, like the pointed arches associated with Gothic architecture. Some are simpler, like the segmental and flat arches that were used extensively in the seventeenth and eighteenth centuries.

Arches can be used as an attractive feature in the elevation of a building, but often their primary function is to span an opening and carry the weight of masonry above them.

The curved shape of an arch evenly distributes the compressive forces from above through the **abutments** to the foundations. Because brickwork is weak in tension (pulling or twisting forces) but strong in compression (squeezing forces), an arch can form an effective structural element to span an opening.

KEY TERM

Abutments: the parts of a structure that directly receive thrust or pressure

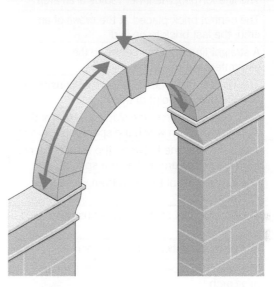

▲ Figure 9.1 Compressional forces from above are transmitted to the abutments

Arch terminology

To be able to set out and build arches, you need to know the terminology that's used. Some terminology describing the parts of an arch is unusual. Study Figure 9.2 and Table 9.1 to begin familiarising yourself with the terms you need to know when setting out arches. You may find you need to refer back to the illustration and table when reading later sections of this chapter.

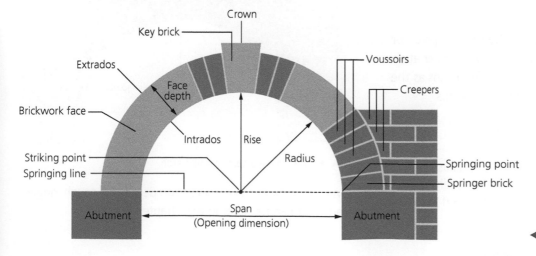

◀ Figure 9.2 Arch terminology

▼ Table 9.1 Arch key terms and definitions

Term	Definition
Collar joint	The joint that runs between the two header courses in a double-ringed arch
Crown	The highest point of the extrados
Extrados	The line forming the outer radius of an arch
Face depth	The distance between the extrados and the intrados
Intrados	The line forming the inner radius of an arch
Key brick	The central brick placed at the crown of an arch; the last brick to be laid
Radius	A straight line from the centre to the circumference of a circle
Rise	The vertical distance between the springing line and the highest point of the intrados
Skewback	On a segmental arch, the sloping surface at the springing point on which the springer bricks rest
Springing line	The horizontal line between the two abutments from where the arc of the arch springs
Springing point	The lowest part of the arch from where the radius begins
Striking point	The centre point of the arch radius
Span	The distance between the abutments
Voussoirs	The wedge-shaped bricks used to form an **axed** arch

KEY TERM

Axed: a type of arch using bricks cut to a tapered or wedge shape around the radius

Structural design of arches

There are two arch structural arrangements from which a range of different arch designs can be produced:

- rough ringed
- axed.

Rough-ringed arch

Rough-ringed arches are formed by using tapered or wedge-shaped mortar joints which are thinner at the intrados and get thicker towards the extrados as the brickwork progresses around the arch radius.

INDUSTRY TIP

This type of arch is called 'rough-ringed' because the wide joints can be difficult to finish acceptably, leading to a rough appearance. This is especially the case if the arch bricks are laid as soldier bricks.

While a rough appearance may not be important if the arch brickwork is hidden under a surface finish such as sand/cement render, it is unlikely to be acceptable where appearance is important. To create an improved appearance, most rough-ringed arches are constructed by laying bricks as headers across the arch radius, creating shorter wedge-shaped joints.

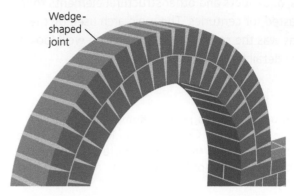

Wedge-shaped joint

▲ Figure 9.3 Rough-ringed arch with wedge-shaped joints around the radius

Axed arch

To produce arches that give a more acceptable appearance for higher quality work, an arch can be constructed using tapered or wedge-shaped bricks, called voussoirs. The shaped bricks for the arch are carefully measured and cut to the same size, so that the joints are the same thickness with parallel edges throughout their length from the intrados to the extrados. The line through each joint radiates from the striking point, giving a neat and attractive appearance.

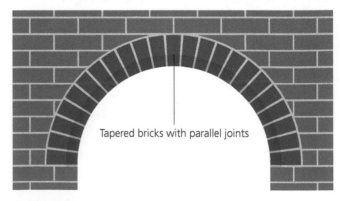

Tapered bricks with parallel joints

▲ Figure 9.4 Axed semi-circular arch

We will look at the construction details for rough-ringed and axed arches later in this chapter.

Types of arches and how they are set out

There are many different types of arch, ranging from simple rough-ringed arches that are relatively easy to build, to complex geometrical shapes which require a great deal of planning and diligence to produce a high-quality result.

The parts of a circle

To set out arches (and curved walls in brick) you need to know the parts of a circle. Study Figure 9.5 to familiarise yourself with the different terms.

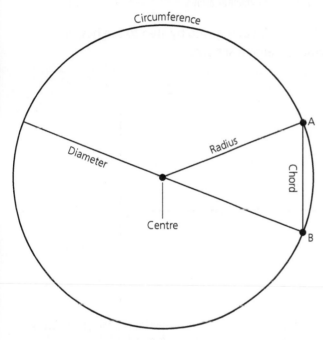

▲ Figure 9.5 Main parts of a circle

Let's look at some of the different types of arch, their features and how to set them out.

Setting out a semi-circular arch step by step

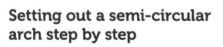

STEP 1 Draw span A–B.

Semi-circular arch

A semi-circular arch is created from half a complete circle. It is sometimes known as a Roman arch because it was frequently used in Roman architecture.

This type of arch is very strong and able to support substantial loadings imposed by masonry constructed above it. Setting out a semi-circular arch is fairly simple since the striking point is easy to locate at the centre of the radius.

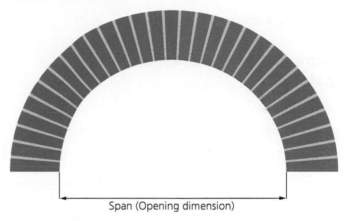

▲ Figure 9.6 Semi-circular arch

Look at the following step by step to see how a semi-circular arch is set out.

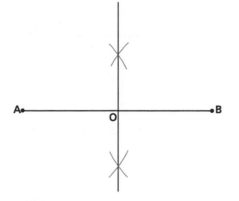

STEP 2 Bisect line A–B to produce the radius centre O.

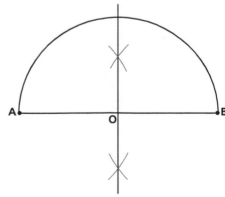

STEP 3 Draw radius O–A to produce the semi-circular shape.

Segmental arch

A segmental arch is formed from a segment of a circle and will have a lower rise than a semi-circular arch.

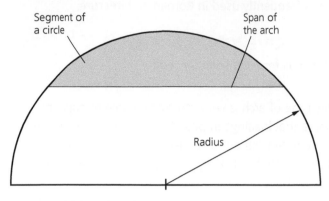

▲ Figure 9.7 A semi-circle and a segment

The rise will be determined and set out so that when the face brickwork crossing the crown of the arch is built, it will form a neat 10 mm mortar joint above the arch.

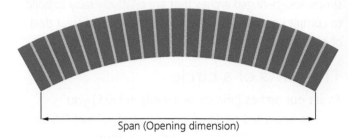

▲ Figure 9.8 Segmental arch

Look at the following step by step to see how a segmental arch is set out.

Setting out a segmental arch step by step

A•————————————————————•B

STEP 1 Draw span A–B.

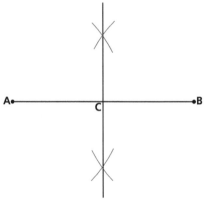

STEP 2 Bisect line A–B to establish the centre point C. Extend the bisecting line above and below the bisecting arcs.

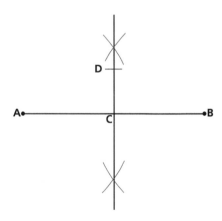

STEP 3 Measure the rise along the bisecting line above C to establish D.

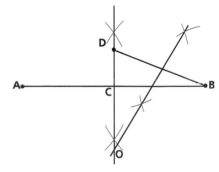

STEP 4 Draw chord B–D and then bisect it. Extend this bisection line to cross the first bisection line to establish O.

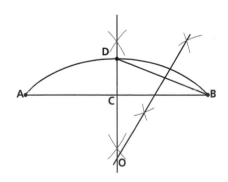

STEP 5 Using O as the centre and O–D as the radius, strike the arc A–D–B to produce the segmental arch radius.

Bulls-eye arch

Bulls-eye arches are sometimes known as wheel arches. This arch is set out as a full circle and can be used to form a decorative window opening. There are four key bricks in this type of arch, located on the vertical and horizontal **axis lines** with the striking point at the centre of the circle.

KEY TERM

Axis line: a straight line about which an object may be divided into symmetrical halves

When a bulls-eye arch is constructed as an axed arch, all the voussoirs around the circular form are cut to the same dimensions and angle of taper. They are carefully set out to radiate from the central striking point.

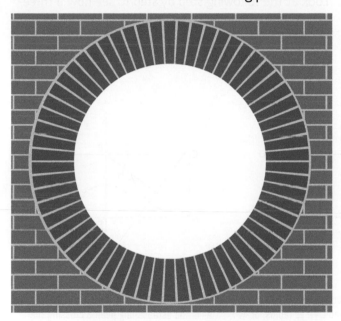

▲ Figure 9.9 Bulls-eye arch formed using tapered bricks (voussoirs)

Setting out a bulls-eye arch step by step

Sometimes, because window frames for a circular opening are an expensive item, a bulls-eye arch may be constructed simply as a decorative feature in an elevation of a building and the open section inside the radius filled in with face brickwork. This is then known as a 'blind' arch. (This practice can also be applied to other types of arch.)

▲ Figure 9.10 Blind arch

Look at the following step by step to see how a bulls-eye arch is set out.

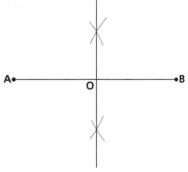

STEP 1 Draw span A–B.

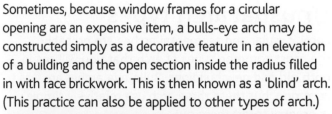

STEP 2 Bisect line A–B to establish the radius centre O.

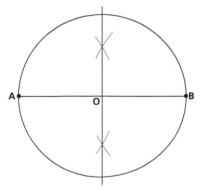

STEP 3 Draw radius O–A to produce the circular shape.

Unusual arch types

Semi-circular and segmental arches are probably the most common types of arch that you will see in both old and new buildings. Bulls-eye arches are less common, but these three arch use relatively simple geometry to set them out.

There are many other types of arch design that can be constructed in masonry. Some of the more complex designs require the use of setting-out geometry that is more demanding. Let's look at some of them.

Three-centred arch

A three-centred arch is formed in the shape of half an oval, sometimes referred to as 'semi-elliptical'. This is a less common arch which you may see in older buildings spanning larger openings. The name 'three-centred arch' comes from the fact that there are three separate striking points to form the shape.

Setting out a three-centred arch step by step

Small semi-circular segments are set out on each side of the face and a segmental shape joins them together across the span. This is a more complicated arch to set out and, if constructed as an axed arch, it has two different-sized voussoirs making up the span.

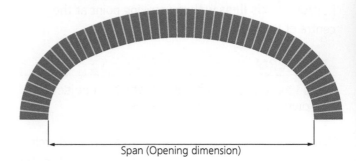

▲ Figure 9.11 Three-centred arch

Look at the following step by step to see how a three-centred arch is set out.

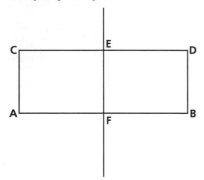

STEP 1 Draw a rectangle A–B–C–D, where A–B is the span and A–C is the rise of the arch. Next bisect A–B to establish points E and F.

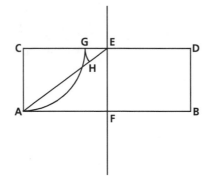

STEP 2 Draw a line between points A and E. Using point C as centre and C–A as radius, draw an arc to cut C–D at G. Using point E as centre and E–G as radius, draw an arc to cut A–E at H.

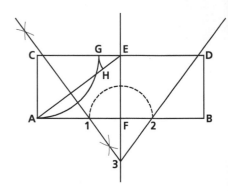

STEP 3 Bisect A–H to produce point 1 cutting A–B and establishing point 3 crossing the extended line E–F. Using F as centre, draw radius F–1 to produce point 2 on line A–B.

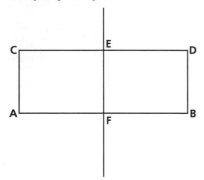

STEP 4 Using point 1 as centre, draw radius 1–A to cut the bisector at J. Using point 2 as centre, draw radius 2–B to cut the bisector at K. Using point 3 as centre, draw radius 3–J through E to point K.

> ### KEY POINT
> In step 4, lines 3–J and 3–K are called 'common normals' because they provide the centre for both the large and small radius.

Gothic arches

Gothic arches are formed from two segmental arches leaning together to form a point. They are always pointed at the crown of the arch.

▲ Figure 9.12 Gothic arch

There are three main types of Gothic arch set out in different ways and to different heights and spans characterised by the position of their striking points:

- Equilateral Gothic arches – always have striking points *on* the springing points located at the ends of the span.
- Dropped Gothic arches – have striking points positioned along the springing line *inside* the span.
- Lancet Gothic arches – have striking points positioned along the springing line *outside* the span.

KEY POINT

Take careful note of the position of the striking points in each of these arches. It is the position of the striking point that will help you identify the type of Gothic arch.

Setting out an equilateral Gothic arch step by step

Look at the following step by steps to see how three types of Gothic arch are set out. In each case, note the position of the striking points.

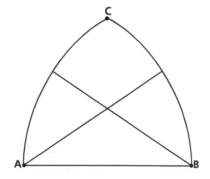

STEP 1 Draw span A–B.

STEP 2 Using the span as the radius, draw arcs A–B and B–A to intersect at C.

Notice that an equilateral Gothic arch uses the end points of the span as striking points for each side of the arch. Now look at the following step by step and note the different position of the striking points.

Setting out a dropped Gothic arch step by step

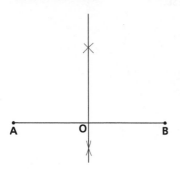

STEP 1 Draw span A–B. Bisect A–B to establish point O.

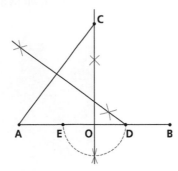

STEP 2 Draw chord A–C then bisect it. Where the bisecting line meets O–B, this establishes the first radius point D. With point O as centre, use radius O–D to establish point E.

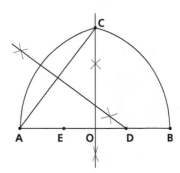

STEP 3 Using radius D–A and radius E–B, draw arcs to point C.

Note that the striking points for a dropped Gothic arch are inside the span of the arch. Now look at the following step by step to see how to set out a lancet Gothic arch. Note the different striking points.

Setting out a lancet Gothic arch step by step

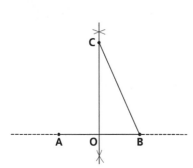

STEP 1 Draw span A–B and extend along the springing line beyond the span. Bisect line A–B to establish point O. Measure rise upwards from point O along bisecting line. Draw chord B–C.

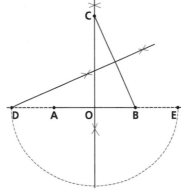

STEP 2 Bisect chord B–C. The bisecting line cuts the springing line past point A to establish point D. Using O as centre, draw radius O–D to establish second centre point at E.

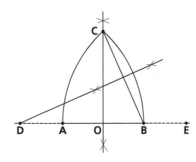

STEP 3 Using radius E–A and radius D–B, draw arcs to point C.

Notice that for a lancet Gothic arch the striking points are outside the span.

ACTIVITY

Many other arch designs have been created as architectural features. Search online and make a list of arch types you discover that are different from the ones we've looked at so far in this chapter.

Make a free-hand sketch of the shape of each new one you find.

Hint: You could start by searching for 'Tudor arch'.

② METHODS AND TECHNIQUES USED TO BUILD BRICK ARCHES

When constructing arches, there are methods and techniques used that are unique to this type of work. You need to think ahead when applying these methods to ensure that the finished work is accurate for gauge, plumb and face plane.

The design of a structure may result in the arch being fully enclosed in an elevation of masonry, with brickwork bearing on the arch from above. When this is the case, it will usually be possible to raise corners at each end of the elevation, allowing you to position a string line from end to end crossing the face of the arch as it is built.

This will keep the face of the arch plumb and maintain accurate face plane. Look at Figure 9.14 to see how this works. (We'll discuss the timber items shown, such as the turning piece, later in this chapter.)

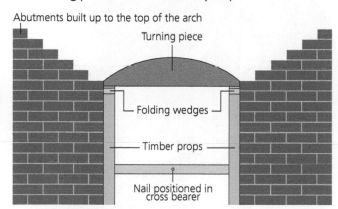

Abutments built up to the top of the arch
Turning piece
Folding wedges
Timber props
Nail positioned in cross bearer

▲ Figure 9.14 Corners raised at each end of a wall to carry a string line

If the arch stands above adjoining masonry, the technique used to maintain accuracy will be different. As an example, think about a semi-circular arch used as a decorative feature above a gate through a garden wall. The masonry at each side of the arch may only be raised to the springing line of the arch.

To maintain accurate plumb and face plane of the face of the arch, one method that can be used is to raise a **deadman** to either side of the arch.

KEY TERM

Deadman: temporary brick pillar built to carry a string line as a guide for masonry under construction

▲ Figure 9.15 Deadman constructed at each end of a wall to carry a string line as a temporary guide

The deadman brickwork is dismantled when the arch brickwork is complete. An alternative and quicker method could be to use timber profiles positioned at the ends of the wall to carry the string line.

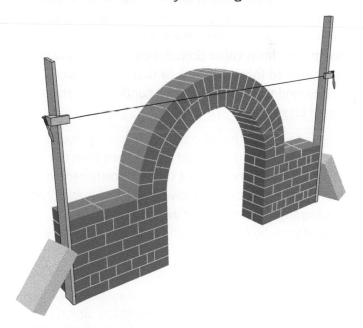

▲ Figure 9.16 Timber profiles at each end of a wall to carry a string line as a temporary guide

Arch support during construction

To span an opening with an arch, we obviously need a way to support the arch bricks until the radius is complete from one side of the opening to the other. This is provided by a temporary timber support.

The support needs to be strongly constructed since there will be considerable weight placed upon it during the building of the arch. A poorly constructed temporary support could lead to distortion of the arch during construction or even a total collapse.

IMPROVE YOUR ENGLISH

Use the internet to find out what a 'relieving arch' is and how it is used. Are there any other names for a relieving arch? Write a short report about what you find.

Let's look at some methods and techniques used to provide temporary support during construction of arches.

Arch centre

Temporary support known as an arch centre is often used when constructing a semi-circular arch. An arch centre is constructed from separate timber sections, with the surface on which the bricks will be laid constructed from either timber slats or laggings with gaps between (open lagging), or thin sheet plywood bent around the radius to form a continuous surface (closed lagging).

Open lagging has the advantage of allowing mortar from the **soffit** of the arch to be removed from the mortar joints where they are suitably positioned, to minimise staining of the brickwork. Closed lagging has an advantage when building axed arches, since the voussoirs can be positioned and adjusted on a more uniform surface.

KEY TERM

Soffit: the underside of an architectural structure such as an arch

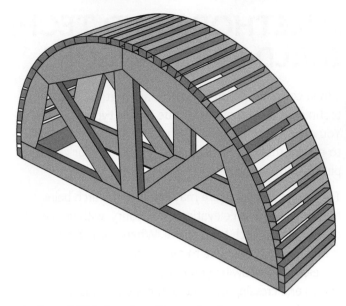

▲ Figure 9.17 Arch centre with open lagging

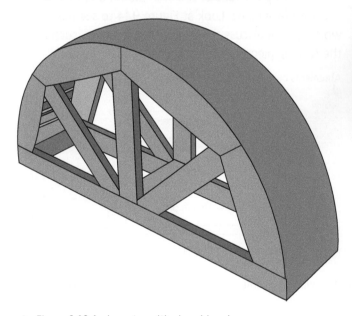

▲ Figure 9.18 Arch centre with closed lagging

Turning piece

Turning pieces can be constructed from solid timber or separate timber components and are often chosen for use in construction of segmental arches or arches with smaller spans. Like arch centres, they must be strongly built and can have open or closed lagging to provide support for the arch brickwork around the radius.

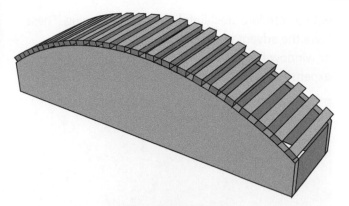

▲ Figure 9.19 Turning piece with open lagging

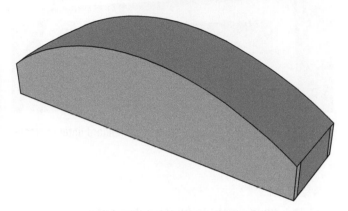

▲ Figure 9.20 Turning piece with closed lagging

Setting up the temporary support

To allow the arch centre or turning piece to be easily adjusted for height prior to construction of the arch, it should be supported on folding wedges placed on props or struts carefully wedged against the walls or abutments either side of the opening. The term 'folding wedges' simply refers to a pair of wedges that are placed on top of each other, with their shallow taper in opposite directions.

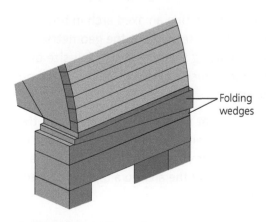

▲ Figure 9.21 Folding wedges

KEY POINT

By using folding wedges, the temporary support can be slightly lowered or eased prior to its removal so that the arch brickwork settles and tightens on itself and bonds securely together. When all the temporary support components are removed and the arch stands unsupported, this is known as 'striking' the arch.

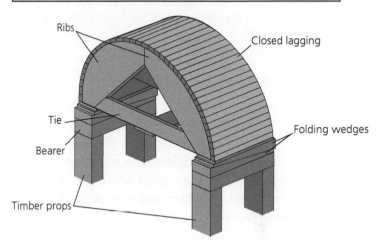

▲ Figure 9.22 Arch centre set up on props and folding wedges

Once the arch centre or turning piece has been set up correctly, the bricklayer marks the centre point of the opening on the timber support and then marks a plumb vertical line to the top or crown of the radius. This will indicate the position of the centre of the key brick. Pencil markings can then be made around the radius of the arch centre or turning piece that will indicate the evenly spaced position of the arch bricks from the springing point to the key brick.

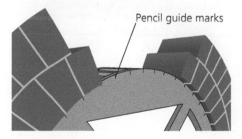

▲ Figure 9.23 Pencil markings on the arch centre to show brick positions around the radius

The spacing of the pencil marks may have to be adjusted by trial and error to make sure the appearance is even across the radius. In a rough-ringed arch, the joint size is slightly reduced from the standard 10 mm at the bottom of the joint, to avoid an overly-large dimension at the top of the joint along the extrados.

Using permanent support

The term 'permanent support' refers to systems that have been developed to speed up arch construction and which remain an integral part of the structure when construction is complete. There are several proprietary components available which take the form of a radiused lintel manufactured in steel which is galvanised to prevent corrosion.

Permanent support products are commonly used in situations where arches are built into a cavity wall or solid wall elevation to span openings for windows or doors. In a cavity wall application, they would require a tray damp proof course to be installed above the crown level of the arch. Using this type of system will require a door or window frame with an arched head (top section) to be installed.

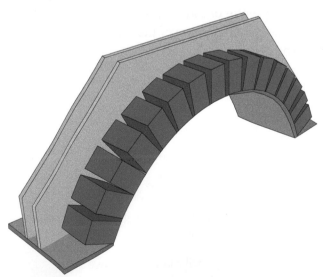

▲ Figure 9.24 Permanent support component

To simplify construction of an arch even further, there are formers available which can be used in conjunction with a standard steel lintel to span an opening. These have the advantage of allowing a standard door or window frame to be installed instead of a more expensive arched frame.

▲ Figure 9.25 Arched former used with a standard lintel to create a segmental arch

Setting out axed arch geometry

Rough-ringed arches are relatively simple to construct since there is minimal cutting of the bricks laid around the radius. By contrast, axed arches are much more complex to construct and require careful preparation and setting out to produce a high-quality job.

The process of constructing an axed arch in brickwork begins with accurately setting out the geometry. This is done by drawing the arch design full-size on a convenient flat surface such as a sheet of plywood or plasterboard.

Let's start by looking at the process for setting out the geometry of an axed semi-circular arch. Remember, the term 'axed' means that you will cut the bricks that are laid around the radius of the arch to a tapered or wedge shape (voussoirs).

Setting out a semi-circular axed arch

Study the step by step to become familiar with the sequence for setting out a semi-circular axed arch.

The end result of this sequence of work is to produce a template which will allow you to accurately mark the voussoir bricks ready for cutting.

Geometric setting out for an axed semi-circular arch step by step

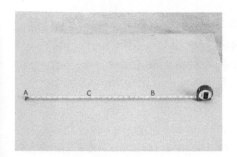

STEP 1 Using a straight edge, draw a horizontal line across the board. This is the springing line. Then mark the span on this line. Mark the two ends A and B. Next, find the centre point between A and B and mark this as point C.

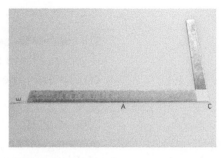

STEP 2 Place a builder's square with its corner at point C and draw a perpendicular line at right angles to the springing line.

STEP 3 Set up the trammel heads and place the compass point at point C and the pencilled end on point A. Scribe an arc from point A round to point B. This is the intrados.

STEP 4 Now find the outer curved line: the extrados. Increase the radius by increasing the distance between the trammel points. In this example the ring is to be one brick thick, so place a brick on the sheet as shown and mark its end. Alternatively, simply open up the trammel points by 215 mm.

STEP 5 Place the trammel point on centre point C and then extend the pencil past point A on the springing line and draw the second curved line around the arch, forming the extrados. The arch is now drawn.

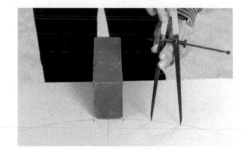

STEP 6 Set the dividers to the size of a standard brick and place the key brick centrally on the extrados to correspond with the centre line of the arch. The maximum spacing should be 75 mm (the same as standard gauge) if using standard-sized bricks. Open or close the dividers to ensure an even spacing of bricks.

STEP 7 When you are satisfied with the spacing, use a straight edge to draw several lines from the striking point through the spaced marks on the extrados.

STEP 8 These lines will allow you to mark out and create a template ready for marking and cutting the voussoirs.

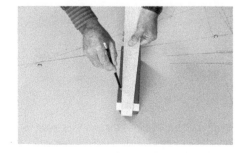

STEP 9 The template is used to mark each brick ready for cutting. The number of bricks needed can be determined from the spacing marks around the extrados on the setting out board.

INDUSTRY TIP

The radiating lines produced in step 7 will mark out a wedge-shaped joint between the voussoirs. To avoid this, you will need to adjust the joint spacing slightly at the intrados to create a truly parallel joint along the arris of the voussoirs. Look at Figure 9.26 to see how this works.

ACTIVITY

Using a scale of 1:10 and a span of 900 mm, set out and draw a semi-circular axed arch on a piece of paper. Show radiating lines for the position of the voussoirs. You will need a compass, pencil, set square, scale rule and eraser.

Setting out a segmental axed arch

The process for setting out an axed segmental arch is very similar to the process used to set out a semi-circular arch. The difference is in how we establish the striking point.

Look at the step by step to see how we begin setting out the geometry for a segmental arch. The abutments of the segmental arch are shown to give some context to the setting out lines.

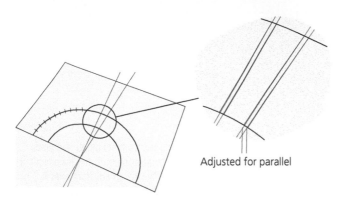

▲ Figure 9.26 Adjust the joint lines to make them parallel

Geometric setting out for an axed segmental arch step by step

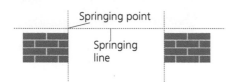

STEP 1 Establish the springing line and the springing point from the working drawing. The springing line is shown above the abutment brickwork here because we want the crown of the arch to correspond with a bed joint in the masonry above the arch.

STEP 2 Establish the rise. This is the measurement from the springing line to the highest point of the intrados on the working drawing.

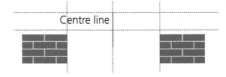

STEP 3 Find the centre line. For this example, we'll say that the distance between the abutments is 910 mm, so the centre line will be 455 mm from one side abutment.

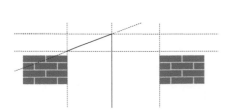

STEP 4 Draw a line from the springing point to the intersection between the centre line and the rise.

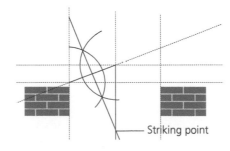

STEP 5 Bisect this new line. Where the bisection cuts the centre line, this will be the striking point for the segmental arch. We've now produced the geometry needed to draw a segmental arch.

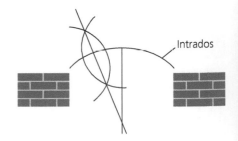

STEP 6 By describing an arc using the striking point as the centre of the radius and the springing point to give us the measurement of the radius, we can draw the intrados for our segmental arch.

Notice that the striking point is well below the springing line which is different to the position of the striking point in a semi-circular arch. This geometry will also allow us to establish the angle of the skewback at the abutments.

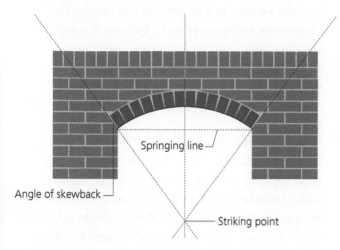

Springing line

Angle of skewback

Striking point

▲ Figure 9.27 Establishing the skewback angle

Cutting the voussoirs

Accurately cutting voussoir bricks into a tapered or wedge shape takes patience and skill. The first step in the process is to assess how easily the selected bricks can be cut and shaped. If an engineering brick is specified, it will be difficult to cut by hand because of the density of the material. Using a masonry bench saw would be the most efficient means of cutting the voussoirs.

INDUSTRY TIP

If a construction project has a large number of axed arches in the design, it may be more efficient to have brick special voussoirs manufactured to the specification if cost considerations allow.

If the specified bricks are suitable for cutting by hand, select the appropriate tools and follow the step by step below.

Marking and cutting the voussoirs step by step

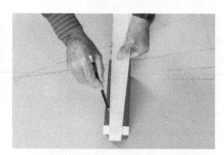

STEP 1 Using the prepared template, carefully mark the outline of the voussoir onto the face of the brick ready for cutting.

STEP 2 Using a lump hammer and a sharp bolster, place the blade of the bolster just to the waste side of the marked line and cut downwards into the face of the brick.

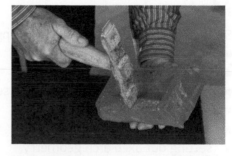

STEP 3 Carefully trim away any excess material from the sides of the brick with a scutch hammer.

HEALTH AND SAFETY

Never cut or trim bricks without using the correct PPE. Eye protection is especially important when cutting bricks by hand since your face will be close to the work you are holding. Think about the possible effects of your cutting activities on those who may be working nearby, since brick dust and fragments that are generated can travel quite a distance.

A traditional practice when cutting accurate shapes in brick was to smooth the cut edges with a carborundum stone if the brick was soft enough. Some arch designs include voussoirs that are bonded across the face of the arch, which increases the skill level required to build them.

▲ Figure 9.28 Bonded voussoirs in an arch face

Methods of work

So far, you have looked at the following:

- arch terminology
- structural design of arches
- types of arches and how they are set out
- some unusual arch shapes
- arch support during construction
- setting out axed arch geometry.

Let's draw all these points together by looking at methods of work that will contribute to consistent and efficient production of high-quality results.

Following the correct sequence

Working in a structured and orderly way is always important when undertaking construction tasks. This is especially the case when creating an arch as a structural element of a building. Once the abutments are built and the temporary support is accurately and securely in place, construction of the actual arch can begin. We'll consider the sequence of work for a segmental arch.

1   The first step is to mark out the skewback bricks. The skewback bricks in a segmental arch form the angle built into the abutments from which the arch will spring. In the first step by step illustration, the cross brace for the timber props holding up the turning piece has been positioned to allow a nail to be driven into it at the striking point. A string line can be attached to the nail and held so that it crosses the springing point to indicate the angle of the skewback.

2   After laying the cut bricks that form the skewback, set up a string line across the face of the arch to assist in maintaining accurate face plane as the arch is built.

3   Now you can start laying the arch bricks. Never lay the arch bricks across the span from one side only. Lay two or three at a time alternating from side to side to distribute the increasing load on the temporary support. The line that has been attached to the striking point can be used to align the arch bricks (or voussoirs in an axed arch) as they are laid around the radius. Hold the line along the edges and through the centre of each brick face to make sure it radiates accurately from the centre of the arc. In a two-ring arch, complete one ring at a time, laying the key brick last and making sure the mortar joints are full. Check the face plane for accuracy as the work progresses using the horizontal string line that you set up from end to end.

4   When the arch brickwork is complete, use the folding wedges under the arch turning piece (or arch centre, depending on the type of arch) to very slightly lower the temporary support. This will cause the mortar joints to compress together slightly, binding the arch bricks together. The process of lowering the arch in this way is called 'easing'.

5   When the arch masonry has hardened, the temporary support components can be completely removed. Remember, this process is called 'striking' the arch.

6   The last stage is to provide a joint finish to the soffit of the arch. This may require some mortar to be raked out and repointed to provide a satisfactory joint finish.

▲ Figure 9.29 A sliding bevel

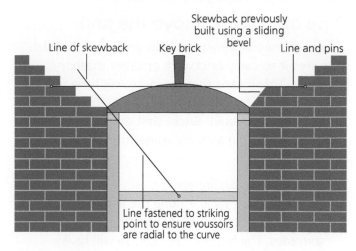

Skewback previously built using a sliding bevel

Line of skewback

Key brick

Line and pins

Line fastened to striking point to ensure voussoirs are radial to the curve

STEP 1 Using the striking point to set the angle of skewback.

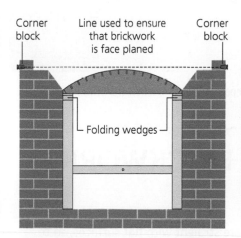

Corner block

Line used to ensure that brickwork is face planed

Corner block

Folding wedges

STEP 2 With the skewback bricks in position, set up a string line across the face of the arch.

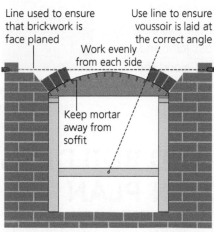

Line used to ensure that brickwork is face planed

Use line to ensure voussoir is laid at the correct angle

Work evenly from each side

Keep mortar away from soffit

STEP 3 Lay the arch bricks alternating from each side and make sure they line up with the centre of the arc.

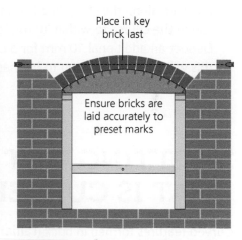

Place in key brick last

Ensure bricks are laid accurately to preset marks

STEP 4 In an arch with multiple rings, complete one ring at a time.

Carefully ease wedges to cause arch to settle slightly

STEP 5 Easing the arch brickwork.

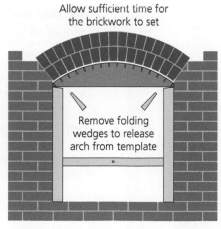

Allow sufficient time for the brickwork to set

Remove folding wedges to release arch from template

STEP 6 Striking the arch support.

Lay facing brick to underside of soffit to be seen

Joint soffit

STEP 7 Provide the specified joint finish to the soffit.

The brickwork above the arch

The arch may be constructed in an elevation which requires it to carry brickwork or other masonry above it. This means that masonry will have to be accurately cut to follow the radius of whatever type of arch has been built. Bricks that are cut to follow the extrados of an arch are sometimes referred to as 'creepers'.

To mark the cuts accurately:

● Position the string line along the line of the first course of brick that will intersect with the arch extrados. The bricks in that course will have already been laid as far along the course as possible approaching the extrados.

● Measure along the string line from the last brick laid in the course to within 10 mm of the extrados. Deduct an additional 10 mm for a cross joint and this will be the dimension to be marked along the top arris of the creeper brick.

● Follow this pattern to establish the marked position on the lower arris of the brick. Joining the marks will show the section of brick to be removed.

Some artistry may be needed to form a curve between the upper and lower marks that will match the curve of the extrados as closely as possible. Hold the cut brick in position to make sure the cut is suitable and use a scutch hammer to trim where required.

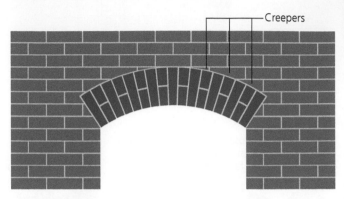

▲ Figure 9.30 Creeper bricks cut to follow the extrados of a segmental arch

③ SETTING OUT AND BUILDING BRICKWORK THAT IS CURVED ON PLAN

Curved features formed in brick (**radial brickwork**) can be used to enhance the appearance of a building. You often see curved walls used in hard landscaping, creating attractive flower beds and boundary walls. Curved walls can be used to construct a range of architecturally interesting features.

Standard bricks are straight with flat surfaces, so to create a curved wall with straight components means using techniques to create the illusion of a curve. **Faceting** is a term used to describe the unsightly overhang between courses of standard bricks used in a wall with too tight a radius. The effect appears more pronounced in strong lighting conditions.

KEY TERMS

Faceting: the effect of straight components overhanging each other on a curve

Radial brickwork: brick features and masonry structures that are curved or rounded, such as arches and curved or radiused walls

▲ Figure 9.31 Header bond boundary wall curved feature

Faceting can be removed altogether if brick specials called radials are used. These are manufactured to precisely follow the specified curve but can be expensive.

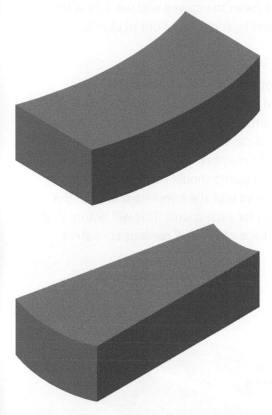

▲ Figure 9.32 Radial stretcher/radial header

There are methods using standard bricks that can be used to minimise the effect of faceting using different bonding arrangements. Let's look at some of them.

Choosing a suitable bond

The bonding pattern selected for a curved wall will depend on the tightness of the curve. If the curve is a tight radius, utilising the shorter header face of a standard brick can be a useful technique to produce an acceptable curved appearance. For example, if we increase the number of header bricks in a curved elevation by using Flemish bond, a smoother curve can be achieved in a tight radius.

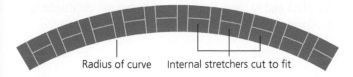

Radius of curve   Internal stretchers cut to fit

▲ Figure 9.33 Flemish bond curved wall

Alternatively, if the radius is suitable, header bond could be used to reduce the faceting effect across the curved elevation.

▲ Figure 9.34 Header bond curved wall

In some cases, if a curved solid wall requires an aesthetically acceptable finish on both faces, **snap headers** can be introduced. This will allow even-sized joints to be formed on both faces. When using this method, it is important to incorporate a means of tying the two faces together.

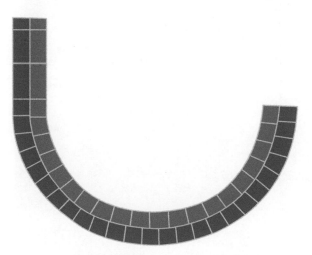

▲ Figure 9.35 Snap headers used in a curved wall

If the radius is large enough and forms a shallow curve, then Stretcher bond can be used to produce a curved wall of satisfactory appearance by forming the perp (or cross) joints in a V-shape.

How a curved wall is set out will depend on the position of the centre of the radius in relation to the wall. You must have clear access from the centre of the radius to the circumference to be able to measure accurately to all points around the curve.

In some cases, straight sections of wall will have been built first from which measurements can be taken to establish the centre point of the radius. Details about the specified radius will be given on working drawings.

Producing accurate work

Care and attention to detail are essential when building walls that are curved on plan. There are a number of methods that can be used to set out and build curved or radiused walls, each of which has advantages and disadvantages depending on how tight or shallow the radius is and the ease of access to the work.

Let's look at some of the methods you can use to build walls that are curved on plan.

Using a trammel

A trammel is a length of timber fixed to the centre point of the curve in such a way that it can sweep through the full arc of the circumference of the curved wall and can be raised as the work progresses. Its length corresponds to the specified radius.

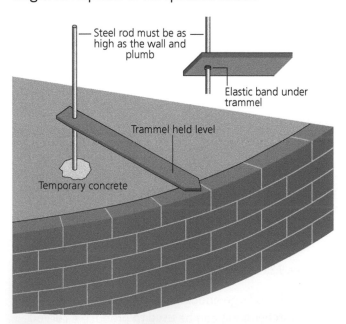

▲ Figure 9.36 Trammel rotating around a fixed centre point

The trammel swings around a steel rod fixed in the required position, which is temporarily secured in concrete. The rod must be carefully plumbed and must be as high as the proposed wall. Any inaccuracy in plumb will mean the curved wall will follow the inaccuracy and be constructed out of plumb.

Using a template

If there is not enough space on site to set up a trammel and rod, or if the curved section of wall is relatively small, then a timber template can be made which is cut to suit the radiused face of the curved wall. To maintain the accuracy of the curve, clearly marked points should be marked along the circumference so that the template is used in the same position for each course. This will ensure that the curved shape of the wall remains consistent throughout.

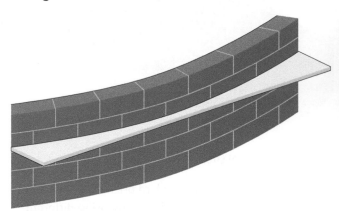

▲ Figure 9.37 Using a timber template to maintain the correct radius

To maintain level around the curve, individual bricks can be laid at marked points along each course and the level between them checked with a straight edge and spirit level. These marked points can also be used to check plumb.

INDUSTRY TIP

Using the same points on each course to position the template and to check the level and plumb will contribute to producing consistent and accurate work.

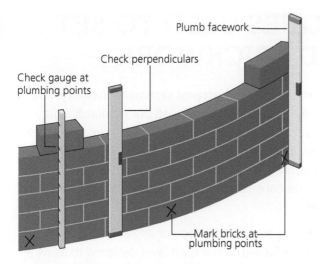

▲ Figure 9.38 Using marked plumbing points

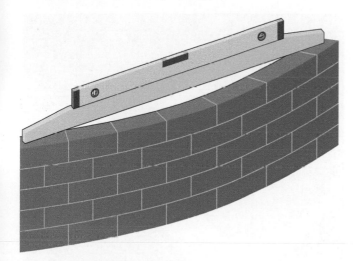

▲ Figure 9.39 Levelling between marked plumbing points

Serpentine walls

Curved walls can be built as a convex curve on plan (curving outwards) or a concave curve on plan (curving inwards). A serpentine wall uses both convex and concave curves.

▲ Figure 9.40 A serpentine wall

A serpentine wall curves in and out along its length with a snaking appearance. If the degree of the curve is large enough, this will add strength and stability to the structure. Serpentine walls are often built without any attached piers.

Setting out a serpentine wall is carried out in much the same way as constructing other walls that are curved on plan, using trammels or templates to form the curve. A line can be stretched through the centre points of the curves of the wall, with lines drawn at right angles from the centres to establish each individual radius.

ACTIVITY

Do some internet research into 'serpentine wall'. Make a note of other names used to describe serpentine walls and list the sort of applications they have been used for.

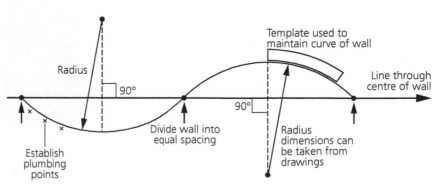

▲ Figure 9.41 Setting out a serpentine wall

4 METHODS AND TECHNIQUES USED TO SET OUT AND BUILD BATTERED BRICKWORK

Battered brickwork is brick masonry that is constructed to a specified angle away from plumb. This method of building can be used to construct retaining walls, to build a supporting pier or **buttress** and in locations where a mass of brickwork must be reduced in size, such as when a large external **chimney breast** is reduced in width.

Retaining walls

A retaining wall must be designed to resist the pressures created by the weight of the material behind it. To achieve this, the wall can be thickened at its base so that it can resist the tendency to slide or overturn. Since the pressure behind the wall decreases as it gets nearer the top, the thickness of the wall can be progressively reduced.

Creating the batter

The reduction in thickness over the height of a wall can be achieved by sloping (or battering) the face of the wall.

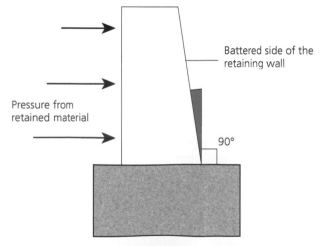

Pressure from retained material

Battered side of the retaining wall

90°

▲ Figure 9.42 Retaining wall with one side built to a batter

When you build a battered section of masonry, you are deliberately building a wall out of plumb. This can seem to work against the instruction a bricklayer receives during training regarding accurate plumbing of masonry. It requires concentration and care to produce consistent results when building battered work.

Producing an accurate slope or angle in a battered wall can be achieved with a spirit level designed to allow a set angle to be followed.

▲ Figure 9.43 Spirit level with an adjustable vial

A simple way to achieve the same result without the need for a special spirit level is to create a **batter board**. The angled edge of the board is set to the specified batter and a standard spirit level held against the remaining edge. When the spirit level and batter board are held against the work and indicate that plumb has been achieved, the face of the wall will be following the desired angle.

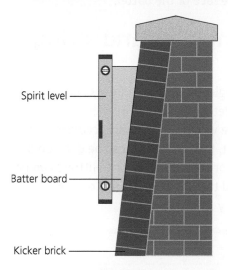

▲ Figure 9.44 Batter board used with a standard spirit level. Note that the first angled brick is called a 'kicker'

Spirit level

Batter board

Kicker brick

To find the specified angle to create a batter board requires a protractor or angle finder. Specialist tools are available to check angles for construction activities which make the task easier.

▲ Figure 9.45 Equipment for measuring angles

Sometimes a retaining wall may be designed with a batter in conjunction with stepped brickwork to reduce the thickness over the height of the wall. The battered face provides a more attractive appearance and a weatherproof finish to the wall, provided a suitable hard-wearing brick is selected.

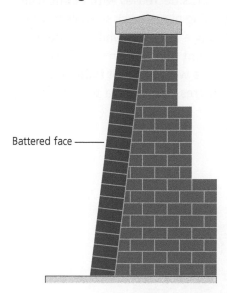

Battered face

▲ Figure 9.46 Stepped brickwork in a retaining wall with a battered face

Buttresses

Where a wall is likely to be subjected to forces which could make it bend or buckle, such as exposed locations where persistent high winds occur, supporting piers called buttresses can be built at intervals along its length. These are bonded into the main wall brickwork to add strength and stability. To resist lateral movement, the buttress can be constructed at an angle or battered into the vertical face of the wall.

▲ Figure 9.47 Battered buttresses supporting a wall

Tumbling-in

The term 'tumbling-in' refers to the method of bonding the buttress masonry to the main wall as the projection of the buttress decreases with height. This can be achieved by using plinth brick specials to give a relatively smooth angle of batter while maintaining the bonding arrangement of the main wall. This is a method which gives a weatherproof finish to the angled face of the buttress, provided a suitable quality brick is used.

▲ Figure 9.48 Plinth bricks used to reduce the thickness of a buttressed pier

A more complex and time-consuming method is to create the specified angle of a buttress by laying brick courses at right angles to the angle of batter, which are cut into the horizontal brickwork bonded with the main wall.

This method involves a lot of setting out and cutting work and may be specified where a more decorative finish is desirable as a feature in the battered buttress. The brickwork can be bonded decoratively across the face of the angled work.

When laying the first section of an angled or battered buttress, it is good practice to create an overhang at the bottom of the angle to act as a drip to shed rainwater from the face of the batter.

Using templates and string lines

Setting out the angle of batter requires good planning and attention to detail. A range of methods can be used to set out the work and maintain accuracy. For a small-scale project, the setting out can be done on a sheet of plywood or plaster board. The buttress can be drawn full size and the bricks positioned on the drawing to be marked ready for cutting.

Templates are useful pieces of equipment that can be easily assembled to set out and accurately construct a battered buttress.

A 'gun template' can be made from two lengths of timber joined to accurately create the specified angle, which can then be followed when setting out and building the batter.

Gauge marks can be added along the angled timber section to maintain accuracy of gauge along the battered face of the buttress. If the battered face extends beyond the length of the gun template, a string line can be attached to extend the line of angle as required.

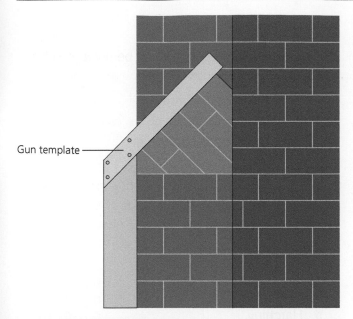

▲ Figure 9.49 Gun template used to create the specified angle

If the angle to be created is not critical, an alternative method using just a string line can be used. The line is fixed to a batten posstioned at the bottom of the batter.

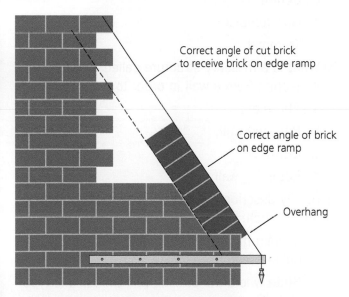

Correct angle of cut brick to receive brick on edge ramp

Correct angle of brick on edge ramp

Overhang

▲ Figure 9.50 A string line at the correct angle of batter fixed to a timber batten

If the angle of batter is minimal, the use of a batter board as previously described for a retaining wall may be appropriate for construction of a battered buttress.

Summary

In this chapter, we've looked at three bricklaying activities that involve complex setting out and construction methods:

- brick arches
- brickwork that is curved on plan
- battered brickwork.

Each of these construction activities requires an understanding of certain principles of geometry and knowledge of special tools, equipment and methods of work.

These bricklaying activities may not be a part of a bricklayer's regular work on site. It is therefore sensible to review the content of this chapter before starting work on any of them, in order to be fully prepared to produce work to the highest standards possible.

Test your knowledge

1 What name is given to a tapered brick used to construct an axed arch?

   a Radial

   b Segment

   c Splay

   d Voussoir

2 What name is given to the angle that the arch bricks spring from in a segmental arch?

   a Screwback

   b Skewback

   c Slantback

   d Slewback

3 What term is used to describe the process of slightly loosening the folding wedges when the construction of a brick arch is completed?

   a Easing

   b Lowering

   c Striking

   d Tightening

4 Why is a correctly constructed arch an efficient means of spanning an opening?

   a An arch is quicker to install than a concrete or steel lintel.

   b An arch transmits compressive forces to the abutments.

   c Arches are strongest when subjected to tension.

   d Arches are very effective at resisting lateral forces.

5 On which curved line should setting out marks for an axed arch cutting template be made?

   a Extrados

   b Intrados

   c Radiating

   d Segment

6 What name is given to brick specials that are manufactured for use in a curved wall?

   a Bullnose

   b Headers

   c Radials

   d Squints

7 What items of equipment can be used to set out and build a battered wall?

   a Batter board and spirit level

   b Plumb bob and gauge rod

   c Protractor and sliding bevel

   d Straight edge and string line

8 What is the name of the unsightly effect in a curved wall, caused by the ends of straight bricks overhanging the course below?

   a Faceting

   b Faulting

   c Grinning

   d Hatching

9 What is the name given to the special wooden template that can be made to set out and build tumbling-in brickwork?

   a Batter template

   b Centre

   c Gun template

   d Trammel

10 What is a masonry structure called that is built projecting from a wall in order to support it?

   a Abutment

   b Battered wall

   c Buttress

   d Retaining wall

11 Briefly describe the characteristics of a 'rough-ringed' arch.

12 a Explain the term 'key brick' in relation to a bulls-eye arch.

   b State where the striking point is for this type of arch.

13 Name three types of Gothic arch.

14 Explain what a trammel is and how it is used to set out and build a curved wall on plan.

15 Explain two ways that tumbling-in can be constructed for a battered buttress.

CHAPTER 10

CHIMNEYS, FLUES AND FIREPLACES

INTRODUCTION

The design and construction of chimneys, flues and fireplaces is carefully controlled by laws and regulations. This is because there is a potential fire risk within a structure when fuel-burning appliances are used to provide heating. The gases that are produced pose a significant risk to health and life if not correctly conducted away from an internal living or working area.

The bricklayer has the responsibility to carefully and conscientiously construct chimneys, flues and fireplaces to the design specification, so that they are safe in operation and long lasting. In this chapter, we will consider the skills and knowledge that must be developed and maintained to be able to confidently build chimneys, construct flues and install fireplaces.

By the end of this chapter, you will have an understanding of:
- the regulations and guidance that control construction of chimneys, flues and fireplaces
- components and materials used in chimney, flue and fireplace construction
- methods of setting out and building chimneys, flues and fireplaces.

This table shows how the main headings in this chapter cover the learning outcomes for each qualification specification.

Chapter section	Level 2 Technical Certificate in Bricklaying (7905-20)	Level 3 Advanced Technical Diploma in Bricklaying (7905-30) Unit 305	Level 2 Diploma in Bricklaying (6705-23)	Level 3 Diploma in Bricklaying (6705-33) Unit 305	Level 2 Bricklayer Trailblazer Apprenticeship (9077) Module 8
1. The regulations and guidance that control construction of chimneys, flues and fireplaces	N/A	1.1	N/A	1.1, 2.1	2.2, 3.1
2. Components and materials used in chimney, flue and fireplace construction	N/A	1.2	N/A	1.2, 1.3, 1.4, 1.5, 1.6, 2.4, 2.5, 2.6, 2.7, 4.2	1.1, 2.1, 3.1
3. Methods of setting out and building chimneys, flues and fireplaces	N/A	2.1, 2.2	N/A	3.1, 3.2, 3.3, 3.4, 3.5, 3.6, 3.7, 3.8, 4.1, 4.3, 4.4, 4.5, 4.6	3.1, 4.1

1 THE REGULATIONS AND GUIDANCE THAT CONTROL CONSTRUCTION OF CHIMNEYS, FLUES AND FIREPLACES

There are extensive regulations that apply to the construction and maintenance of chimneys, flues and fireplaces. While the bricklayer cannot be expected to have in mind every detail of the laws and regulations, it is important to develop and maintain an awareness of the main regulations and control documents and where they can be found for reference.

> **KEY POINT**
>
> - When we refer to chimneys, we mean the masonry structures you can usually see above the roof line of a building.
> - When we refer to flues, we mean the ducts or tubes that take the gases produced during **combustion** safely outside a building.
> - When we refer to fireplaces, we mean the locations inside a building where the combustion process takes place.

> **KEY TERM**
>
> **Combustion:** the chemical process in which substances mix with oxygen in the air to produce heat

▲ Figure 10.1 A fireplace must be constructed to comply with regulations

Information sources

Recognised information sources for chimney, flue and fireplace construction can be categorised as regulation and advice.

Regulatory documents

First, we'll look at documents that contain information in the form of regulations. Keep in mind that the term 'regulatory' means that the directions and instructions given *must* be followed.

Building Regulations

The design and construction of fireplaces and flues is controlled by the Building Regulations Part J. This is called an 'Approved Document' and is issued by the government to ensure that buildings and specific parts of them are constructed to a minimum standard. There are many individual components that form chimneys, flues and fireplaces that are defined and described in the Building Regulations.

> **INDUSTRY TIP**
>
> You can access Approved Document J at: www.gov.uk/government/publications/combustion-appliances-and-fuel-storage-systems-approved-document-j

Codes of practice

An Approved Code of Practice (ACOP) is a set of written directions and methods of work issued by an official body or professional association to its members and approved by the Health and Safety Executive. There are many codes of practice that give direction when constructing chimneys, flues and fireplaces, provided by a range of professional bodies.

British Standards

A BS (British Standard) document applies in the UK and some British territories. A BS EN (British Standard European Norm) document applies to many European countries and includes the UK while it remains a member of the European Union. These standards define good practice. They are agreed ways of doing something, written down as precise criteria so they can be used as rules or definitions. An example is: **BS EN 998-2**, a specification for mortar for masonry.

Drawings

Drawings produced by an architect or architectural technician are specific to the task and form part of the contract documentation. This means that they are legally binding. Drawings will provide the design and details of the work to be carried out, including measurements and dimensions which are critical and comply with regulations.

INDUSTRY TIP

If you find a mistake in a drawing that you've been given, always bring it to the attention of your line manager or supervisor. They will refer back to the architect. Because the drawings are contract documents, the architect must approve alterations to drawings. He or she will inform site personnel about alterations with a document called a 'variation order'.

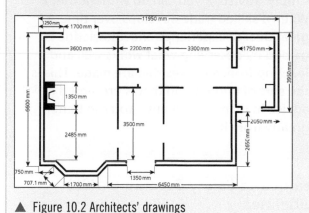

▲ Figure 10.2 Architects' drawings

Specifications

These link with the drawings and specify the types, colours, quality and other specific details of the materials to be used in the construction process. Like a drawing, the specification is a contract document and so is legally binding. It may state which manufacturer or contractor should be used.

Specification		
102 External cavity walling		
• Walling below ground:		
- Type:	Cavity wall, concrete filled.	▼
- Masonry units:	Common bricks.	▼
- Mortar:	Class M6 mortar.	▼
• DPC at ground floor:	Flexible cavity trays.	▼
• Walling above ground:		
- External leaf above ground:		
Masonry units:	Facing bricks.	▼
Bond or coursing:	Flemish bond.	▼
- Internal leaf above ground:		
Masonry units:	Aerated concrete blocks.	▼
- Mortar:		
Type:	Class M4 mortar.	▼
Joint profile to		
external faces:	Bucket handle.	▼
- Wall ties:	Insulation retaining wall ties.	▼
- Cavity insulation:	Full fill cavity insulation.	▼
- Ventilation components:	Air bricks and sub-floor ventilation ducts.	▼
• Openings:		
- Lintels:		
Type:	Manufactured stone lintels.	▼
Cavity tray cover:	Flexible cavity trays.	▼
- Cavity closers:	Flexible insulated DPCs.	▼
- Sills:		
Type:		▼
DPC below:	Manufactured stone sills.	▼
• Abutments:	Natural stone sills.	
Cavity trays and DPCs:	Precast concrete sills.	▼
Flashings built into masonry:	As drawings.	▼

▲ Figure 10.3 A specification

Safety regulations

Health and safety issues are a major consideration when constructing chimneys, flues and fireplaces, as outlined below.

- Safe installation of a fireplace includes applying specific minimum dimensions to the setting out and construction process.
- Safety regulations will be applied to the design of the flue to ensure the safe and efficient removal of dangerous gases.
- During installation, safety regulations must be applied when moving heavy components.
- Safety regulations apply to working at height when building a chimney.

An information source that is specific to chimney, flue and fireplace construction is the Heating Equipment Testing and Approval Scheme (HETAS), which provides regulatory information in the UK.

Advisory documents

Documents categorised as 'advisory' form an important part of the range of documents you will use to provide information on safe and effective construction methods and practices. Consider them carefully.

Schedules

Schedules can advise on sequence and timing when selecting components for chimneys, flues and fireplaces. If a project has construction features that are repeated frequently throughout its duration, a schedule cuts down on repetitive documentation and contributes to efficiency.

Good practice guides

Good practice guides can advise on proven methods and techniques to ensure successful completion of the work task. Drawing on past experience can be an asset when undertaking technically demanding construction tasks.

▲ Figure 10.4 Check the relevant documentation before you start work

Manufacturer's instructions

Manufacturer's instructions give information on the correct way to assemble or install an item. They can also provide valuable information and advice on interactions of components with other materials or substances such as adhesives. Following manufacturer's instructions ensures that the item will wear better, be safe in use and function properly over time.

▲ Figure 10.5 Manufacturer's instructions

VALUES AND BEHAVIOURS

Avoid thinking that because documents are simply 'advisory' you can do without them. When a skilled trade develops over time, a great deal of experience and knowledge is gained by observing what works well and learning from mistakes that are made. This is the sort of information that is presented as advice in many documents. You can benefit greatly from it.

Health and Safety Executive (HSE) guidelines

HSE guidelines are developed for all industries and there is a specific set for the construction industry. These provide information, regulation and guidance on work in construction and help to prevent accidents in the workplace. These guidelines provide useful information regarding the risks involved with the work you are carrying out, such as moving materials and working at height.

Work on chimneys is likely to involve work at height so reference to these guidelines will be essential.

Regulations and standards in materials

The materials for the construction of various parts of a chimney, flue and fireplace are closely regulated to ensure that the completed chimney, flue and fireplace will work effectively and be safe to use over the life of the building.

This means that the materials must be of the right quality and strength to cope with the demands that will be placed on them. The components specified will obviously need to be manufactured from materials that are not affected by heat and combustion; very high temperatures are generated within a solid fuel fireplace or appliance.

Components that must function in high temperature situations are manufactured from materials that are described as **refractory**. Many of these materials are made from fireclay, which is a natural material that is extracted from deeper excavations than other clay types.

KEY TERM

Refractory: difficult to melt; resistant to heat

ACTIVITY

Research the different uses of refractory bricks. Select one use and find out what temperature your choice operates at.

▲ Figure 10.6 Refractory bricks are resistant to high temperatures

▲ Figure 10.7 Refractory brick

The way the regulated components for chimneys, flues and fireplaces are assembled and installed by the bricklayer can have a critical effect on the proper functioning of the finished work. We'll look in greater detail at how regulations and advice are applied when measuring, setting out and constructing chimneys, flues and fireplaces later in this chapter.

First, let's get familiar with the names and descriptions of some of the regulated components that will be used.

② COMPONENTS AND MATERIALS USED IN CHIMNEY, FLUE AND FIREPLACE CONSTRUCTION

There is an extensive range of components and materials used in the construction of chimneys, flues and fireplaces. Within groups of components there can be variations in design or appearance, but the component will still be required to conform to regulations and standards.

Component descriptions

▼ Table 10.1 Main components of a fireplace, flue and chimney

Component	Description
Throat unit	Sometimes referred to as a 'starter unit', this item serves several purposes: ● provides support for the masonry to be constructed above it, serving as a lintel ● allows the flue to be correctly positioned above the combustion area ● reduces (or 'gathers') the dimensions of the fireplace opening to suit the size of the flue.
Flue liners	Non-combustible liners are assembled to form a duct or tube to conduct fumes and gases safely outside a structure. They can be manufactured from clay, concrete or metal. They can be specified as round, square or rectangular cross-sections.
Fireback	The fireback (sometimes called a 'burr') sits at the back of the combustion area. It must withstand very high temperatures because of its position within the fireplace. Some designs are manufactured in two sections to make installation easier.
Chimney pot	Chimney pots are fitted at the top of the flue. They are often designed with a tapered shape which can increase the updraft through the chimney for greater efficiency. There are many different designs of chimney pot.

▼ Table 10.1 Main components of a fireplace, flue and chimney (continued)

Component	Description
Chimney cowl 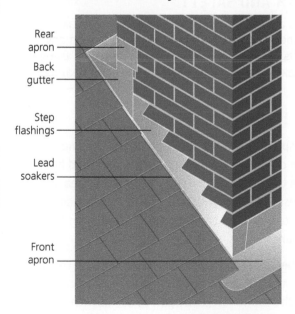	A cowl is a component added to the top of a chimney pot to prevent the entry of rain. A cowl can also be used to prevent persistent strong winds affecting the efficient flow of flue gases away from the chimney.

Components used for weathering

Remember, 'weathering' refers to the methods we use to make a masonry structure resistant to bad weather. A chimney is usually constructed in an exposed position and projects through the roof structure, so it will need to include components that are designed to effectively weather it and prevent water **ingress**.

KEY TERM

Ingress: the act or process of entering something

A traditional material used for waterproofing and weathering the location of a chimney is lead. (For a reminder about faults that can develop in lead components used in chimney design, refer back to page 217.) Lead is a soft metal that can be formed into complex shapes to produce components that follow the angle of pitch of a roof and corners of a chimney.

Rear apron
Back gutter
Step flashings
Lead soakers
Front apron

▲ Figure 10.8 Chimney waterproofing components

INDUSTRY TIP

Shaping (or 'dressing') lead is a skilled activity. The bricklayer may be required to undertake simple dressing of lead, but more complex shaped components may require lead soldering to produce them. Unless you have the skills required, this sort of work should be done by a plumber or fabricator.

Where the chimney projects through the roof structure, a suitable system of waterproofing must be provided in two key areas:

- Within the body of a chimney – a lead tray is inserted across a bed joint covering the whole area of the stack. The tray is a lead component that can require a high level of skill to produce. In the illustration, the cylindrical upstand is formed to sit in the joint of the flue liners so that water travelling down the inside of the flue is directed onto the body of the tray. Water collected on the body of the tray can flow out of the chimney masonry through weep holes.
- Where the chimney emerges through the roof – lead is used to form an apron at the lower point and an apron and gutter at the upper point where the chimney meets the roof. At each side, lead flashings are stepped along the angle of pitch. Where the roof material is slate, lead or plastic angled sheets called soakers are inserted under each course of slate to correspond with each lead flashing.

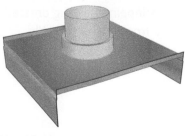

▲ Figure 10.9 Lead tray

297

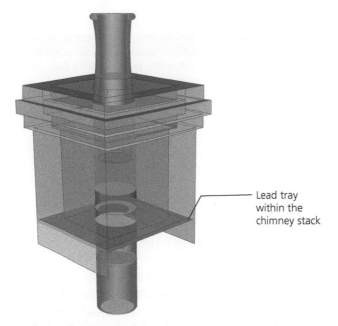

▲ Figure 10.10 Lead tray positioned within the chimney

Lead tray within the chimney stack

▲ Figure 10.11 Placing soakers

If the roof material is a concrete or clay tile, a lead flashing can be dressed (or shaped) under each course, with an overlapping flashing dressed above it, extending over the front edge of the tile.

▲ Figure 10.12 Flashings, apron and gutter on a chimney

The fireplace

The complete structure of the chimney, flue and fireplace combines to become a significant and heavy part of a building. That is why the fireplace area, as the lowest point of the arrangement, must be solidly constructed to carry the weight of masonry and components above it.

Remember, when we refer to the fireplace, we mean the area inside a building where the fire is contained. Look at the sectional illustration to see how the various parts are arranged in relation to each other.

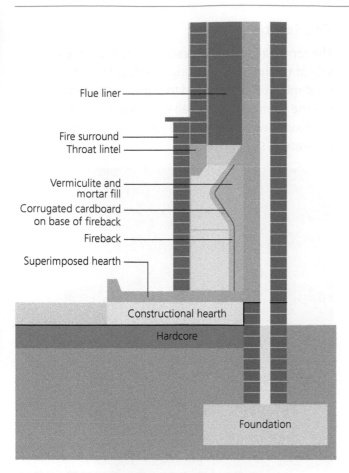

▲ Figure 10.13 Section through a fireplace (solid floor)

Fireplace terminology

The bricklayer needs to become familiar with the terminology used for each part of the fireplace to be able to construct it in accordance with the regulations. We'll look at how regulations apply when setting out and building the work later in the chapter. Here we'll just look at the description and some details of each part.

The hearth

The **hearth** is an area in front of the fireplace that gives protection from heat and sparks to the floor and floor covering. It is described as two sections:

- The 'constructional hearth' – this is usually formed during construction of the main structure. This provides structural support for the area around the chimney breast and prevents damaging heat transfer to surrounding components, so it is usually constructed from concrete.
- The 'superimposed hearth' – this is laid on top of the constructional hearth and can provide a

decorative finish which obviously must also be heat resistant. It can be constructed in brick, ceramic tiles or slate, or more expensive decorative materials such as marble.

KEY TERM

Hearth: an area in front of a fireplace that gives protection from heat and sparks to the floor and floor covering

INDUSTRY TIP

A constructional hearth can be installed on the ground floor of a building as part of the structural design of the solid floor slab. It can also be installed as part of a suspended floor at ground level or even at upper levels of a structure in timber floor designs. The dimensional regulations must be followed closely to reduce fire risk.

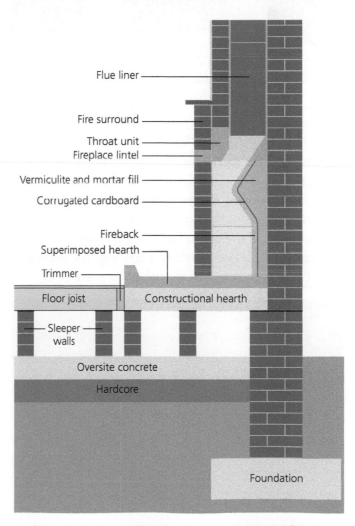

▲ Figure 10.14 Section through a fireplace (suspended floor)

▲ Figure 10.15 Superimposed hearth finished in ceramic tiles

The chimney breast

The chimney breast can be constructed internally as a projection into a room or externally projecting from an outside wall, leaving more uninterrupted space inside the building. It contains the fireplace opening and the flue. As the height of the breast increases, the width is often reduced at upper floor levels of a building. The chimney breast can be constructed in brick or concrete block.

▲ Figure 10.16 Chimney breast above a fireplace

The jambs

The term '**jamb**' simply describes the masonry either side of the fireplace opening. The fireplace opening itself must be constructed to regulated dimensions, but the chimney breast can vary in size depending on the room design, so that the jambs can increase or decrease in size.

▲ Figure 10.17 Chimney breast jamb

Back boiler

Back boilers were common in the 1950s and 1960s as a means of utilising heat produced in the fireplace to heat domestic hot water. A back boiler is a simple device which uses **convection** and gravity feed to create a flow of heated water through pipes connected to a cylinder which stores the hot water.

KEY TERMS

Back boiler: a simple device which uses convection and gravity feed to create a flow of heated water through pipes connected to a cylinder which stores the hot water

Convection: the flow of heat through a gas or a liquid as the hotter part rises and the cooler part falls

Jamb: the masonry either side of a fireplace opening; also describes a vertical inside face of an opening in a wall (see Chapter 5)

The flow of hot gases from the combustion process is directed through a duct in the boiler unit to heat the water contained in the jacket. A damper plate can be fitted at the top of the duct to control the flow of hot gas. The gases then join the main flue to be drawn up the chimney.

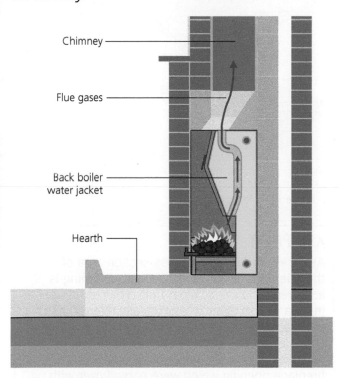

Chimney

Flue gases

Back boiler water jacket

Hearth

▲ Figure 10.18 Fireplace with a back boiler

This arrangement is a useful way of utilising heat that would otherwise be lost up the chimney and is often employed in modern wood-burning stoves.

The flue

The flue must be carefully constructed to prevent gases created by the combustion process escaping into the living or working area of a building. The gases are toxic and can have potentially fatal results if not effectively contained within the flue and conveyed to the open air.

HEALTH AND SAFETY

The gases from the combustion process are often undetectable by the occupants of a room. A carbon monoxide detector should be fitted in a room with a fireplace, to alert the occupants to possible leaks from a flue.

There are flue components manufactured in metal that form a continuous tube and provide a sealed duct that are relatively easy to install. Flues that are constructed by the bricklayer will most likely use individual liner components.

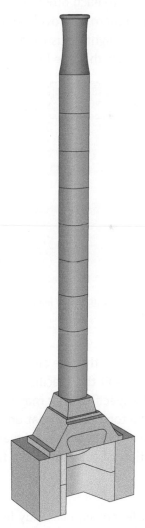

▲ Figure 10.19 Flue liners assembled to form a flue

While a flue is essentially a tube, it must function in conditions that can include extreme heat at the lower end of the flue and damp cooler conditions that can initiate damaging chemical reactions at the top end of the flue. It must cope with expansion and contraction throughout its length as it heats up and cools down.

Flue terminology

Becoming familiar with the terminology relating to flue components and how they fit together will assist the bricklayer in constructing a flue that continues to work safely and efficiently over time.

Liner spigot and socket

Liners are assembled in a defined way to achieve a gas-tight flue. One end of a liner has a projection called a spigot and the other end has a recessed socket. These are designed to fit into each other with an adhesive or mortar between them to provide an effective seal and also to allow to slight adjustments as construction proceeds to make sure that the flue is correctly aligned.

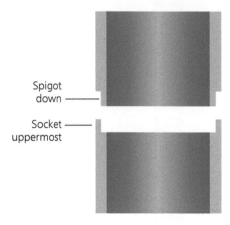

▲ Figure 10.20 Section through flue liners

Reducers

The liners used to form the flue are manufactured in a variety of sizes to suit the specified dimension and cross-section. The specified liner may be smaller than the opening at the top of the throat unit and will therefore require a component known as a 'reducer' to allow a smooth flow of gases from the fireplace.

ACTIVITY

Choose a website that supplies clay flue liners. How many different diameter flue liners can you find? Discuss with another person why you might need different diameter flue liners.

The joint between the throat unit and the reducer is in a location that will be subjected to very high temperatures, so fireclay may be specified as the jointing material to maintain a gas-tight seal in such harsh conditions. A standard sand and cement mortar is unlikely to be able to withstand the high temperatures generated in the fireplace area and over time will become prone to cracking.

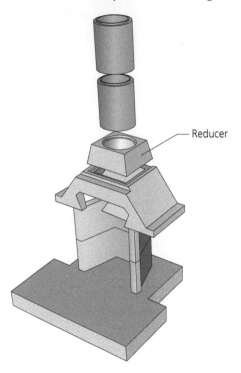

▲ Figure 10.21 Reducer

KEY POINT

As noted, reducing the cross-section area of the flue directly above the fireplace opening is achieved by using a throat unit. The reduction in area at this point is traditionally referred to as the 'gather'. Before the introduction of concrete throat units, the gather was formed by stepping out bricks cut to an angle on each side of the fireplace opening which were parged over with mortar to give a smooth flow to the flue gases.

Off-set

The most efficient design for a flue is a straight vertical arrangement. However, there may be situations which require the flue to change direction to pass obstructions that occur as the flue passes through the floor levels of a structure.

Curved sections of flue liner referred to as 'off-sets' are manufactured in a range of lengths, diameters and angles to allow the direction of the flue to be adjusted as required.

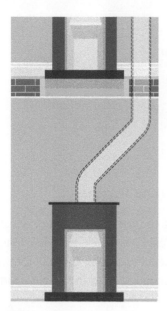

▲ Figure 10.22 An off-set used to avoid an obstruction

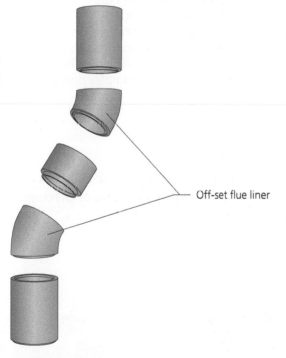

Off-set flue liner

▲ Figure 10.23 Off-set flue liner

The chimney stack

The chimney stack is the most exposed part of the construction arrangement that allows an open fire or other solid fuel appliance to be used safely in a living or working space. A chimney stack constructed in masonry must be strong enough to resist high winds, rain, icy conditions and heat from the sun.

> **KEY POINT**
> Because the chimney stack is in such an exposed position, the masonry can remain saturated over extended periods. This can lead to the occurrence of damaging chemical reactions in the materials from which the bricks are manufactured and the mortar. To reduce the potential for attack by chemical reaction, sulphate-resisting cement can be used in the mortar mix.

▲ Figure 10.24 A chimney stack is often in an exposed position

Some terminology used in chimney design is unusual and many terms relate to features that offer protection from the elements. You should become familiar with them to completely understand the purpose and function of a chimney.

Chimney terminology

Flaunching

'**Flaunching**' refers to the weatherproofing measures applied to the exposed top of a chimney stack. There are two main methods of flaunching a chimney:

- A strong cement/sand mix is formed on top of the chimney, shaped around the chimney pot to direct water away from the surface. This also serves to securely anchor the chimney pot in position.
- A pre-formed concrete unit is laid on top of the chimney stack, which includes an opening for the pot to project through. The concrete unit can be designed with an overhang to give greater protection from rain.

▲ Figure 10.25 A concrete flaunching unit. Note the overhang to increase protection from the rain

Oversailing course

Oversailing courses can be used as a decorative feature (for more details, see page 248.) The projecting course or courses of brickwork have the additional function of forming a 'drip' in the elevation of masonry to direct water away from the face. This is a design feature that can be applied as a protection measure in a chimney stack.

▲ Figure 10.26 Oversailing courses

The oversailing courses enlarge the dimensional cross-section of the chimney stack, allowing space for a pot to be bedded securely within them before the brick courses are stepped back to the original dimensions.

Necking course

A 'necking course' is constructed as oversailing brickwork part-way down the height of the stack as an additional decorative feature and a means of directing water away from the face of the chimney stack.

▲ Figure 10.27 Necking course

Mid-feathers

Mid-feathers are sometimes referred to as 'withes'. These terms refer to the masonry walls that separate multiple flues in a chimney stack. The mid-feathers or withes should be tied into the outer masonry of the stack, either by appropriately setting out the brick bonding arrangement of each course or by including suitable reinforcement in each course.

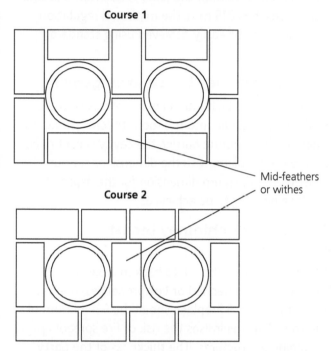

▲ Figure 10.28 Mid-feathers bonded into the outer brickwork

Parging

This is a more traditional practice used to finish the inside surface of a flue. Before flue liners were introduced, the inside of a brick-built flue was coated in a thin coat of mortar to seal it and provide a smooth surface to encourage a good flow of gases.

▲ Figure 10.29 Parging

③ METHODS OF SETTING OUT AND BUILDING CHIMNEYS, FLUES AND FIREPLACES

A bricklayer setting out chimneys, flues and fireplaces must carefully follow the working drawings and other contract documentation to make sure that the work fully complies with regulations. Remember, regulatory information *must* be followed.

Let's look at some dimensional regulations that apply to the construction process.

The fireplace

The layout of the fireplace design will determine which regulation dimensions apply. Remember, the fireplace can be constructed within a chimney breast that's internal, projecting into a room, or external projecting from an outside wall. Two fireplaces can be constructed back to back in a party wall between semi-detached dwellings. Let's look at the options.

Fireplace regulation dimensions

INDUSTRY TIP

Further information on fireplace dimensions can be found in 'Combustion appliances and fuel storage systems: Approved Document J', pages 37–38: www.gov.uk/government/publications/combustion-appliances-and-fuel-storage-systems-approved-document-j

There are minimum dimensional requirements for many of the elements of fireplaces and chimneys to prevent the spread of fire in the location of the fireplace. For example, combustible material other than a skirting board (such as a timber floor joist) must be positioned a minimum of 200 mm from the inside of the flue.

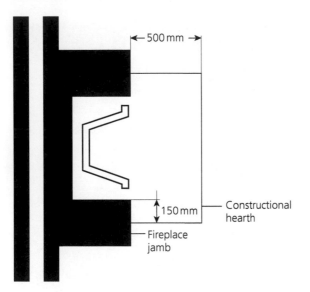

▲ Figure 10.30 Minimum dimensional requirements for a hearth

Internal chimney breast arrangement

The fireplace in an internal chimney breast arrangement will often be constructed against an external cavity wall. The cavity will be maintained behind the fireplace with the jambs either side of the fireplace extending into the living or working area from the inner leaf.

The minimum width of the jamb is 200 mm and since a brick length is 215 mm, the minimum regulation dimension can be easily achieved using standard bricks.

External chimney breast arrangement

If the fireplace is built into an external breast, there will be no projection into the room to form jambs. To maintain the moisture barrier, the cavity is continued around the fireplace, with the dimensions set out to allow the minimum dimension for the depth of fireplace opening to be achieved.

Back-to-back chimney breast arrangement

If two flues are built back to back in a party wall between semi-detached or link properties, the backing masonry for *each* fireplace should be a minimum of 200 mm. This minimises the risk of fire spreading to adjoining structures. The thickness of the party wall itself can be reduced either side of the chimney breast.

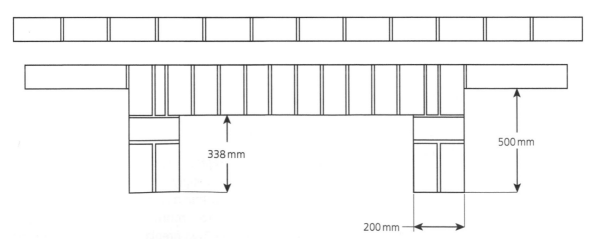

▲ Figure 10.31 Fireplace arrangement for an internal chimney breast

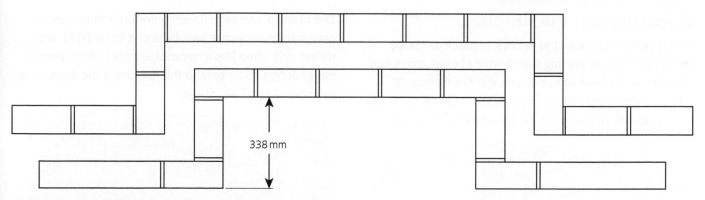

▲ Figure 10.32 Fireplace arrangement for an external chimney breast

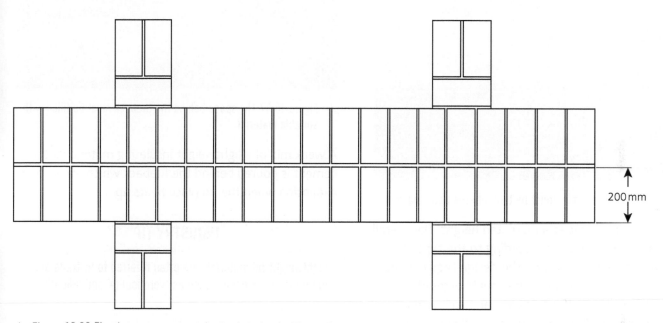

▲ Figure 10.33 Fireplace arrangement for back-to-back chimney breasts

The hearth

The minimum thickness of the constructional hearth is 125 mm to give it the required load-bearing qualities. The minimum projection from the face of the fireplace is 500 mm to ensure adequate heat and spark protection for floor materials and floor coverings. The distance from the fireplace opening to the side edge of the hearth must be a minimum of 150 mm, to give protection from heat and sparks.

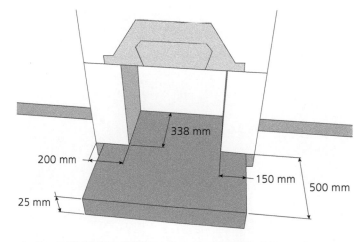

▲ Figure 10.34 Constructional hearth minimum dimensions

Constructing the fireplace

The jambs are constructed in brick or block and serve the purpose of supporting the chimney breast above and containing the heat source to prevent fire damage to surrounding components. They should be bonded into the adjoining wall to increase structural strength and rigidity.

▲ Figure 10.35 The jambs must be tied into the adjoining wall

The throat unit is set at a regulated height. The overall size of the opening is set according to the type of appliance, flue diameter and the design requirements for the room's appearance. The usual height of the opening is 550 mm, but this can vary, so it is important to check the specification and the working drawing before commencing construction.

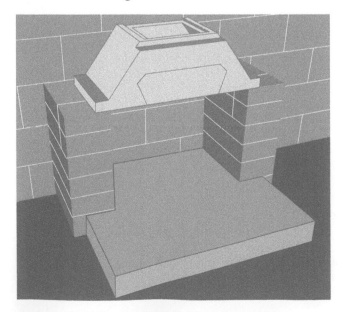

▲ Figure 10.36 Throat unit set on the masonry jambs

The fireback can be installed before the throat unit if preferred to allow the area behind it to be filled. The throat unit often has a removable insert which allows easier access to fill behind the fireback if the decision is to install it at a later stage.

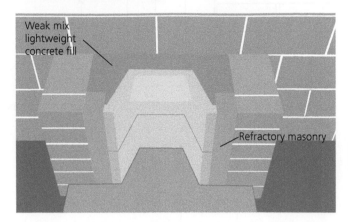

▲ Figure 10.37 The space behind the fireback must be filled with suitable material

A weak mix of a lightweight insulating material and cement is poured behind the fireback which allows for expansion when the fireplace heats up.

INDUSTRY TIP

Lightweight fill materials are often referred to by trade and manufacturer's names such as 'Vermiculite' and 'Micafil'.

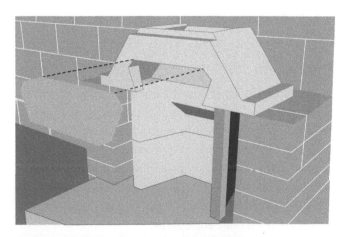

▲ Figure 10.38 Insert in throat unit removed for access

A sheet of corrugated cardboard can be placed directly behind the fireback before the fill is put in place. The first time the fire is lit, the cardboard will char away to leave a small permanent expansion gap.

If the depth of the fireplace opening is greater than the minimum required measurement, it may be beneficial to use an additional special-shaped lintel just below the front edge of the throat unit to improve the flow of gases up the flue and prevent them from entering the room. The specification and drawings will give information on whether this is needed.

▲ Figure 10.39 Fireplace lintel

The flue

As noted, traditionally flues were constructed entirely in masonry with the internal faces parged in mortar to smooth and seal them. However, corrosive deposits often formed on the flue masonry, due to the presence of condensed moisture which combined with chemicals and carbon in the soot created by the combustion process. These deposits caused the masonry to deteriorate, which could lead to extensive repair or rebuilding.

With the introduction of flue liners manufactured from more durable materials, the durability of flues was greatly improved. Solid fuel fires require a flue rated as Class 1 which can be met by using liners manufactured from clay or pumice concrete. These materials can resist the high temperatures and chemical substances created when burning solid fuels such as coal or wood.

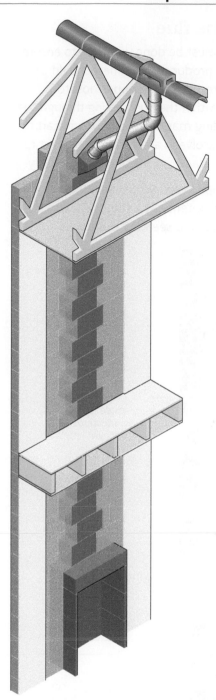

▲ Figure 10.40 Flue for a gas fire

> ### INDUSTRY TIP
>
> A Class 1 flue has a diameter greater than 175 mm.

Flue components for a gas-fired appliance cannot be used for a Class 1 flue since they are not durable or heat resistant enough. They are often produced in a rectangular shape with dimensions that match standard blocks to allow them to be integrated into the bonding arrangement of a block wall. There are a number of modern flue component designs that are manufactured for heating appliances that do not burn solid fuels.

IMPROVE YOUR ENGLISH

Research 'balanced flues' and write an explanation of how they differ from a standard flue arrangement.

Constructing the flue

Assembling flue liners must be done with care to ensure that a gas-tight seal is produced. The liners should always be installed with the socket uppermost to prevent any moisture within the flue passing through the joint into the surrounding masonry. Sulphur and carbon in the flue gases can dissolve in any moisture present to form acids which can attack the surrounding masonry.

Understanding this simple principle protects the chimney from deterioration and potentially extends its working life. Study the illustrations to see how this works.

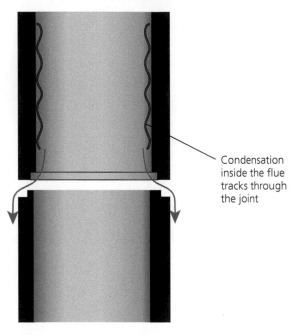

Condensation inside the flue tracks through the joint

▲ Figure 10.41 Spigot uppermost WRONG!

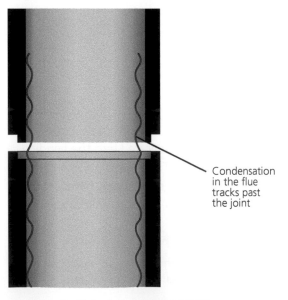

Condensation in the flue tracks past the joint

▲ Figure 10.42 Socket uppermost RIGHT!

Liner joints should be carefully sealed with fire cement or refractory mortar. Mortar produced for bricklaying is not suitable since it will not accommodate the expansion and contraction caused by the high temperatures present in a solid-fuel appliance. Adhesives and sealants are available in ready-mixed form which can be applied using a sealant gun to simplify the process.

The space between the flue liners and the chimney breast masonry should not be filled with ordinary mortar or concrete. The liner manufacturer will provide instructions on a suitable mix of weak concrete using a lightweight insulating material. This will reduce condensation within the flue and stabilise the internal temperature to improve the flow of gases to the outside air.

ACTIVITY

Look at 'Combustion appliances and fuel storage systems': Approved Document J: www.gov.uk/government/publications/combustion-appliances-and-fuel-storage-systems-approved-document-j

Locate Section 1.28 on page 22. Write down two suitable mix ratios for a weak insulating concrete for filling the space between flue liners and the chimney breast masonry.

The chimney stack

Building a chimney stack involves handling and lifting heavy components into position and working at height, so the first consideration must be the relevant health and safety requirements. Method statements and risk assessments must always be consulted before starting work.

▲ Figure 10.43 Be safe when working at height

Constructing the chimney stack

The stack will be constructed around the flue liners to the height at which the lead tray is to be installed. To accommodate lead flashings below the height of the lead tray, the appropriate bed joints can be raked out to a depth of about 25 mm to allow the flashings to be installed later. (Flashings *can* be installed as the masonry is built but they will be prone to damage and mortar staining.)

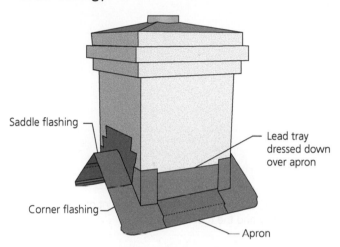

▲ Figure 10.44 A lead tray and lead flashings work together to weather a chimney stack

Saddle flashing

Lead tray dressed down over apron

Corner flashing

Apron

Once the lead tray is in position and formed tightly around the flue liners, the masonry work can continue to the level of the pot. At a point below the base of the pot, the masonry can be oversailed or corbelled (stepped) out a number of courses to allow the base of the pot to sit comfortably within the outline of the stack. This masonry overhang arrangement also serves to direct water away from the faces of the chimney stack.

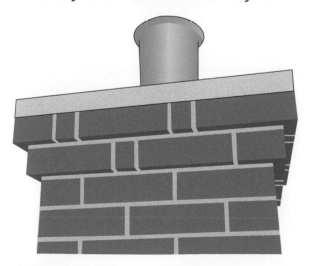

▲ Figure 10.45 Oversailed or corbelled courses

When the pot has been positioned, the top of the stack can be completed with a weatherproof finish. If flaunching is specified using a strong sand/cement mortar mix, a measure of skill and care is required to produce a durable and effective finish by hand.

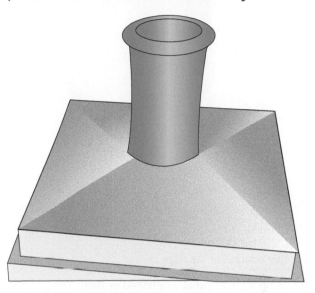

▲ Figure 10.46 Skill is needed to produce a smooth uniform finish to sand/cement flaunching

Use of a flaunching system such as a pre-formed concrete cap will require care in positioning to ensure any overhang projects evenly on all sides of the stack.

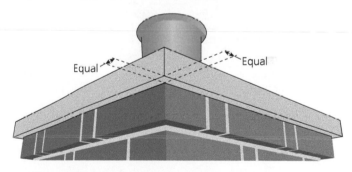

Equal   Equal

▲ Figure 10.47 Create an even overhang on all sides

Keep in mind that pre-formed flaunching systems are usually heavy components that must be handled and manoeuvred into position with care.

Regulation heights

The position of a chimney in relation to the roof pitch and the roof ridge line has an effect on the efficiency with which it conducts flue gases to the open air. There are specific regulations regarding the height to which a chimney stack must be built in different positions.

▼ Table 10.2 Regulation height measurements for chimneys

Position of chimney	Height of chimney stack
At or within 600 mm of the ridge ▲ Chimney emerges at the ridge	Minimum 600 mm above the ridge
Elsewhere on the roof (whether pitched or flat) ▲ Chimney emerges below the ridge	Minimum 1000 mm above the highest point of intersection of the chimney and the roof surface

▼ Table 10.2 Regulation height measurements for chimneys (continued)

Position of chimney	Height of chimney stack
Within 2300 mm horizontally of an openable rooflight, dormer window or other opening ▲ Chimney emerges near a dormer window	Minimum 1000 mm above the top of the opening
Within 2300 mm of an adjoining or adjacent building, whether or not beyond the boundary ▲ Chimney emerges near an adjacent building	Minimum 600 mm above any part of the adjacent building within 2300 mm

Source: Adapted from the Building Regulations 2010 Approved Document J: Combustion appliances and fuel storage systems

KEY POINT

The regulation measurement refers to the dimension from the weather surface of the roof to the top of the chimney pot.

ACTIVITY

The design of a chimney can often include decorative features that are striking in appearance. Be observant as you travel around and look out for chimneys that feature decorative designs in brick. Take photographs and discuss with others how the designs might have been built.

Installing a fireplace surround

When the fireplace area and the room it is in are complete, the fireplace surround and the superimposed hearth can be installed. Fitting these components earlier would risk causing damage to them. It may be the case that the whole building is completed before the fireplace surround and superimposed hearth are fitted to avoid damage to them during other construction operations.

▲ Figure 10.48 A fireplace surround

Fitting the fireplace surround and superimposed hearth step by step

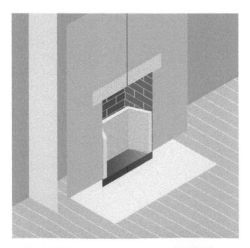

STEP 1 Establish the centre point of the opening and check that the face of each jamb is plumb. If the face is out of plumb, you will need to take this into account when sealing the fireplace surround at the fireplace opening.

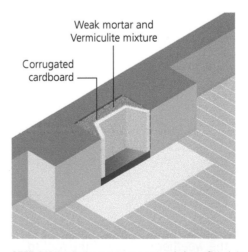

STEP 2 If the fireback hasn't previously been fitted, it should now be positioned and filled at its rear with a weak mix of cement and lightweight insulating material. Remember, corrugated cardboard should be placed against the rear of the fireback before the fill is introduced. If the fireplace opening is wider than the fireback, the open sides of the fireback will need to be filled with refractory masonry.

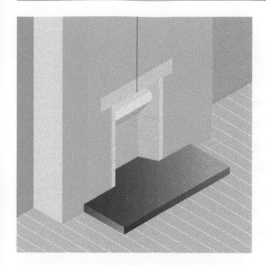

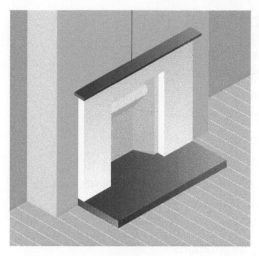

STEP 3 Depending on the design of the fireplace surround, the superimposed hearth will be installed next. (Some designs of hearth have cut-out sections to allow the hearth to be laid last.) The superimposed hearth should be bedded evenly on suitable mortar and levelled carefully in both directions.

STEP 4 The fire surround can now be placed in position very carefully to avoid causing any damage to the hearth or the fire surround itself. The joints around the fire opening will need to be sealed between the fireback and the surround using fireproof rope. Once the surround is in place, it can be securely fixed to the jambs and chimney breast and any damaged plasterwork around the fireplace restored.

HEALTH AND SAFETY

Keep in mind that some fire surrounds are very heavy. Enlist the help of an assistant to move the fire surround into place and make sure that it is supported during fixing. If it falls on someone it could cause serious injury.

Summary

This chapter has emphasised how construction of chimneys, fireplaces and flues is carefully controlled by laws and regulations so that the occupants of a building are protected from potential hazards. When working on any part of a structure that includes a chimney, fireplace or flue, always remind yourself that your work must be completed to the standards required to maintain safety.

Remember that the bricklayer's work when constructing chimneys involves protecting the building by correctly installing weatherproofing measures. Poorly installed flashings, trays and flaunchings can allow water penetration that may go unnoticed for some time, leading to a need for expensive repairs to the building at a later stage. Review the material in this chapter periodically to remind yourself about the complexities of constructing chimneys, flues and fireplaces.

Test your knowledge

1 What is the minimum thickness of a constructional hearth?

   a 100 mm

   b 125 mm

   c 150 mm

   d 175 mm

2 Which one of the following documents can be described as 'advisory'?

   a British Standards

   b Building Regulations

   c Schedule

   d Specification

3 What is the minimum height to which a chimney emerging at the ridge of a roof must be constructed?

   a 300 mm

   b 450 mm

   c 525 mm

   d 600 mm

4 What method was used to seal brick flues before flue liners were introduced?

   a Flashing

   b Haunching

   c Painting

   d Parging

5 The construction of chimneys, flues and fireplaces is controlled by which part of the Building Regulations?

   a Part A

   b Part C

   c Part F

   d Part J

6 What is the name of the concrete unit positioned above the fireplace opening that supports the flue liners?

   a Base

   b Head

   c Step

   d Throat

7 What is the name of the weathering arrangement on the top of a chimney stack?

   a Flashing

   b Flaunching

   c Pointing

   d Projecting

8 What is one name for the dividing walls between multiple flues in a chimney stack?

   a Indents

   b Links

   c Spacers

   d Withes

9 What is the minimum projection of the constructional hearth from the face of the fireplace?

   a 150 mm

   b 338 mm

   c 450 mm

   d 500 mm

10 What is the name of the masonry built either side of the fireplace opening?

   a Abutment

   b Head

   c Jamb

   d Surround

11 Explain the purpose of oversailing courses in relation to chimney stack construction.

12 Describe the differences between an internal chimney breast and an external chimney breast.

13 Name three materials that can be used as a heatproof finish on a superimposed hearth.

14 Explain how a back boiler works.

15 Explain why flue liners must be installed with their socket uppermost.

Glossary

Abutments the parts of a structure that directly receive thrust or pressure.

Accelerant a substance added to speed up a process.

Acute an angle less than 90°.

Aesthetics related to attractiveness and appeal.

Agenda a written document that sets out the points for discussion in sequence.

Aggregate the coarse mineral material, such as sharp sand and graded, crushed stone (gravel), used in making mortar and concrete.

Alternate interchanging repeatedly.

Approved Code of Practice (ACOP) a set of written directions and methods of work, issued by an official body or professional association and approved by the HSE, that provides practical advice on how to comply with the law.

Area the measurement of a two-dimensional surface such as the face of a wall.

Arris any straight sharp edge of a brick formed by the junction of two faces.

Attached in this context, bonded into the main section of masonry following the specified bonding arrangement.

Axed a type of arch using bricks cut to a tapered or wedge shape around the radius.

Axis line a straight line about which an object may be divided into symmetrical halves.

Back boiler a simple device which uses convection and gravity feed to create a flow of heated water through pipes connected to a cylinder which stores the hot water.

Batter board a board with one edge cut to a slope or batter.

Battered brickwork masonry that is deliberately built leaning inwards from its base.

Bearing the portion of the lintel that sits on the wall and transmits structural weight.

Bisection in geometry, the division of something into two equal parts, usually by a line, which is then called a bisector.

Broken bond the use of part bricks to establish a bonding pattern where full bricks will not fit in.

Brownfield site this type of land will have been previously used for dwellings or industry.

Building information modelling (BIM) a method of planning and managing a construction project throughout the building's lifecycle from the design and planning stages through to demolition.

Building line a boundary line set by the Local Authority beyond which the front of a building must not project.

Bulging when the face of a wall is forced out by pressure or loadings from above or behind.

Bulking an increase in volume of soil or earth, caused by introducing air.

Buttress a masonry structure that projects from and supports a wall.

Cant brick a special brick with one or two corners removed to form an angle (chamfer) at 45°.

Carbon emissions result in the release of carbon dioxide into the atmosphere; also referred to as greenhouse gas emissions contributing to climate change.

Chimney breast contains the fireplace opening and the flue; can be constructed internally as a projection into a room or externally projecting from an outside wall.

Cladding the application of one material over another to provide a skin or layer.

Client a person or company that receives a service in return for payment.

Climate change a large-scale, long-term change in the Earth's weather patterns and average temperatures.

Cold bridge low temperatures conducted to the interior of a cavity wall where the masonry of the outer leaf touches the masonry of the inner leaf at a reveal.

Combustion the chemical process in which substances mix with oxygen in the air to produce heat.

Commercial when referring to buildings, those used for business activities to generate a profit, e.g. an office block.

Concave curving or hollowed inwards.

Convection the flow of heat through a gas or a liquid as the hotter part rises and the cooler part falls.

Convex curving or bulging outwards.

Coping the masonry covering of the top of a wall, usually with a slope.

Corrosion the gradual destruction of a material by chemical reaction with its environment.

Courses continuous rows or layers of bricks or blocks on top of one another.

Critical path analysis (CPA) a network analysis technique of planning complex working procedures.

Cure to cause to set hard, often using heat or pressure.

Damp proof course (DPC) a barrier designed to prevent moisture travelling through a structure.

Damp proof membrane (DPM) a sheet of strong waterproof material.

Datum a fixed point or height from which reference levels can be taken.

Dead load the weight of all the materials used to construct the building.

Deadman temporary brick pillar built to carry a string line as a guide for masonry under construction.

Deflection the degree to which a structural element is displaced under a load.

Demolition the process of destruction, when a structure is torn down and destroyed.

Dentil course arrangement of alternating set-back and projecting bricks to give a castellated effect.

Detached when referring to a house, a stand-alone free-standing residential building.

Dog-tooth type of string course with bricks laid diagonally to the face line of the wall to give a serrated effect.

Duct an enclosed passage or channel for conveying air.

Durability ability to withstand wear and tear from elements that cause decay.

Duty holder the person in control of any potential danger.

Economical using the minimum of resources necessary for cost effectiveness.

Efflorescence a powdery white crystal deposit that can form on the face of the masonry.

Era a period of time distinguished by particular characteristics.

Estimate essentially a 'best guess' of how much specific work will cost, or how long it will take to complete.

Ettringite a hydrous calcium aluminium sulphate mineral that causes cracking in joints.

Fabric in this context, the structure or framework of the building.

Face plane the alignment of all the bricks in the face of a wall to give a uniform flat appearance.

Faceting the effect of straight components overhanging each other on a curve.

Fall a downward slope or decline.

First fix in simple terms, this refers to all work prior to plastering.

Flaunching the weatherproofing measures applied to the exposed top of a chimney stack.

Float in critical path analysis, the difference between the earliest start time (EST) and the latest finish time (LFT).

Footings the masonry constructed from the top of the foundations up to finished floor level (FFL).

Foul water includes sewage from toilets and waste water from sinks, baths and appliances such as washing machines.

Friction the resistance that one surface or object encounters when moving over another.

Frog the indentation in a brick.

Frontage line the front wall of a building.

Fungal caused by a fungus (a type of organism that obtains food from decaying material).

Gable the triangular wall at the end of a ridged roof.

Galvanise to cover iron or steel with a protective zinc coating.

Gauge the process of establishing measured uniform spacing between brick or block courses including horizontal mortar joints.

Gauging the method used to accurately and consistently measure quantities of material for mortar or concrete.

Greenfield site land that has not previously been used for construction, such as a countryside location or an open area such as a park or school field that has not previously been built on.

Hardcore solid materials of low absorbency used to create a base for load-bearing concrete floors, for example, crushed stone.

Hearth an area in front of a fireplace that gives protection from heat and sparks to the floor and floor covering.

Helical having the shape or form of a helix (a spiral).

Hierarchy a system in which roles are arranged according to their importance or level of responsibility.

Hydrated caused to heat and crumble by treatment with water.

Hydration chemical reactions in which products of hydration bond individual sand and gravel particles together to form a solid mass.

Imposed load additional loads that may be placed on the structure, such as the weight of people and furniture or the effects of wind and snow; also referred to as live load.

Improvement notice issued by an HSE or local authority inspector to formally notify a company that improvements are needed to the way it is working.

Industrial when referring to buildings, those used for processing materials and manufacturing goods, e.g. an insulation manufacturer.

Inert not chemically reactive.

Infestation the presence of a large number of insects or animals causing damage.

Ingress the act or process of entering something.

Inspection chamber a masonry structure that allows inspection of services below ground.

Insulation materials used to reduce heat transfer in a building. Can also be used to control sound transmission.

Integrity in a bricklaying context, being whole and complete.

Isometric projection a pictorial method of presenting information on a drawing. The structure is drawn at specified angles with one corner represented as closer to the person viewing the drawing.

Jamb a vertical inside face of an opening in a wall (Chapter 5); the masonry either side of a fireplace opening (Chapter 10).

Joists parallel timber beams spanning the walls of a structure to support a floor or ceiling.

Kiln a type of large oven.

Kinetic relating to, caused by, or producing movement.

King closer a brick with one corner cut away, making the header at that end half the width of the brick.

Lateral movement movement or pressure from the side.

Legislation a law or set of laws made by a government, for example, the Health and Safety at Work etc. Act (HASAWA) 1974.

Leptospirosis also known as Weil's disease, this is a serious disease spread by contact with urine from rats and cattle.

Live load *see* imposed load.

Load bearing relating to the carrying of a load.

Local exhaust ventilation (LEV) an engineered control system that reduces exposure to airborne contaminants by sucking the dust and fumes away from the workplace.

Mid-feathers sometimes referred to as 'withes'; refer to the masonry walls that separate multiple flues in a chimney stack.

Non-ferrous containing little or no iron; resistant to corrosion.

Oblique projection a drawing method where the front elevation is drawn to its actual size and shape, with the third dimension (or depth) shown drawn back to either the left or right as required. The lines are drawn horizontally, vertically or at a 45° axis to the left or right.

Obtuse an angle between 90° and 180°.

Orthographic projection a two-dimensional method of laying out a drawing of a structure. The front elevation has the plan view below it; the side or end elevations are shown directly either side of the front elevation.

Oversailing projecting from the main face of a wall.

Parapet a low wall along the edge of a roof or balcony projecting above the roof surface.

Percentage parts of a hundred.

Perforated having lines of small holes.

Perp short for 'perpendicular', this is the vertical mortar joint which joins two bricks or blocks together at right angles (or perpendicular) to the bed joint.

Personal Protective Equipment (PPE) this is defined as 'all equipment (including clothing affording protection against the weather) which is intended to be worn or held by a person at work and which protects against one or more risks to a person's health or safety.'

Perspective in drawing, a way of portraying three dimensions on a flat, two-dimensional surface by suggesting depth or distance.

Pinch rod a piece of timber cut to the width of an opening and used to check that the opening size stays the same width throughout the construction of the panel.

Plasticiser an additive that makes a material more pliable.

Potable suitable to drink.

Prefabricated factory-made units or components transported to site for easy assembly.

Private sector projects that are independently financed, as well as public projects with private finance.

Profiles an assembly of timber boards and pegs set up at the corners and other points of a building to allow string lines to be accurately positioned.

Prohibition notice issued by an HSE or local authority inspector when there is an immediate risk of personal injury. A prohibition notice will only be issued when there is a clear breach of health and safety legislation and must be taken seriously.

Proprietary an item manufactured and distributed under a trade name.

Public sector projects that are funded by local and central government.

Queen closer a brick split vertically along its length to a specified dimension.

Quoins the vertical external angles (corners) in walling.

Radial brickwork brick features and masonry structures that are curved or rounded, such as arches and curved or radiused walls.

Rafter a beam forming part of the internal framework of a roof.

Ratio the amount or proportion of one thing compared to another.

Recycled manufactured from used or waste materials that have been reprocessed, for example, bench seating made from recycled carrier bags.

Refractory difficult to melt; resistant to heat.

Reinforce in masonry and concrete, strengthen by adding steel.

Renovation the process of renewing outdated or damaged building structures to make them useful again; sometimes termed 'remodelling'.

Requisition a written request or order for supplies.

Residential when referring to buildings, means suitable for use as a dwelling.

Respiratory relating to breathing.

Retaining wall a wall built to hold back a mass of earth or other material.

Retarder a substance added to slow down the rate of a chemical change.

Reveals the masonry forming the side of a window or door.

Reverse bond in the same course, starting with a stretcher and ending with a header.

Ridge the highest horizontal line on a pitched roof where sloping surfaces meet.

Risk assessment a systematic examination of a job or process to identify significant hazards and risks and evaluate what control measures can be taken to reduce the risk to an acceptable level.

Scale when accurate sizes of an object are reduced or enlarged by a certain amount.

Screed a levelled layer of material (often sand and cement) applied to a floor or other surface.

Second fix in simple terms, this refers to all work after plastering.

Semi-detached sometimes referred to as a 'semi', a house that is joined to another house on one side by a shared (party) wall.

Services systems installed in buildings to make them comfortable, functional, efficient and safe.

Setting-up costs includes costs of hoarding, temporary services and temporary site accommodation.

Sharp in this context, sand that has pointed or angular grains.

Snap headers full bricks cut in half with the header face showing.

Soakaway a hole dug in the ground, filled with rubble and coarse stone which allows liquid to slowly drain into the surrounding earth. A soakaway can also be used to drain storm water where mains drainage is not available.

Soffit the underside of an architectural structure such as an arch.

Solid walls masonry that has no cavity in the structural design.

Spalling when the face of a brick or block crumbles away.

Stability resistance to movement or pressure.

Statutory undertakings the various services that are brought to the site such as water and electricity.

Storm (or surface) water rainwater falling on the roof and surrounding grounds of a building.

Striking point the centre point of the radius from which an arc is described.

Substructure all brickwork, blockwork and structural materials below finished floor level (FFL).

Subsidence the sudden sinking or gradual downward settling of the ground's surface with little or no horizontal motion.

Subsoil the soil lying immediately under the surface soil.

Sulphate a salt of sulphuric acid.

Superstructure the upwards part of a building that begins where the substructure ends.

Tender the process where a contractor will make calculations of resource requirements and their profit margins for a project, which they will submit to a client for consideration.

Terraced when referring to houses, a row of similar dwellings joined together by the side (or party) walls.

Thermal movement changes in dimension of masonry or concrete because of fluctuations in temperature over time.

Thermoplastic becoming soft when heated and hard when cooled.

Tolerance allowable variation between the specified measurement and the actual measurement.

Toolbox talks short talks given at regular intervals onsite at a work location.

Tooled or ironed the procedure using steel tools to create a specified type of joint.

Topsoil the upper, outermost layer of soil, usually the top 5–10 inches (13–25 cm).

Underpinning to prop up or support from below; to replace or strengthen the foundation of a building.

Undulation a gentle rising and falling.

Uniformly evenly and consistently.

Verbal communication talking to others to transmit information.

VOC (volatile organic compounds) the measure that shows how much pollution a product will emit into the air when in use.

Void an open space or gap.

Walkthrough an animated sequence as seen at the eye level of a person walking through a virtual structure.

Wall plate a timber plate bedded in mortar on the inner skin of the cavity wall, to which the timber roof structure can be securely fixed.

Water suppression system a system where water is sprayed onto the cutting disc to keep dust to a minimum when cutting bricks, blocks and stone.

Watt the unit measurement for power.

Working at height work in any place where, if there were no precautions in place, a person could fall a distance liable to cause personal injury.

Test your knowledge and activities answers

CHAPTER 1

Test your knowledge

1 d – Risk assessment
2 a – Blue circle
3 d – Oxygen
4 c – Powder
5 b – Control of Substances Hazardous to Health Regulations
6 b – 75°
7 d – Volatile organic compound
8 c – 85 dB(A)
9 b – More than 7 days
10 c – 410 V
11 A dangerous occurrence which has not caused an injury (a near miss) should still be reported because if it were to happen again, the consequences could be more serious. Steps can be taken to prevent it happening again.
12 Independent scaffold transmits the loadings carried by it to the ground. Putlog scaffold transmits the loadings partly to the ground and partly to the building under construction.
13 Back straight
   Elbows in
   Knees bent
   Feet slightly apart
14 These signs are usually square or rectangular and coloured green. They tell you the safe way to go, or what to do in an emergency.
15 The layout of the site
   Any site-specific hazards of which operatives need to be aware
   The location of welfare facilities
   The assembly areas in case of emergency
   Site rules

CHAPTER 2

Activities

Answers to 'Improve your maths' activity, page 39

a 500 mm
b 6 mm
c 10,000 mm

Answers to Gantt chart activity, page 58

1 4.5 weeks
2 8 weeks

Answers to 'Improve your maths' activity, page 65

1 $29 + 51 = 80$
2 $79 - 23 = 56$
3 $54 \times 76 = 4104$
4 $23 \div 4 = 5.75$
5 Three quarters as a decimal fraction = 0.75

Answer to 'Improve your maths' activity, page 66

20.2 m

Answers to 'Improve your maths' activity, page 66

1 5.04 m²
2 2.43 m²
3 0.85 m²

Answer to 'Improve your maths' activity, page 68

Area = 11.04 m²
Total number of bricks required = 663

Answer to 'Improve your maths' activity, page 69

3.5 × 0.6 × 0.3 = 0.63 m³ of concrete

Answers to 'Improve your maths' activity, page 69

1 53.9 m
2 32.4 m
3 20.9 m

Test your knowledge

1   d – Variation order
2   d – Toolbox talk
3   b – Drawings can be amended quickly
4   d – 1:2500
5   a – A0
6   a – Isometric
7   d – Quote
8   b – Preliminaries
9   d – Quantity surveyor
10 d – Skirting boards
11 Written communication can form a permanent record of information passed to others. If it is read and interpreted carefully, it can reduce the chance of misunderstandings and allow reference to relevant information later.
12 Any three of the following:
   - Rolling your eyes (perhaps meaning 'here we go again!')
   - Yawning (indicating boredom)
   - Hands in pockets (indicating lack of interest)
   - Crossed arms (indicating discontent)
   - Smiling (indicating happiness)
   - Frowning (indicating disagreement).
13 The minutes of a meeting form a record of decisions made and actions planned by those in attendance. The minutes from past meetings are used to make sure that actions decided on previously have been carried out.
14 Orthographic projection is a two-dimensional method of laying out a drawing, where the front elevation of a structure has the plan view directly below it. The side or end elevations are shown directly each side of the front elevation.
15 Standardised symbols are used on drawings as a means of passing on information in a simpler, more easily understood way. If all the parts of a building were labelled in writing, the drawing would soon become very crowded with text and potentially confusing.

CHAPTER 3

Test your knowledge

1   c – Funded by government
2   b – Dwellings
3   b – Professional
4   b – Modular
5   d – The section below finished floor level
6   a – End-bearing
7   a – Coarse aggregate
8   b – Fast construction
9   c – Dry-lining
10 b – 10°
11 This refers to the method of understanding the arrangement of responsibilities and relationships between each member of the team. The greater the responsibility, the higher up the 'line' the role will be placed. Each person on the line of management is responsible for managing those below his or her stated position.
12 A raft foundation is often laid over an area of softer soil that would be unsuitable for a strip foundation or where a pile foundation would be too expensive. It consists of a reinforced slab of concrete covering the entire base of the building, spreading the weight over a wide area.
13 It is important that adequate ventilation is provided to keep the underfloor space dry. This reduces the chance of rot in the timber components.
14 Solid timber joists can be prone to movement and bending due to changes in moisture levels in a structure. Timber engineered components are more stable in use and lighter to use by the installer.
15 U-values show the rate of transfer of heat through a structure in watts. Well insulated parts of a building have a low thermal transmittance whereas poorly insulated parts of a building have a high thermal transmittance. The lower the U-value, the better the insulation properties of the structure.

CHAPTER 4

Activities

Answers to 'Improve your maths' activity, page 112

1 325 m³
2 33 lorries

Test your knowledge

1 a – Land in countryside never built on before
2 c – Because it has low loadbearing capacity
3 c – Bulking
4 a – Ranging
5 d – Site
6 c – Longest
7 b – Local authority
8 b – Temporary bench mark
9 a – Optical level
10 a – Gauge
11 It may be necessary to consult with the relevant authorities to obtain permission for tree removal. Sometimes trees are the subject of preservation orders, in order to protect the environment. If existing trees and other vegetation are to be preserved as part of the development, they will need to be clearly identified at the site preparation stage of the project.
12 This type of land will have been previously used for homes or industry and so existing structures may need to be demolished and removed, which will increase development costs. Soil contamination as a result of industrial use will need to be removed before new development starts.
13 Timber profiles are an assembly of timber boards and pegs set up at the corners and other points of a building to allow ranging lines to be accurately positioned. They consist of timber rails attached to timber pegs and are assembled on site to suit the requirements of the job.
14 Using this method will cancel out any defects or distortions in the straight edge. This is especially important if the straight edge is made from timber as this can warp or twist over time.
15 Once the self-levelling instrument has been set up on its tripod, the operator can move around the site carrying a staff mounted receiver.

CHAPTER 5

Activities

Answers to 'Improve your maths' activity, page 136

2.700 m × 0.600 m =1.62 m²
1.62 × 60 bricks per m² =
97.2 bricks (round up to 98)
97.2 × 5% = 4.86 (round up to 5)
98 + 5 = 103 bricks required

Answers to 'Improve your maths' activity, page 157

Standard brick depth = 65 mm
Standard bed joint = 10 mm
65 + 10 = 75
75 × 12 = 900 mm

Test your knowledge

1 a – Firing
2 b – Compressive
3 a – Gauging
4 b – Plasticiser
5 a – Damp
6 c – Entrainment
7 c – Penetrating damp
8 a – Broken bond
9 a – Drip
10 b – Jointing
11 Full fill insulation sheets can push the two skins of a cavity wall apart.
12 Forming a joint finish gives a good appearance and weatherproofs the joint.
13 It is lighter than a concrete lintel to work with; it has insulation built in; it doesn't show in the face of a cavity wall.
14 Concrete performs well in compression but not in tension. The steel strengthens the concrete to prevent cracking.

15 A special type of steel tie is used which has a de-bonding sleeve bedded in the mortar joint.

CHAPTER 6

Activities

Answer to 'Improve your maths' activity, page 183

The width of a standard brick is 102.5 mm. Divide this in half to give 51.25 mm. A standard mortar joint is 10 mm. Divide this in half to give 5 mm. Deduct 5 mm from 51.25 to give 46.25 mm – the *exact* size of a queen closer.

Test your knowledge

1 a – English
2 c – Dry
3 c – Quarter of a brick
4 b – 102.5 mm
5 b – 3 or 5
6 b – 46 mm
7 b – Engineering brick
8 d – Weather struck
9 c – Eight
10 a – 1 brick
11 A queen closer is laid next to the header brick in each course of a quoin in a quarter bond solid wall.
12 Flemish garden wall bond or English garden wall bond because there are fewer header bricks in the design. This means varying projections in the back of the wall are reduced. (Or similar explanation.)
13 Engineering brick, because the brick on edge forms the weathering of the wall and must resist the effects of rain and frost. (Or similar explanation.)
14 The three factors are: i) the durability of the chosen joint; ii) the time required to produce a satisfactory result; iii) the appearance required as a design feature.
15 A tile creasing is formed by laying two courses of concrete or clay tiles, half-bonded and bedded in mortar. Since the tiles will form an overhang on either side of the wall, they allow rainwater to run off the coping away from the face of the wall. (Or similar explanation.)

CHAPTER 7

Activities

Answer to 'Improve your maths' activity, page 224

4.5 m³ ÷ 0.5 m³ = 9 batches of gauged material

Answers to 'Improve your maths' activity, page 234

Depth of timber = 100 mm
Multiply 100 × 40 (fall ratio of 1:40) = 4000
Length of timber required = 4000 mm or 4 m

Test your knowledge

1 d – Water freezing in the body of the brick
2 d – Structural
3 d – A chemical reaction with the cement in the mortar
4 c – To measure movement either side of a crack in a wall
5 a – Capillary attraction
6 d – Stainless steel
7 a – Gloves, ear defenders, safety boots, dust mask and a combined helmet/visor
8 b – Needle
9 d – Underpinning
10 c – Step iron
11 Preventative work is carried out to avoid a building falling into disrepair. Responsive work is carried out after a fault has occurred.
12 Spalling can be caused by one of the following:
   ● Frost damage caused by the repeated cycle of freezing and thawing which creates stresses in exposed concrete and breaks it down over time.
   ● Internal reinforcing steel expands if it rusts or corrodes, creating pressures that can lead to the concrete cracking and spalling.
13 Vertical cracking can occur when mortar is stronger than the individual bricks or blocks; any movement will pass through the bricks or blocks and so crack them. (Or similar answer.)
   Horizontal cracking can be caused by sulphate attack or wall tie failure. (Or similar answer.)

Diagonal cracking often follows the brick joints and is usually caused by structural movement such as settlement or subsidence. (Or similar answer.)

14 The three methods are: i) adjustable steel props and needles; ii) offset props (known as 'Strongboys'); iii) dead shoring using timber components.

15 A combined system has foul and storm water running together in a single pipe. A separate system has storm and foul drainage running in separate pipes to different destinations.

CHAPTER 8

Activities

Answer to 'Improve your maths' activity, page 253

675 mm

Test your knowledge

1 b – Cant
2 a – Acute
3 c – Squint
4 c – Quetta
5 b – Pinch rod
6 c – Horizontal
7 b – Dentil
8 b – Above the plinth course
9 b – 45°
10 a – An acute-angled quoin using standard bricks
11 Hard hats are to be worn where there is a risk of falling objects.
Safety glasses must be worn to prevent any flying brick fragments or debris causing eye injury.
A suitable dust mask must be worn when dust is generated. If fumes and smoke are present, make sure that suitable respiratory protection equipment (RPE) is worn.
Ear defenders must be worn, as brick cutting can be noisy, especially when disc cutters are being used.
12 Materials such as bricks, concrete and stone can contain large amounts of crystalline silica. Cutting these materials produces airborne dust particles that can penetrate deep into the lungs, causing serious damage, including lung cancer and silicosis.

13 a Dimensional consistency and colour consistency
   b Dimensional consistency is important because bricks of varying sizes will affect the straight appearance of any joint lines that extend through the panel.
   Colour consistency is important because bricks tend to have multiple colours within them and one colour can merge into another, so the bricklayer must develop an 'eye' for selection of bricks that are consistent in colour.
14 When the work is at high level above the eye line of an observer. Bricks can vary slightly in size and shape so the part of the brick that is more easily seen should be more carefully aligned horizontally.
15 a Straight; concave curve; convex curve
   b Ramped work can be used where there is a change of height along the length of a wall. It can include a string course to add visual interest.

CHAPTER 9

Test your knowledge

1 d – Voussoir
2 b – Skewback
3 a – Easing
4 b – An arch transmits compressive forces to the abutments.
5 a – Extrados
6 c – Radials
7 a – Batter board and spirit level
8 a – Faceting
9 c – Gun template
10 c – Buttress
11 Rough-ringed arches are formed by using tapered or wedge-shaped mortar joints which are thinner at the intrados and get thicker towards the extrados as the brickwork progresses around the arch radius.
12 a There are four key bricks in this type of arch, located on the vertical and horizontal axis lines.
   b The striking point is at the centre of the circle.
13 i) Equilateral; ii) dropped; iii) lancet
14 A trammel is a length of timber fixed to the centre point of the curve in such a way that it can sweep through the full arc of the circumference of the curved wall and can be raised as the work progresses. Its length corresponds to the specified radius.

15 Plinth special bricks can be used to create the slope of the batter by reducing the thickness of the buttress as it gains height.

Bricks laid at right angles to the angle of batter are cut into the horizontal brickwork bonded to the main wall.

CHAPTER 10

Test your knowledge

1 b – 125 mm
2 c – Schedule
3 d – 600 mm
4 d – Parging
5 d – Part J
6 d – Throat
7 b – Flaunching
8 d – Withes
9 d – 500 mm
10 c – Jamb
11 The oversailed or corbelled masonry allows the base of the pot to sit comfortably within the outline of the stack. The pot is secured by filling around it with strong mix mortar.
12 An internal chimney breast projects into a room, while an external chimney breast projects from an outside wall, leaving more uninterrupted space inside the building.
13 Any three of the following:
   ● Brick
   ● Ceramic tile
   ● Slate
   ● Marble
14 A back boiler is a simple device which uses convection and gravity feed to create a flow of heated water through pipes connected to a cylinder which stores the hot water. Hot gases from the combustion process are directed through a duct in the boiler unit to heat water contained in a vessel known as a jacket.
15 To prevent any moisture within the flue such as condensation passing through the joint into the surrounding masonry. Sulphur and carbon in the flue gases can dissolve in any moisture present to form acids, which can attack the surrounding masonry.

Index

A

abutments   265, 317
accelerant   83, 317
access
   equipment   27–30, 219–22
      see also working at height
   onto site   55–6
accidents
   recording and reporting   16–17
   see also injuries
acid rain   201
acute angle   254
acute angled quoins   258–60
addition   64
adjustable steel props   209–10
aesthetics   317
agenda   37, 317
aggregate   82, 317
air brick   128
air conditioning   97
air source heat pumps   106
angles   254–6
   see also arches; battered
   brickwork
apartment block   80
Approved Codes of Practice (ACOPs)
2, 292, 317
arches
   axed   266, 276–9
   bulls-eye   269
   construction methods   273–82
   creeper bricks   282
   deadman   273
   gothic   270–1
   rough-ringed   266
   segmental   268, 278–9
   semi-circular   267, 276–8
   setting out   267–72, 276–9
   structure   265
   terminology   265–6
   three-centred   270
   voussoir bricks   279–80
architectural drawings   38
architectural periods

Edwardian   80
Elizabethan   79
Georgian   79
Victorian   80
architect   74
area   317
   measuring   66–8
arris   155, 317
asbestos   4–5
assembly drawings   41
attached piers   187–8, 317
autoclave   139
autoclaved aerated concrete blocks
100
axed arches   266, 276–9, 317
axis line   269, 317

B

back boilers   300–1, 317
bargeboard   91
basket-weave patterns   244–7
bath, symbol for   46
batter board   317
battered brickwork   286–9, 317
battery power   24
bearing   142, 317
benched drain   231
bench saw   204–5
bill of quantities   49, 62–3
biomass heating   105
bisection   254–5, 317
bitumen felt   130
'block and beam' method   86
block plan   40
blocks
   dense concrete   140, 147, 182
   hollow   140, 177–8
   lightweight insulation blocks
   140
   symbol for   46
   thin-joint masonry   167–9
   see also brickwork
blown render   198–9
blue jean insulation   98

body language   36
bolster   203
BREEAM (Building Research
Establishment Environmental
Assessment Methodology)   97
brick autoclave   139
brick bolster   137
brick hammer   137, 203
brick kiln   137–8
brick-on-edge (BOE) coping   190–2
bricks
   characteristics   139–40
   clay   137–9
   common   138–9
   concrete   139–40
   engineering   138–9
   facing   137–8
   moving   147–8
   sand/lime   139
   stacking   147–8
   storing   146–7
   see also blocks; brickwork
brick tongs   21
brickwork
   basket-weave patterns   244–7
   battered   286–9
   bonding junctions   184
   broken bond   153
   buttresses   287–8
   curved   282–5
   decorative   238–53
   English bond   174–5, 180,
   183–6
   English garden wall bond   181
   faults in   196–201
   Flemish bond   174, 181, 183–6,
   283
   Flemish garden wall bond
   181–2
   forming openings in   162
   good practice   155–6
   herringbone patterns   240–4
   joint finishes   169–70, 189
   laying to the line   158–9

lining in   158
one-brick-thick   179–82
panel surrounds   247–8
plinth courses   250–1
plumbing   158
quarter bond   180
queen closer   183
ramps   252–3
reinforcing   260–2
removing a damaged brick
206–7
repairing   202
reverse bond   152–3
setting out   152–3
stepped   287
Stretcher bond   152–3, 179,
283–4
string courses   248–50
symbol for   46
tumbling-in   288
voids in   156
weather protection   190–2
see also arches; quoins; wall ties
British Standards
chimney stacks   292–3
wall height   155
wall ties   159–60
broken bond   153, 317
brownfield site   111, 112–13, 317
builder's square   119
building control officer   76
building information modelling (BIM)
60–1, 317
building lines   119–20, 121, 317
Building Regulations
compliance   40–2
and sustainability   97
building regulations, chimney stacks
292
Building Research Establishment
(BRE)   97
building technician   74–5
bulging   200–1, 208, 317
bulking   111–12, 317
bulls-eye arch   269
buttresses   286, 287–8, 317

C
calculation methods   64–9, 254–6
cant bricks   191, 251, 317
capillary attraction   179
carbon dioxide   96, 128
carbon emissions   128, 317
carbon footprint   105
career progression   78
carpenter's saw   114
caution signs   20
cavity walls
below ground   138
closing at reveals   163–4
components of   87, 128
design of   128–35
forming junctions in   165
forming openings in   162
gables in   165–6
insulation   131–4
lateral movement   135
materials
See blocks; bricks; mortar
thermal efficiency   102–4
thermal movement   134
thin-joint masonry   167–9
timber frame   87–8, 166–7
wall ties   87, 88, 130–1,
159–60
see also brickwork; damp proof
course (DPC)
CDM coordinator   7
cement   82
mortar   143
storing   144
symbol for   46
ceramic tiles   90
cesspits   229
changing rooms   9
chemicals   82–3
see also hazardous substances
chimney breast   286, 300, 317
chimney pot   218, 296
chimney stack   303–5
British Standards   292–3
building regulations   292
components of   296–7

construction methods   311
faults in   217–18
flue   296, 301–3, 309–10
materials   295
mid-feathers   305
necking course   304
over-sailing course   304
parging   305
regulation dimensions   312–13
repairing   217–18
safety regulations   293–5
setting out   305–7
weatherproofing   297–8
see also fireplace; hearth;
working at height
chisels   203
circle
parts of   267
see also angles
civil engineer   74
cladding   88, 317
claw hammer   116
cleaning masonry   205
clerk of works   77
client   317
communicating with   36–7
responsibilities of   7
climate change   96, 317
club hammer   114, 137
coarse aggregate   82
coatings, for walls   89
cold bridge   163, 317
combustion   292, 317
commercial buildings   72, 317
common bricks   138–9
communication
body language   36
with customers   36–7
email   35
letters   34–5
at meetings   37–8
verbal   34, 35–6
written   34–5
company structure   72
component/range drawings   42
computer aided design (CAD)   42–3

concave ramps   252–3
concrete   82
   faults in   223
   mixing   224–5
   mix ratios   82–3, 224
   moving and placing   225–6
   producing   223
   repairing   223–4
   symbol for   46
concrete bricks   139–40
concrete lintels   140–1
conservation projects   75
Construction, Design and
Management (CDM) Regulations
2015   5–6
construction drawings   40–2, 117
construction methods
   modular   80–1
   monolithic   81
construction sequence
   first fix   94
   second fix   94–5
   services   95–6
contractor   7
contract planning   53–4
contracts
   penalty clause   53
   planning stages   53–4
contracts manager   76
Control of Noise at Work Regulations
2005   9–10
Control of Substances Hazardous to
Health (COSHH) Regulations 2002
4–5, 238
Control of Vibration at Work
Regulations 2005   13–14
convection   317
conversion tables   64
convex ramp   253
coping   129, 190–2, 317
copper pipe   95
corbelling   249
corners see quoins
corrosion   198, 317
costing process   62–3
   see also estimating

courses   125, 317
   see also brickwork
cowl   297
cracking
   blown render   198–9
   horizontal   200
   in masonry walls   199–200
   repairing   206–7
   vertical   200
cradles   222
creeper bricks   282
critical path analysis (CPA)   58, 60,
317
cure   317
curved brickwork   282–5
cutting
   equipment   137, 154, 204–5,
   238–9
   by hand   154–5, 239
   step-by-step   239–40

D
damp   129, 179, 196–7
damp proof course (DPC)   83, 128,
317
   closing cavity at reveals   163–4
   installing   156–7
   lintels and sills   164
   materials   129–30, 142
   replacing   213–15
   solid walls   178–9
   symbol for   46
damp proof membrane (DPM)   86,
317
datum   122–3, 125, 318
dead loads   81, 318
deadman   273, 318
dead shoring   212
debris chute   30–1
decorative brickwork   238–53
deflection   208, 318
delivery note   50
demolition   6, 318
dense concrete blocks   140, 147,
182
dentil course   249, 318

dermatitis   16
designer   7
detached building   79–80, 318
disc cutter   204–5, 206, 238
division   65
documents
   bill of quantities   49, 62–3
   delivery note   50
   invoice   50
   job sheet   51
   manufacturer's instructions   52
   method statement   47
   permit to work   52
   requisition order   48
   schedule   48
   site diary   52
   specification   47
   time sheet   51
   variation order   48
   see also drawings
dog-tooth course   249–50, 318
door
   replacing   212–13
   symbol for   46
   see also lintels
double glazing   99, 101
drainage systems
   benched drain   228, 231
   cesspits   229
   domestic   95, 228–34
   excavation   230–2
   falls   232–3
   foul water   228
   inspection chambers   230–2
   leaking   208
   materials   229–30
   rainwater   90, 95–6, 228
   septic tanks   229
   sewage treatment plant   228
   site preparation   110
   soakaway   229
draught proofing   99
drawings
   architectural   38–9
   construction   40–2, 117
   floor plans   117–18

format and layout of 43–4
isometric projection 45
location 39–40
oblique projection 45
orthographic projection 43–4
producing 42–3
setting out 116–18
solid walls 176
specification 176
standardised symbols 45–6
standard paper sizes 43
walkthroughs 42
dredged sand 144
drills 204
drinking water, on site 8
drum mixer 151
dry-lining 89–90
dry rot 201
duct 128, 318
dummy frame 162–3
durability 318
dust masks 11
duty holders 6–7, 318

E

ear defenders 10, 11
earth (subsoil), symbol for 46
Edwardian buildings 80
efflorescence 146, 197, 205, 318
electricity
    dangers of 24
    extension cables 25
    legislation 23–4
    Portable Appliance Test (PAT)
    25
    power tool checks 24–5
    safety checks 25
    services 95, 109–10
    voltage 24
    wiring a plug 24
Electricity at Work Regulations 1989
23–4
elevations 41
Elizabethan buildings 79
email 35
emergency procedures 19
employment
    career progression 78

documents 50–2
employee responsibilities 4,
9–10
employer responsibilities 3–4,
9–10
job roles 73–7
    price work (piecework) 50
    sub-contracting 50
Energy Performance Certificate
(EPC) 101–2
engineering bricks 130, 138–9
English bond 174–5, 180, 183–6
English garden wall bond 181
equipment
    access 27–30, 219–22
    cutting 137, 154, 204–5,
    238–9
    electric 23–5
    finishing 136
    hand tools 203
    laying 136–7, 167
    legislation 22–5
    levelling 123–5
    lifting and handling 20–3
    for maintenance 203–4
    mixing 151
    power tools 24–5, 204–5
    setting out 113–16, 119
estimating 61–3, 318
estimator, role of 63
ettringite 198, 318
excavation
    drainage systems 230–2
    preparing for 121
expanded foam insulation 100
expanded metal lathing (EML) 184
expanded polystyrene (EPS) 133
expanded polystyrene (EPS) beads
99–100
extension cables 25
external wall insulation 89
eye protection 11

F

face plane 158, 318
faceting 282, 318
facilities, on site 8–9, 56
facing bricks 137–8

fall 318
fall ratio, drainage systems 232–3
fascia 91
fibre cement slate 93
fibreglass insulation 98
finished floor level (FFL) 129
finishing, equipment 136
fireback 296
fireplace 298–302
    construction methods 308–9
    jambs 163–4, 300, 308
    regulation dimensions 306
    surround 314–15
    see also chimney stack; hearth
fire safety 17–19
first aid 15–16
first fix 94, 318
flashings 94
flat roofs 90
    thermal efficiency 105
flaunching 218, 304, 318
Flemish bond 174, 181, 183–6,
283
Flemish garden wall bond 181–2
floating concrete 226
floor plans 40, 117–18
floors 85–7
flue 296, 301–3, 309–10
    for gas-fired appliance 309
    see also chimney stack
flue liners 296, 302–3
flush joint 189
footings 83, 121, 318
    see also foundations
footwear 11
Forest Stewardship Council (FSC)
105
foul water 228, 318
foundations 81–5, 134
    excavation 121
frame fixings 162
free-standing walls 174, 176
friction 85, 318
frog 139, 318
frontage line 119–20, 122, 318
frost damage 83, 198, 201, 208
full fill insulation 132, 161

fungus   196–7, 318

G

gabled roof   91
gables   318
   in cavity walls   165–6
galvanised steel   130–1
galvanising   184, 318
Gantt chart   57–9
gas, services   95, 109–10
gates   185
   see also isolated piers
gauging   125, 151, 157, 318
gauntlets   12
general arrangement drawing   40
geometry   254–6
Georgian buildings   79
gloves   12
goggles   11
gothic arches   270–1
greenfield site   111–12, 318
grey water   95
grinder   204
ground floors   86–7
ground source heat pumps   106
gun template   289

H

half-bond *see* Stretcher bond
half-round joint   170
hammer drill   204
hammers   114, 116, 137, 203
hand-arm vibration syndrome (HAVS)
13
hands, protecting   12
hand saw   114, 203
hardcore   86, 318
   symbol for   46
hard hat   11
hazardous substances   4–5
hazards
   on site   13–16
   see also risk
health and safety
   emergency procedures   19
   employee responsibilities   4,
   9–10
   employer responsibilities   3–4,
   9–10

fire safety   17–19
first aid   15–16
handling equipment   20–5
legislation   2–13
method statement   15
personal protective equipment
(PPE)   10–13
risk assessments   9, 14–15
site welfare facilities   8–9
see also electricity; hazards
Health and Safety at Work etc. Act
(HASAWA) 1974   3–4
Health and Safety Executive (HSE)
2
   notifying   6
   powers of   2
hearth   299–300, 307, 318
   minimum dimensions   307
heat loss, sources of   97
heaving   199
helical bars   206, 318
herringbone patterns   240–4
hierarchy   77, 318
high-rise buildings   85
high visibility (hi-vis) jacket   11
hipped roof   79, 91
hollow blocks   140, 177–8
hollow piers   262
Home Quality Mark   97
horizontal cracking   200
horizontal reinforcement   260–1
hydrated calcium silicate   139
hydrated lime   145
hydration   82, 143, 318

I

i-beam floor joists   87
imposed loads   81, 318
improvement notice   2–3, 318
independent scaffold   219, 222
inductions   15
industrial buildings   72, 318
inert filler   143, 318
infestation   201–2, 318
ingress   318
injection insulation   132, 162
injuries, recording and reporting
16–17

inner leaf   87
inspection chambers   175, 177,
230–2, 318
insulation   75, 87, 88, 318
   cavity walls   131–4
   external wall   89, 213
   installing   132, 160–1
   internal wall   213
   materials   98–100, 105, 133–4,
   142
   and sustainability   97
   symbol for   46
   thermal efficiency   102–5
   thermal transmittance   101
integrity   319
intermediate profiles   120
internal walls   88–90
   setting out   120–1
invoice   50
irregular angles   255
isolated piers   177, 184–7
isometric projection   45, 319

J

jambs   163–4, 300, 308, 319
job roles   73–7
job sheet   51
jointer   136
jointing   169–70, 189
jointing chisel   203
joists   86–7, 319
junctions, in solid walls   184

K

kiln   319
kinetic   319
kinetic lifting   21–2
king closer   188, 319
knee pads   12

L

ladders   27, 220
lambswool insulation   98
laser level   124–5
lateral movement   135, 319
laying
   blocks 'on flat'   182
   equipment   136–7, 167
   to the line   158–9

see also brickwork

lead flashings   217

lead sheet   94

lead tray   297–8

leaking gutters   208

lean-to roof   91

legislation   319

  codes of practice   2, 292, 317

  construction industry   5–6, 13–14

  employment   22–5

  health and safety   2–13, 23–4, 238

  noise   9–10

  see also Building Regulations

leptospirosis   16, 319

letters   34–5

levelling   123–5, 157

lifting   21–2

Lifting Operations and Lifting Equipment Regulations (LOLER) 1998   23

lightweight insulation blocks   140

lime   145

  see also sand/lime bricks

lime leaching   197

line and pins   137

linear measurement   65–6

line management   77–8

lintels   140–2, 164

  replacing   209

listed buildings   73

load bearing capacity   85

loadings   81

local exhaust ventilation (LEV) 238, 319

location drawings   39–40

loose-fill insulation   99

lump hammer   114, 203

M

maintenance

  bulging   208

  cleaning masonry   205

  concrete   223–4

  cracking in walls   206–7

  projects   73

removing a damaged brick 206–7

repairing a chimney stack 217–18

replacing a lintel   209

replacing failed damp proof course (DPC)   213–15

replacing failed wall ties   215–16

tools for   203–5

management hierarchy   77–8

mandatory signs   20

manual handling   20–5

Manual Handling Operations Regulations 1992 (amended 2002) 21

manufacturer's instructions   52

masonry

  definition   88

  faults in   196–201

  see also brickwork

masonry cement   143

mason's line and pins   115

materials

  chimney stacks   295

  damp proof course (DPC)   129–30, 142

  drainage systems   229–30

  foundations   82

  insulation   98–100, 105, 133–4, 142

  mortar   142–5

  positioning   147–9

  recycled   98

  storing   56

  see also blocks; bricks

mathematics   64–9

measurement

  area   66–8

  linear   65–6

  tape measure   115, 116

  units of   64, 101

  volume   68–9

meetings, communication in   37–8

metal lathing   184

method statement   15, 47

mid-feathers   305, 319

mineral fibre   133

mineral wool insulation   98

minutes   37

mixing mortar   149–52

mobile elevated working platforms (MEWPs)   221

modular method, of construction 80–1

moisture

  efflorescence   146, 197, 205

  points of entry   129, 131

  weep holes   164

  see also damp; damp proof course (DPC)

monolithic structures   81

mortar   142–5

  failure   201

  mixing   149–52

mould   196–7

multifoil insulation   99

multiplication   65

N

near miss   16

necking course   304

new-build projects   72

noise

  ear defenders   10

  impact on public   54

  legislation   9–10

Noise Abatement Order   54

non-ferrous metal   218, 319

north, symbol for   46

O

oblique projection   45, 319

obtuse angle   254

obtuse angled quoins   256–8

offset props   211–12

operatives   7

optical level   124

optical site square   119

Ordinary Portland Cement (OPC) 143

ordnance bench mark (OBM)   123

Ordnance Survey   123

orthographic projection   43–4, 319
outer leaf   87, 88
over-sailing course   192, 248–9, 304, 319

P

pad foundations   84
painting, new plaster   90
panel surrounds   247–8
parapet wall   129, 319
parging   305
partial fill insulation   132, 161–2
pedestrian routes, on site   56
pegs and rails   114
penalty clause   53
penetrating damp   129
percentages, calculating   69
perforated   319
permit to work   52
perp joint   138, 142, 156, 163–5, 169–70, 189, 319
personal protective equipment (PPE)   3–4, 10–13, 319
Personal Protective Equipment (PPE) at Work Regulations 2002   10–11
perspective   45, 319
piers
    attached   187–8
    design of   176–7
    hollow   262
    isolated   177, 184–7
pile foundations   84–5
pinch rod   247, 319
pipework   95
PIR (polyisocyanurate)   98
pitched roofs   91
    thermal efficiency   104
pitch polymer   130
pit sand   144
planning
    building information modelling (BIM)   60–1
    costing process   62
    critical path analysis (CPA)   58, 60
    Gantt chart   57–9
    sequence of work   57

site layout   54–7
    slippage   58
    stages of   53–4
planning officer   76
plasterboard   88, 89–90
plasticiser   145, 319
plastic waste pipe   95
plinth courses   250–1
pneumatic breaker   204
pods   81
pointing   189
pointing trowel   136
pollution   228
polyisocyanurate (PIR) foam   133
polypropylene   130
polystyrene   98
polythene   130
Portable Appliance Test (PAT)   25
potable water   145, 319
power tools   24–5, 204–5
preambles   62
pre-contract planning   53
prefabricated units   88, 319
preliminaries   62
preservation order   112
pre-tender planning   53
price work (piecework)   50
prime cost (PC)   62
principle contractor   7
private sector   72, 319
profiles   113, 118–20, 319
prohibition notice   2–3, 319
prohibition signs   20
proprietary   319
provisional sum   62
Provision and Use of Work Equipment Regulations (PUWER) 1998   22–3
public sector   72, 319

Q

quantities, calculating   63–4
quantity surveyor, role of   62, 74
quarter bond   180
queen closer   183, 319
Quetta bond   175, 261
quoins   148, 157–8, 183, 319

angled   256–60
quotes   61

R

radial brickwork   319
    see also arches; curved brickwork
radiator, symbol for   46
rafters   91–2, 319
raft foundations   85
rainwater, drainage systems   90, 95–6, 228
ramps   252–3
ranging line   113–14, 115
Rapid Hardening Portland Cement (RHPC)   144
rasp   168
ratio   39, 319
recessed joint   170
recycled materials   98, 319
refractory   319
reinforcing brickwork   260–2
render systems   89
    cracking in   198–9
renovation projects   72–3, 319
repairing faults *see* maintenance
Reporting of Injuries, Diseases and Dangerous Occurrences Regulations (RIDDOR) 2013   16–17
requisition order   48, 319
residential buildings   72–3, 319
resources, estimating   61–3
respirators   10, 12
restoration projects   73, 75
restraining straps   166
rest rooms   9
retaining walls   174–6, 320
retarder   83, 151, 320
retention, of payments   53
reveals   129, 320
reverse bond   152–3, 320
ridge   91, 320
right angles, setting out   118–19, 120
rising damp   129, 179
risk
    assessments   9, 14–15, 320
    hazardous substances   5

permit to work   52
roof ladder   27, 221
roofs   90–4
    thermal efficiency   104–5
rot   201–2
rough-ringed arches   266

S
safety
    checks   25
    electrical   24–5
    glasses   11
    helmet   11
    notices   20, 55
    signs   20
    see also health and safety
sand   144–5
    see also aggregate
sand/lime bricks   139
sawn timber, symbol for   46
saws   114, 203
scaffold systems   29–30, 219–22
    stacking bricks on   147–8
    waste chute   218
scale   39, 320
    used in drawings   40–2
schedule   48
scissor lift   23
scoop and hopper   167
screed   75, 320
scutch hammer   137, 203
second fix   94–5, 320
sectional drawings   41
segmental arch   268, 278–9
self-employment   50
semi-circular arch   267, 276–8
semi-detached houses   80, 320
septic tanks   229
serpentine walls   285
services   320
    drainage   110
    electricity   95, 109–10
    gas   95, 109–10
    locating   109
    telecoms   110
    water   110
setting out
    arches   267–72, 276–9

brickwork   152–3
building line   119–20
chimney stacks   305–7
drawings for   116–18
dry   152
equipment   113–16, 119, 123–5
excavation   121
frontage line   119–20
internal walls   120–1
methods   118–25
positions of door and windows   154
profiles   113, 118–20
right angles   118–19, 120
step-by-step   120, 122
transferring levels   122–5
wall positions   121
setting-up costs   62–3, 320
settlement   199–200
sewage treatment plant   228
sharp   82, 144, 320
sheathing board   88
sheep's wool insulation   134
shower tray, symbol for   46
signs, safety   20
sills   164
silt test   144–5
single glazing   101
single leaf masonry   155
sink, symbol for   46
sinktop, symbol for   46
site diary   52
site layout
    access   55–6
    planning   54–7
    stationary plant   57
    storage of materials   56
    temporary accommodation   56
site plan   40
site preparation
    clearance   110–13
    investigation   109
    locating services   109–10
    see also setting out
site welfare
    facilities   8–9, 56

see also health and safety
slate
    as damp proof course (DPC)
    material   130
    as roof covering   93
snap headers   283, 320
soakaway   229, 320
socket, symbol for   46
soffit   91, 320
soil
    contamination   113
    testing   81–2
    topsoil   111
solar panels   96
soldier course   192
solid concrete floors   86
solid walls   174, 320
    design   176–7
    thermal efficiency   102
    see also brickwork; walls
spalling   170, 198, 201, 320
specialist operatives   75
specification   47
specification panel   176
spirit level   123–4, 157
spoil   111
spray foam see expanded foam insulation
spray paint   115, 121
stability   174, 320
stacking bricks   147–8
stack scaffold   222
stairs, symbol for   46
stakes   114
stationary plant   57
statutory undertakings   62–3, 320
steel lintels   141
steel reinforcing mesh   82
stepladders   28, 220
stitch drilling method   206–7
stone erosion   201
stonework   88
    symbol for   46
storm water   320
    see also rainwater
straight edge   123–5
straight ramps   252

Stretcher bond   152–3, 179, 283–4
striking point   252, 320
string courses   248–50
string lines   113–14, 288
strip foundations   83–4, 134
structure
    floors   85–7
    foundations   81–5
    roofs   90–4
    see also construction methods; walls
stud walls   89
sub-contracting   50
subsidence   200, 320
subsoil   199–200, 320
substructure   81, 87, 320
subtraction   65
sulphate attack   198
Sulphate Resisting Portland Cement (SRPC)   144
sunscreen   12
superstructure   87, 320
suppliers   63
surface water see drainage systems; rainwater
surveyor   74
suspended concrete floors   86
suspended timber floors   86
sustainability   96–7
switch, symbol for   46
symbols
    for drawings   45–6
    hazardous substances   5

T

tamping concrete   226
tape measure   115, 116
technicians   74–5
telecoms   110
tell-tale   200
temporary bench mark (TBM)   123
tender   53, 320
terraced houses   79–80, 320
terracotta   192
thermal
    efficiency   102–5
    movement   134

transmittance   101
    see also insulation
thermal movement   320
thermoplastic   142, 320
thin-joint masonry cavity walls   167–9
three-centred arch   270
three-dimensional walkthroughs   42
throat unit   296
tile cutter   204
tiles
    ceramic   90
    roofing   93–4
timber
    infestation   201–2
    pegs and rails   114
    profiles   113
    rot   201–2
    storing   149
    sustainability   105
    symbol for   46
    woodworm   202
timber frame walls   88, 166–7
    see also cavity walls
time sheet   51
toe boards   220
toe-cap boots   11
toilet
    facilities   8
    symbol for   46
tolerance   320
toolbox talks   15, 38, 320
tooled or ironed   169, 320
tools
    hand tools   203
    power tools   204–5
    see also equipment
topsoil   111, 320
tower scaffold   29, 221
trade operatives, job roles   75–6
trammel   252–3, 284
trees, retaining   112
trenches, excavation   121
trestles   28–9, 220
triple glazing   100, 101
trowel   136
tubular scaffold   29–30

tumbling-in   288

U

ultraviolet light   12–13
underpinning   226–7, 320
undulation   320
units of measurement   64
upper floors   87
U-value   101

V

vapour control layer   88
variation order   48
ventilated cavity   88
verbal communication   34, 35–6, 320
vertical cracking   200
vertical reinforcement   261–2
vibrating concrete   226
vibration white finger (VWF)   13
Victorian buildings   80
void   320
volatile organic compounds (VOC)   5, 320
voltage   24
volume, measuring   68–9
voussoir bricks   279–80

W

walkthroughs   320
wall plate   166, 320
walls
    attached piers   187–8
    external finishes   89
    faults in   196–201
    free-standing   174, 176
    internal   88–90, 120–1
    isolated piers   177, 184–7
    positions of   121
    Quetta bond   175
    repairing   202
    retaining   174–6
    solid   174
    stud   89
    timber frame   88
    see also brickwork; cavity walls; maintenance
wall ties   87, 88, 130–1, 148, 159–60, 168

failure of   199
    replacing   215–16
wash basin, symbol for   46
washing facilities   8
water   145
waterproof breather membrane   88
water services   95, 110
water suppression system   238, 320
watts   101, 320
WC, symbol for   46
weatherproofing
    brickwork   190–2

chimney stacks   297–8
    roofs   91, 94
weather struck joint   189
weep holes   164
Weil's disease   16
wet rot   201–2
wheelbarrow   21
whisk   167
window
    frames   212–13
    hinging position of   46
    symbol for   46
    see also lintels

wind turbines   96
wiring a plug   24
woodworm   202
Work at Height Regulations 2005
(amended 2007)   26–7
working at height   26–31, 219, 320
    scaffold systems   29–30,
    219–22
    trestles as platform   28–9
    using ladders and stepladders
    27–8
written communication   34–5